Advances in Spatial Science

Springer

Berlin
Heidelberg
New York
Barcelona
Budapest
Hong Kong
London
Milan
Paris
Santa Clara
Singapore
Tokyo

Titles in the Series

Cristoforo S. Bertuglia · Silvana Lombardo
Peter Nijkamp (Eds.)

Innovative Behaviour in Space and Time

With 87 Figures
and 50 Tables

 Springer

338.064
I5893

Prof. Dr. Cristoforo S. Bertuglia
Politecnico di Torino
Dipartimento di Scienze e Tecniche per i Processi di Insediamento
viale Mattioli 39
I-10125 Torino
Italy

Prof. Dr. Silvana Lombardo
Università di Roma "La Sapienza"
Dipartimento di Pianificazione Territoriale ed Urbanistica
via Flaminia 70
I-00197 Roma
Italy

Prof. Dr. Peter Nijkamp
Free University
Department of Spatial Economics
De Boelelaan 1105
NL-1081 HV Amsterdam
The Netherlands

The editors gratefully acknowledge the financial support for this book provided by the Italian National Research Council as part of the PFT2 Project (Progetto Finalizzato Transporti), contracts no. 94.01344.PF74 and no. 94.01345.PF74, in association with the Department of Sciences and Techniques for Settlement Processes (DINSE) of Turin Polytechnic.

МК

ISBN 3-540-62542-9 Springer-Verlag Berlin Heidelberg New York

Cataloging-in-Publication Data applied for
Die Deutsche Bibliothek – CIP-Einheitsaufnahme
Innovative behaviour in space and time: with 50 tables / Cristofore S. Bertuglia; Silvana Lombardo; Peter Nijkamp. – Berlin; Heidelberg; New York; Barcelona; Budapest; Hong Kong; London; Milan; Paris; Santa Clara; Singapore; Tokyo: Springer, 1997

© Springer-Verlag Berlin · Heidelberg 1997
Printed in Germany

Hardcover design: Erich Kirchner, Heidelberg

SPIN 10546804   42/2202-5 4 3 2 1 0 – Printed on acid-free paper

Preface

In the past decade there has been growing recognition that economic development is not mainly exogenously determined but, to a large extent, is a transformation process induced and governed by economic actors who respond to competitive, institutional and political challenge. This 'challenge and response' model is increasingly accepted as a valid analytical framework in modern growth theory and also explains the popularity of endogenous growth approaches to technological innovation issues.

However, a major and as yet largely under-researched topic is the question of the diffusion and adoption of new technological changes in the context of space-time dynamics. This diffusion and adoption pattern has obviously clear spatial and temporal variations connected with behavioural responses which may vary over time and different locations. This means that a closer analysis of spatio-temporal opportunities and impediments is necessary in order to fully map the complex interactions of technology and economy in space and time.

This volume sets out to bring together a collection of original contributions commissioned by the editors to highlight the spatio-temporal patterns and backgrounds of the diffusion and adoption of new technologies. Some are in the nature of a survey, others have a modelling background and again others are case studies. The contributions originate from different countries and different disciplines. This book is complementary to a previously published volume on technological innovation, *Technological Change, Economic Development and Space*, edited by C.S. Bertuglia, M.M. Fischer and G. Preto, and also published by Springer-Verlag (1995).

We wish to express our gratitude to Marianna Bopp (Springer-Verlag) who - as a charming, contemporary Mrs. Rasputin - has in a forceful but friendly way, guided this project to successful completion. We also wish to thank Angela Spence for her thorough linguistic editing and Dianne Biederberg (Contact Europe) for her uninterrupted efforts to prepare the text of this book in a professional way.

Cristoforo S. Bertuglia
Silvana Lombardo
Peter Nijkamp

Contents

PART A

Overviews

1 An Interpretative Survey of Innovative Behaviour and Diffusion

Cristoforo S. Bertuglia, Silvana Lombardo and Peter Nijkamp

031
032

1.1 Issues in Diffusion Analysis

"Made in Taiwan". The Western world has witnessed an avalanche of consumer goods produced in the Asian Pacific Rim which have flooded the markets in countries which used to have a strong industrial base. A world wide process of industrial restructuring has taken place, leading to a shift of the industrial heartland to traditionally remote areas. At the same time, the pattern of international trade and the geographical distribution of goods and information has gone through a drastic change. Market dynamics and product diffusion are apparently parallel processes which derive from two intriguing research themes:
- the incubation and generation of new ideas and technologies and their translation into commercially successful new products;
- the geographical spread (diffusion) of technologies or new products (or goods, in general) and their market acceptance.

The first issue has extensively been addressed in a recent companion book, where 'technogenesis' received central attention (see Bertuglia et al. 1995). The second topic is the focal point of the present book. It seeks to analyze the forces which govern the process from the production of new goods to their adoption by the market (or by clients). Such phenomena are not only relevant in the context of commodity trade, but also in the transfer or exchange of services, ideas or information. This issue has become of primary importance in the emerging new network society, in which information and communication technology plays a dominant role as the vehicle for the transmission of goods and for spatial interactions.

It is increasingly recognized that nodal points in a network, notably cites, play a strategic role in the creation and diffusion of innovations. Not only does the 'urban milieu' offer favourable seedbed conditions for innovative behaviour (e.g. socio-cultural facilities, knowledge and education infrastructure, financial and venture capital support mechanisms, and institutional/managerial ramifications), but it also acts as a catalyst for transmitting new findings to other places in a network (through its connectivity infrastructure). Clearly, the adoption and diffusion of innovation may take a multiplicity of forms (e.g. satellite

4

communication, teleport networks, physical transportation, logistics and telematics platforms etc.). Cities are essentially focal points of transactions favouring new ways of doing things. Such transformation processes clearly have a geographical and time dimension, as explained in traditional space-time geography.

The analysis of spatial, temporal and spatiotemporal diffusion already has a long history, not only in geography or marketing, but also in biology and medicine. Diffusion theory offers a multidisciplinary framework of analysis for a wide range of phenomena, e.g. migration, distribution of news, pollution, acceptance of innovations, spread of epidemic diseases, transportation and technology transfer. An extensive survey of different models can be found in the historical work by Rogers (1983) and more recently Banks (1993). There has also been in the course of time an analytical shift from simple exponential and logistic distribution functions to time-dependent growth parameters and critical threshold levels (see Nijkamp and Reggiani 1994).

Thus, diffusion analysis has become a major strand in social-science orientated explanatory frameworks for the geographical-temporal spread of phenomena. Clearly, diffusion patterns are not independent of the source from which they originate or of the specific phenomenon to be transmitted. Especially in the area of technological change, diffusion analysis has become a major topic, not only analytically but also politically. Therefore, in the next section we will look briefly at the nature and origin of technological innovation.

1.2 The Process of Technological Innovation[1]

The revival of Schumpeterian views on current economic restructuring phenomena has increasingly induced scientific interest in innovations. Both the behavioural stimuli and the selection environment for the creation and adoption of technological and organizational change in firms have become a subject of intensive study (cf. Suarez-Villa and Fischer 1995). In this context a rich field of economic research has recently been developed, including, for instance, long wave tests, analysis of incubation hypotheses, impact studies on small and medium-sized enterprises, neo-Fordist approaches to industrial organizations and labour markets, and the growth potential of high technology industries (cf. Suarez-Villa and Hasnath 1993). Several studies have been devoted to the seedbed conditions of new technologies, especially in the context of small and medium-size firms. Two particular lines of inquiry have called for much attention in the recent past, viz. the urban incubator hypothesis and the product life-cycle model. Both approaches focus on dynamic competition of firms.

The need for technological innovation starts normally in the competitive environment of the firm. New products and ideas are necessary in order to guarantee survival of the firm, leading to the intrinsic need for dynamic behaviour,

---

[1]The authors wish to thank Mariëlle Damman for her contribution to this section.

risk management and new patterns of management and organisation (see Ciciotti et al. 1990; Damman 1994).

Thus, the company life-cycle is a process aimed at continuity. To maintain this process, the company must be permanently innovative. In this context, Schumpeter (1934) developed his dynamic entrepreneurial concept in which innovations play a critical role. There appear to be two distinct moments favourable to the development of innovations (Mouwen and Nijkamp 1992). The first is in periods of an upswing of the economy, the second is a phase of downswing. A major reason for the importance of these two phases is the product life-cycle (see Figure 1.1). In an upswing period companies have the money available to develop new products. It is then likely that a variety of new products is introduced on the consumers' market. The companies are willing to take the risk of more uncertainty with regard to their business activities. In Schumpeter terms this is the 'demand-pull' hypothesis, where the market absorbs innovative products. Competitive companies will imitate these new products in order to prevent a gap in terms of their market share. After a phase of growth, the product will enter a phase of maturity. During the next phase of the cycle, moving towards a crisis, companies have to innovate in order to survive in such a turbulent period. Such innovation will - for risk avoidance reasons - mainly be focused on new processes rather than on new products. Process innovations, however, will lengthen the phase of maturity and the phase of decline of existing products. In Schumpeter terms this the 'depression-trigger' hypothesis. An extreme policy to create and favour an innovative business is to pursue an expensive wage policy through which the innovative companies can maintain their business activities while weaker companies have to leave the market: wage restraint keeps old-fashioned non-innovative companies alive in an artificial way (Kleinknecht 1994).

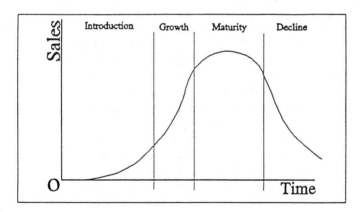

Fig. 1.1.   Product life-cycle

(Source: Van Geenhuizen 1993)

The above Schumpeterian model with regard to the phases of innovation is summarized in Figure 1.2. This model is predominantly concerned with small firms. In the case of large companies, exogenous science and invention is replaced by endogenous science and technology (mainly in-house R&D) (cf. Nijkamp and

6

Poot 1993). After the endogenous phase, the subsequent phases can be distinguished as follows: management of innovative investment, new patterns of production, changed market structures and eventually profits (or losses) from innovation. The basic difference in the model for smaller companies is the way technology becomes institutionalized within existing firms.

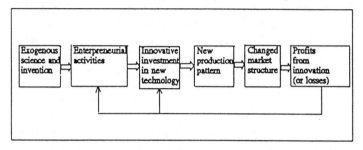

Fig. 1.2.   Schumpeter's model of 'Entrepreneurial Innovation'

(Source: Davelaar 1989)

When an innovation is successful, companies have a competitive advantage for a certain period. Competition may emerge from various sources, and this focuses attention on the firm's environment, which we now discuss.

In the context of competitive behaviour, Porter (1990) has emphasized five forces of competition (Figure 1.3). These five forces determine industrial profitability because they determine the prices firms can charge, the costs they have to bear, and the investment required to compete in the industry. The threat of new entrants (incumbent firms) limits the overall profit potential in the industry because new entrants bring new capacity and seek a new market share, while pushing the margins down. Powerful buyers or suppliers bargain away their own profits. Fierce competitive rivalry erodes profits since it brings higher costs (e.g., as for advertising, sales, or R&D) or by passing on profits to customers in the form of lower prices. The presence of close substitute products limits the price competitors can charge without inducing substitution and eroding the industry's volume. Companies must defend themselves against competition, but use the opportunities offered by competition. In this context, Porter has formulated generic strategies to develop a situation of competitive advantage. These strategies are cost leadership and product differentiation. Both cost leadership and differentiation can be achieved by following the strategy advocated by Schumpeter.

Porter's five forces are clearly present in the economic environment of companies, especially regarding the market and the technology. Environment can also be understood here in a more narrow, *spatial* sense. According to the latter interpretation, the environment refers for example to infrastructure, transport possibilities, the labour market, the capital market, energy, raw materials and the local government.

The latter observation calls for more analytical attention to the impact of geographical locations on the process of technological change. In an urban context, the existence of forward and backward linkages between industrial and service sectors means that new innovations may have substantial multiplier effects on the whole urban economy, especially if these effects are of a cumulative and circular nature (the 'urban breeding place' hypothesis). In this way, innovations may also stimulate a recovery from urban decline processes. Therefore, cities may offer a diversity of valuable facilities, making it easier for companies to develop innovations.

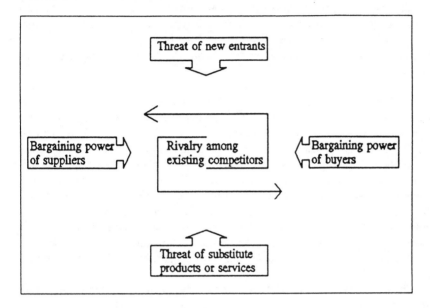

Fig. 1.3.   Porter's five competitive forces

(Source: Porter 1990)

Cities, or urban areas in general, are agents or incubators of economic, social, scientific, technological and cultural change. They are sources of socio-economic mutation through their cultural opportunities and their geographical connectivity. As a result, cities have become complex entities, with a multidimensional portfolio of economic, social, political and cultural features which offer a competitive advantage in terms of labour skills, locational possibilities, capital assets, infrastructural provisions, and socio-cultural amenities. Such a heterogeneous configuration creates the conditions for survival and continuity, even though urban developments may exhibit fluctuating growth patterns (see Nijkamp 1990). At the same time, cities are also early adopters of innovations generated elsewhere. Thus, modern cities seem to offer the most favourable selection environment for new ideas and new ways of doing things.

Davelaar (1989) mentions various conditions why large cities (incubation areas) favour the rise of new firms and innovation in established firms:

- Agglomeration economies caused by spatial clustering of (dis)similar firms offer significant scale advantages. Generation of innovations needs flexibility, which can be provided by agglomeration economies, especially in metropolitan or central areas.

- As regards the demography or population structure, the availability of skilled labour will be highly relevant, especially in the initial phases of innovation projects.
- The availability of specialized information and intensive communication patterns provides risk reduction to enterprises.
- The diffusion process of innovation often involves new types of demand regarding social overhead capital. Such new types of social overhead capital will normally first be developed in metropolitan or central areas. The density of economic activities in such areas will favour the profitability of investments in these areas.

Innovation means the incorporation of new knowledge into products and processes. This new knowledge normally also involves new technology. If a city more or less satisfies the above mentioned conditions, various pathways can be distinguished to develop new technology. According to Townroe (1990) technology comes to the city along various pathways: (i) new plant and equipment, (ii) migrant company investment, (iii) intracompany transfers, (iv) local research and development expenditure, and (v) spin-offs (e.g. via company networks, technopoles, science parks etc.) (Segal et al. 1985). Each pathway may favour the technological potential or the innovation capacity of a city. To reap the fruits of this potential, the city should offer appropriate host conditions, in particular: knowledge-based skills supported by diagnostic, flexible and entrepreneurial attitudes and abilities.

Not only the innovations themselves, but also the diffusion of innovation is an important element, as innovation diffusion leads to spatial spill-over effects (cf. Masser and Onsrud 1993). Spatial diffusion normally exhibits a hierarchical and distance-decay pattern (also Pred 1977). In this respect, large urban agglomerations act as the nodes in a spatial network. A rise (decline) in innovation efforts usually starts first in these nodes, and then affects smaller cities.

It should be added that cities or regions often exhibit bottlenecks which may hamper their growth. In general, threshold values prevent the start of urban growth, while congestion hampers a continuation of urban growth. Such bottlenecks may be removed by more investments in public overhead capital or infrastructure capital. A bottleneck analysis also provides an explanation for the shift in innovation tendencies from large agglomerations to medium-size agglomerations. Large agglomerations often exhibit too many bottleneck factors for a dynamic and flexible innovation strategy; as a result, smaller agglomerations sometimes become more favourable locations for innovative firms. Clearly, a certain minimum stock of social overhead capital has to be available. Thus small-size cities often do not provide an appropriate breeding ground for new activities, unless they have a direct accessibility to public overhead capital in large agglomerations (Damman 1994).

Analytical and modelling contributions to the phenomena of innovation and urban dynamics can be found in earlier works by Forrester (1969), who applied concepts from system analysis to analyze trajectories of urban development phenomena. These theories describe the long-run trajectory of a closed urban system. Such urban systems may lead to self-organizing equilibrium paths in the long run. The

Forrester model starts off from an empty area with a favourable growth potential, good locational conditions and a large labour force. These attraction factors induce urban development in all fields. This is a self-regenerating process through which the city is permanently expanding, until it loses its attractiveness. After this period, the rings surrounding the old city become attractive, so growth is focused in the periphery. This phase is followed by the next stage of urban decay in the rings surrounding the old city.

The above mentioned process will come to a standstill if the limits of the urban territory are reached. A final equilibrium state is characterized by too large a housing stock compared to the industrial land use needs, too many obsolete houses and plants compared to new ones, and too large a volume of unskilled labour compared to the desired urban labour supply. Differences in the attractiveness of various urban systems may have different impacts on in-migration, overcrowding, unemployment and urban decay, though within the limited setting of a closed urban system, the market forces will sooner or later generate a new equilibrium state. Davelaar (1989) has made interesting attempts such as testing a blend of the above mentioned incubation and bottleneck theories.

The next question to be dealt with is how innovative processes spread to other actors or areas.

1.3 Innovation Networks

The spread of innovations does not take place in a random way, but follows organized structures based on interaction and communications networks. It is increasingly recognized that most modern companies operate in an uncertain dynamic environment. The world is more and more characterized by a complex web of relationships. Network relationships seem to become an important feature of industrial economies nowadays (Kamann and Strijker 1991). According to Damman (1994), there are two reasons for the emergence of this phenomenon: first, since the early 1980s, companies have increasingly withdrawn from non-core activities, leaving these activities to various suppliers. Second, related to this is the increased flexibility in production based on the decline of demand for mass products.

In our examination of the relevance of diffusion patterns, we will focus in particular on network use and knowledge in a modern industrial economic system. Uncertainty, and the related presence of imperfect information, prevents a pure price mechanism from allocating resources in an optimal way and driving economic activities to a stable competitive equilibrium. Uncertainty can only be incorporated in a competitive equilibrium system by assuming an equal (imperfect) access of all individuals to the same information (Ouwersloot 1994); a condition which, in the presence of highly differentiated firm sizes, market structures and spatial situations can be considered highly unrealistic. In their economic behaviour and decision-making processes, companies are usually facing different kinds of uncertainty. Uncertainty can be divided into static and dynamic uncertainty. Static uncertainty can be caused by an 'information gap', an 'assessment gap' and a 'competence gap'. By 'information gap' is meant the access to specific information. The access can be hampered, for example, by costs, time and distance. When other characteristics, like for example inputs, components, production factors and technical equipment do exist, but are hidden in the results or the performance, an 'assessment gap' will emerge. The 'competence gap' will arise as a result of limited ability to process and understand available information. Dynamic uncertainty can be formed by a 'competence-decision gap' or a 'control gap'. In that case, uncertainty involves the impossibility of precisely assessing the outcomes of alternative actions. By 'control gap' we mean that the outcomes of present actions depend in fact on the dynamic interaction among different decisions of many actors over which the company has by definition a minimum control (Camagni 1991).

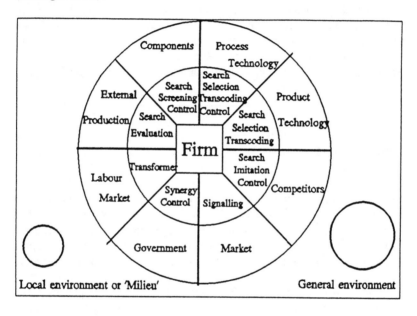

Fig. 1.4.   Main uncertainty reducing functions performed by the 'milieu'

(Source: Camagni 1991)

While complexity will cause uncertainty (Lombardo 1996), the local environment of a company may be considered one of the most important uncertainty-reducing factors (Figure 1.4). The local environment may be defined as a set of territorial relationships encompassing in a coherent way a production system, different economic and social actors, a specific culture and representation system, which generate a dynamic collective learning process. Different functions of the regional environment with regard to uncertainty reduction can be identified. First, a collective information-gathering and screening function. Second, the signalling function toward the direction of the market of local companies. A third element is a collective learning process, mainly through skilled labour mobility within the local labour market, customer-supplier technical and organizational interchange, imitation process and reverse engineering, or exhibition of successful technologies. A fourth function is a collective process of definition of managerial styles and decision routines. The last function is an informal process of decision coordination. All these functions contribute to greater effectiveness and innovativeness of companies in a local knowledge environment.

In analyzing the diffusion behaviour of a firm, we have to see the company as an open actor influenced by different factors. These factors can be subdivided into company characteristics and environmental characteristics.

Examples of company characteristics are age, company activity, R&D, competitive edge and company size. The influence of the age of a company is manifested in the phase of the company life-cycle. The phase of the company life-cycle influences the degree of uncertainty of the actor and thus the dependency on its environment; uncertainty is often found among young, newly established companies. Companies may also differ with regard to the kind of product manufactured. Companies which manufacture traditional products deal with a different stage of product life-cycle from companies which manufacture (very) modern products. The product life-cycle is shorter for (very) modern products than traditional products. This life-cycle also influences the degree of uncertainty: products which can be found in the introductory phase are risky, whereas products in the growth phase rarely deal with uncertainty. If the product life-cycle is relatively short, the introductory phase involves even more uncertainty than a longer product life-cycle. Here R&D may be seen as an essential influencing factor. Companies have to make choices about their competitive edge or strategy, regarding, for instance, cost leadership or product differentiation. In the case of product differentiation, a good product design plays an important role in the introductory phase of the product life-cycle. This means that companies with such a strategy are confronted with much more uncertainty than companies with cost leadership as a competitive edge. The size of a company may also play a role with regard to the degree of uncertainty. In this respect small companies are often seen as vulnerable actors in the economy.

Examples of environmental characteristics are the city and the region. These characteristics influence the company from outside. The city and the region contain facilities which make it attractive for a company to be located and to innovate successfully (Haag and Lombardo 1991). Can cities therefore be seen as the 'cradle' for innovative behaviour or not? Is the infrastructure sufficient or not? These questions are related to different theories which concern bottlenecks, thresholds or other kinds of constraints which affect an external environment to a firm.

Uncertainty, in all its forms, plays an essential role in the importance attached to the industrial location of firms. This uncertainty can be reduced through knowledge networks connected with the external world. The company and its environment have in this respect an interactive relationship which is full of uncertain elements. Clearly, the company will try to reduce uncertainty by creating different provisions, for example, access to knowledge networks. These networks play a role in the external environment and within the company itself, as information gaps can be reduced by knowledge networks.

Networks are thus a *sine qua non* for innovation diffusion. There is a multiplicity of networks and hence diffusion patterns may take various forms (cf. Alderman 1990). In the present book on innovative behaviour in space and time, two main themes of paramount analytical interest can be distinguished, viz. (i) the methodology for analyzing patterns and effects of innovative processes, and (ii) the general lessons and success factors which can be drawn out of empirical research or case study work. These two themes also underlie the structure of the present volume which will be set out in more detail in the next section.

1.4 Organization of the Book

The first section of the book, Part A, aims to build the conceptual and methodological framework which defines and classifies the areas of interest covered by the papers presented in the second and third part.

In the opening chapter, Davelaar and Nijkamp provide a survey of location theories, followed by an analysis of the location determinants of innovation production and of key spatial factors of innovation diffusion. This is followed by Frenkel and Shefer, who present a review of the literature, focusing on the influence of innovation on economic growth and on the definition of regional innovative capacity. The authors also deal with the process of spatial diffusion, describing the structure of several diffusion models. Karmeshu and Jain, in the next chapter, provide a classification and description of several different approaches to the modelling of innovation diffusion, with particular attention to the temporal aspects of diffusion.

From these reviews, the breadth and the complexity of the issues involved clearly emerges. Among other things, we can see that innovation processes can be better understood if they are analyzed in their different 'stages': (i) creation and generation (connected with R&D activities), (ii) adoption by individual actors (which is often usefully classified as early adoption and late adoption), (iii) diffusion (in space and time).

The characteristics and the effects of these three phases are strongly influenced by (and in their turn influence) the socioeconomic, cultural and infrastructural attributes of the locations in which the innovative processes are born or are introduced, spread and decay. This reciprocal influence raises a series of important issues. One of them is the spatial scale of phenomena, and hence the relevance of different spatial levels of analysis at which the triggers and the bottlenecks of innovative behaviour can be identified. The analysis must therefore range from the micro level, where we are concerned with the development of R&D activities inside a firm, to the meso level of urban settlements viewed as incubators and/or early adopters of new ideas and technologies, and finally the macro level of international diffusion of innovations.

But the most challenging issue is connected with the role of space (Lombardo and Preto 1993) in economic development, in particular at regional and urban level.

If there exists a reciprocal influence between territorial structure and economic and technological change, then the trajectories of spatial development will interact with those of economic and technological development. This implies that the spatial structure cannot be considered merely as a passive support of phenomena which have their explanation in other contexts (Ratti 1995). On the contrary, space assumes the role of one of the active components of development (Rallet 1988); this leads to the concept of 'active space' as a field of force (Van Geenhuizen and Ratti 1995). We will not develop this approach here, except to note that the idea of creative space, which is of course deeply rooted in territorial sciences (such as territorial and urban planning and design), together with the idea of creative time, is developed and applied, at present, only in a relatively small part of the literature on regional and urban economics.

The attempt, then, to carry out an organic treatment of a field of interest which presents such a complex plot of different aspects, led us to build the structure of this book on the basis of an imaginary grid. In such a grid, the rows are identified by three aspects of the innovative process: generation, adoption and diffusion, while the columns are characterized by different spatial levels of analysis: firm level, urban level, regional level, national and international level.

Each chapter, therefore, finds its (prevalent) place in one of the boxes of the grid. Some boxes are relatively 'crowded', while others are empty, in correspondence with the intensity of research efforts or depth of analysis concerning the different issues.

The papers in the two remaining parts of this book are organized following the main themes presented in the first section, i.e. (i) theories and models for describing and simulating the patterns and effects of the different phases of innovative processes and (ii) general lessons and suggestions for policies gleaned from empirical studies.

In Part B, dedicated to theories and models at the micro scale, Campisi et al. deal with the issue of the generation of innovation. They discuss the innovative behaviour of a firm, relating it to the extent of its R&D activity, and its market share, while Lucertini and Telmon examine the adoption aspect, using a benchmarking approach.

At the urban level, most of the papers are devoted to the study of the effects of innovation adoption on the spatial organization of cities and on the structure of mobility. In particular, Azzini et al. present a survey and a comparison among different modelling approaches. Lombardo and Occelli describe the results of experiments carried out by means of a model built to simulate the interacting processes of (i) adoption of new information technologies, (ii) location choice and (iii) mobility choice on the part of firms in an urban area. The simulations concern different scenarios deriving from urban policies and firms' strategies. Martellato presents the formulation of a model in which the spatial effects of innovation are connected with agglomeration economics.

Hewings et al., on the other hand, discuss diffusion and structural change. They derive the processes of structural change from a regional econometric input-output forecasting model for the Chicago economy. At the regional scale and focusing on diffusion aspects, Nijkamp and Poot survey how spatial interdependence in a system of regions can influence technological change and present a model in which technological change is endogenized in macroeconomic theories of growth. And on a global scale, Feldman and Kutay discuss how the effects of new technologies could be incorporated into a new locational theory of the firm, together with the configuration of the new global economy.

Part C relates mainly to the assessment of the results of empirical studies and on their implications for policies.

The paper by Haynes et al. concerns the firm and decisions about new product development. They present an empirical analysis which relates to the behaviour of individual firms bringing innovations to the market. At the same level, Van Geenhuizen and Nijkamp, from an adoption process perspective, explore the strategies which enabled individual companies to adopt new product and process technologies. Their empirical analysis is applied to the textile industry in the Netherlands.

As to the innovation generation aspect, Suarez-Villa assesses the crucial relevance of the capacity to generate inventions endogenously for regional development. In the context of adoption processes at regional level, with specific reference to innovation in the information and telecommunications sector, Capello and Nijkamp develop the concept of network externality. In order to assess the effects of telecommunication network externalities on the performance of firms and regions, they formulate a conceptual model and present the results of a comparative empirical analysis carried out in the north and the south of Italy.

Advanced communication systems play an important role in any business environment. The determinants of the firm's adoption of such systems based on micro data are investigated by Heli Koski, who also gives the results of an empirical application to the Finnish metal working industry.

The potential impacts of new information and telecommunications systems applied to transport are analyzed by Nijkamp et al., who investigate the user side of these technologies by exploring the interaction between human behaviour and transport telematics. The concepts are illustrated by case studies concerning two different information systems: one relating to bus passengers in Southampton, the other relating to motorway drivers in a Dutch region.

The last two papers concern the supraregional level. Geerlings et al. analyze innovative behaviour in the context of environmental technology, relating to the problems caused by the transport sector. They pay special attention to the role of the governments in this context and, by way of example, discuss the Clean Air Act of the State of California.

Finally, Funck and Kowalski survey innovation diffusion and R&D activities in Central and Eastern Europe and draw policy recommendations for countries in transition.

References

Alderman N (1990) Methodology Issues in the Development of Predictive Models of Innovation Diffusion, Technological Change in Spatial Context. In: Ciciotti E, Alderman N and A Thwaites (Eds). Springer Verlag Berlin 148-166

Banks R B (1993) Growth and Diffusion Phenomena. Springer Verlag Berlin

Bertuglia C S, Fischer M M and G Preto (Eds) (1995) Technological Change, Economic Development and Space. Springer-Verlag Berlin

Camagni R (Ed) (1991) Innovation Networks: Spatial Perspectives. Belhaven Press London

Ciciotti E, Alderman N and A Thwaites (Eds) (1990) Technological Change in a Spatial Context. Springer-Verlag Berlin

Damman M M G J L (1994) Individual and Spatial Success Factors in Innovative Firms. Master's Thesis. Department of Economics Free University Amsterdam

Davelaar E J (1989) Incubation and Innovation. PhD Thesis. Free University Amsterdam

Forrester J W (1969) Urban Dynamics. MIT Press Cambridge Massachusetts

Haag G and S Lombardo (1991) Information Technologies and Spatial Organization of Urban Systems: a Dynamic Model. In: Ebeling W, Peschel M and W Weidlich (Eds). Selforganization in Complex Systems. Springer-Verlag Berlin 29-36

Kamann D J and D Strijker (1991) The Network Approach: Concepts and Application, Innovation Networks: Spatial Perspectives. In: Camagni R (Ed). Belhaven Press London 145-174

Kleinknecht A (1994) Heeft Nederland een Loongolf Nodig? Inaugural Address. Free University Amsterdam

Lombardo S (1996) Assessing Uncertainty in Urban Systems. Submitted for publication in Environment and Planning A

Lombardo S and G Preto (Eds) (1993) Innovazione e Trasformazioni delle Città. Teorie, Metodi e Programmi per il Mutamento. Angeli Milan

Masser I and H J Onsrud (Eds) (1993) Diffusion and Use of Geographic Information Technologies. Kluwer Dordrecht

Mouwen A and P Nijkamp (1992) Technology, Innovation and Dynamics of Urban Systems. Indian Journal of Regional Science. 24 1:21-41

Nijkamp P (Ed) (1990) Sustainability of Urban Systems. Gower Aldershot UK

Nijkamp P and J Poot (1993) Technological Progress and Spatial Dynamics. In: Kohno H and P Nijkamp (Eds) Potentials and Bottlenecks in Spatial Economic Development. Springer-Verlag Berlin 196-223

Nijkamp P and A Reggiani (Eds) (1994) Nonlinear Evolution of Spatial Economic Systems. Springer-Verlag Berlin

Ouwersloot H (1994) Information and Communication from an Economic Perspective. PhD Thesis. Free University Amsterdam

Porter M E (1990) The Competitive Advantage of Nations. Free Press New York

Pred A R (1977) City-Systems in Advanced Economics. Hutchinson London

Rallet A (1988) La Région et L'Analyse Économique Contemporaine. Revue d'Economie Régionale et Urbaine. 3:15-23

Ratti R (1995) Lo Spazio Attivo: Una Risposta Paradigmatica dei Regionalisti al Dibattito Locale-globale. Paper presented at XVI Conference of the Italian Regional Science Association. Siena Italy

Rogers E M (1983) Diffusion of Innovations. Free Press New York

Schumpeter J A (1934) The Theory of Economic Development. Harvard University Press Cambridge Massachusetts

Segal N, Quince R and W Wicksteed (1985) The Cambridge Phenomenon. Cambridge University Press Cambridge Massachusetts

Suarez-Villa L and M M Fischer (1995) Technology, Organization and Export-driver Research and Development in Austria's Electronics Industry. Regional Studies. 29 1:19-42

Suarez-Villa L and S A Hasnath (1993) The Effect of Infrastructure on Invention, Technological Forecasting and Social Change. 44 3:333-358

Townroe P (1990) Regional Development Potentials and Innovation Capacities. In: Ewers H J and J Allesch (Eds) Innovation and Regional Development. De Gruyter Berlin 48-63

Van Geenhuizen M S (1993) A Longitudinal Analysis of the Growth of Firms. PhD Thesis Erasmus University Rotterdam

Van Geenhuizen M S and R Ratti (1995) New Network Use and Technologies in Freight Transport. An Active Space Approach. Paper presented at the ESF/FC Euroconference on European Transport and Communication Networks. Espinho Portugal

2 Spatial Dispersion of Technological Innovation: A Review

Evert Jan Davelaar and Peter Nijkamp

2.1 Introduction

In recent years there has been growing interest in the dynamics of existing firms and the formation of new firms. World-wide economic stagnation has called for a thorough analysis of the conditions that favour the offspring of new economic activities (see also Cuadrado-Roura et al. 1994 and Suarez-Villa and Cuadrado-Roura 1994). In this context, much emphasis has been placed on the growth potential of the high-technology industry. Although this sector, through the diffusion and widespread application of high-technology products, may indirectly account for a large share in total employment change, the direct employment base of this sector is relatively small (ranging from 3 to 13 percent of total national employment). Consequently, the broader process of technological progress, the diffusion of technological innovations and the birth of new firms deserves much more attention than the high-technology sector per se.

It is also worth noting that excessive attention to the high-technology sector carries the risk of neglecting the growth potential of other new - sometimes small-scale - activities outside the high-technology sector which might also significantly contribute to a further economic growth.

The geographical location of new firms (both inside and outside the high-technology sector) does not have a uniform pattern. Both concentration and dispersion may occur simultaneously. High-technology industries (such as micro-electronics and computer-based firms) can be found in large geographical concentrations like California's Silicon Valley or Massachusetts' Route 128. Other non directly high-technology based new activities are often found in older districts of larger cities, as these provide seedbeds for the emergence of new - often semi-informal - firms in a dynamic and often uncertain development climate. Consequently, a more thorough analysis of the geographical dimensions of industrial dynamics is warranted (cf. Capello 1994).

In general, innovation is considered as one of the key elements in the dynamics of the industry. Especially in recent years a wide variety of studies have pointed out the close links between (often basic) innovations and the long-term performance of an economy. In this context, structural economic changes and long-run fluctuations (e.g., Kondratieff cycles) have to be given particular mention (see, among others, Bianchi et al. 1985; Davelaar 1992; Kleinknecht 1985; Mensch 1979). Such dynamic processes have a clear impact on regional and urban economic development, as the latter phenomenon is co-determined by the production, diffusion and adoption of innovations (see Nijkamp 1986).

Various investigations have taken place on the employment effects of industrial innovations. If one takes a microscopic view on the individual firm level, it is indeed in many cases difficult to assert the existence of a clearly positive relationship (see Gunning et al. 1986). On the other hand, if one takes a macroscopic viewpoint, it is no doubt true that innovations increase the competitive position of a country or region. It can be claimed that without innovations the country or region would be worse off (see OECD 1984). Industrial innovation can therefore be considered at worst inevitable and at best beneficial. This dilemma is closely related to the well-known technology-push demand-pull discussion in industrial and spatial dynamics.

In this context, it is useful to make a distinction between the *production and the diffusion* of innovative activities (see also Stoneman 1983). The production of innovations has an impact on the competitive position of a region, while the diffusion influences the spatial distribution of the performance or potential of a spatial system. Thus both the production and the diffusion of innovations are complementary processes in the development path of regions.

In recent years, the potential and the success of local or regional initiatives in encouraging innovative firm behaviour has led to increased interest in the so-called urban incubator hypothesis. The main focus of the present paper will be on this hypothesis, which states that urban centres provide the seedbed for the emergence of new - often small-scale - firms. It suggests that older districts in such cities provide the breeding place for the creation of new activities due to their less structured, more flexible and often informal economy (see Hoover and Vernon 1959; Vernon 1960). Later on this hypothesis was broadened, and it was assumed that large-scale agglomerations provided favourable conditions for the start of a range of entrepreneurial activities (Davelaar 1992; Fagg 1980; Davelaar and Nijkamp 1986). More recently, it has also been hypothesized, that urban areas specifically provide the seedbed for new innovative firms.

In the literature it is often stated that certain cities or even parts of cities are particularly fertile environments for the creation of new firms. The term *incubation milieu* is often used to describe such areas (cf. De Ruijter 1983; Steed 1976; Fagg 1980; etc.). This concept can also be related to the *innovation potential* of cities (cf. De Ruijter 1978; Nijkamp 1985). The essence of the innovation incubation hypothesis is that certain cities are especially favourable for the production and adoption of innovations. It is useful at this point to make a clear distinction between the use of this hypothesis for the *production* and the *diffusion* process of innovations. A fertile production environment of innovations does not necessarily imply a fertile adoption environment. The central question in this context is: Which regions are especially favourable for the production (adoption) of innovations? The problem one encounters in trying to answer this question is then: *when* is a region *especially* fertile?

This paper is organized as follows. After a general discussion on the location and formation or new activities (Section 2.2), a concise presentation is given of the main aspects of the innovation production process (Section 2.3) and of the innovation diffusion process (Section 2.4) in the framework of the incubator hypothesis. This concept is then linked to agglomeration economies and social overhead capital, and followed by a discussion of impact assessment of incubation phenomena.

2.2 Formation and Location of New Activities

Firms seek to locate in a certain place for a wide variety of complex reasons. The needs and expectations of a firm (the 'demand profile') on the one hand, and the potential and quality of a certain area (the 'supply profile') on the other jointly determine the locational pattern of new economic activities. From a US study it has become evident that the factors considered important by high-technology and non high-technology firms are very similar, although apparently weighted differently (see Office of Technology Assessment 1984).

The geographical pattern of new economic activities can be analyzed from two different viewpoints, viz. the *macro-economic oriented regional growth theory* and the *micro-oriented industrial location theory*. The following examples of conventional regional *macro* growth theory can be mentioned:

- **export base theory**: this theory presupposes that regional growth rates are dependent on the region's interregional or international export performance. Multiplier effects are thus decisive for the growth potential of a region. In this framework, the main locational question is therefore the selection of appropriate new activities in a given region.
- **factor price equalization theory**: this theory (based on international trade theory) takes for granted that mobility of production factors depends on their regional return. Spatial equilibrium is hence determined by a combination of factor substitution and spatial substitution. Consequently, technological efficiency is decisive for the locational pattern of new activities.
- **unbalanced growth theory**: this theory focuses on spatial divergence as a mechanism of economic growth. Regional stress factors, cumulative causation, and backwash and/or spread effects determine the locational picture of a country, and thus the type of new activities determines the spatial development pattern.
- **growth pole theory**: this theory is based on the notion of centrifugal polarization effects from a central region outwards. The success of a growth pole strategy thus depends on the degree of propulsiveness of new activities (including both the generation and diffusion of promising growth effects).
- **product life cycle theory**: this theory recognizes that economic activities have different locational requirements at different stages in their development (innovations need a different seedbed from mass produced goods, for instance). Consequently, the locational profile of a firm is determined by the type of new activity at some place (including the development stage).

- **diffusion theory**: this theory states that it is the speed and spatial trajectory of new (innovative) activities which determine the performance of entrepreneurial initiatives.

Various alternative approaches can be identified, such as the adoption-adaption model, the hierarchical diffusion model, the interindustry diffusion model and the institutional-management diffusion model. In all these cases, attention is focused on new activities that have maximum performance in terms of diffusion potential.

Examples of *micro*-based industrial location theory cannot easily be classified according to major dynamic growth forces. However, it is possible to make a distinction according to the main determinants of individual location decisions of firms, such as labour costs, transportation and communication costs, capital costs, availability or land, market access, access to information and research centres, quality of life and amenities, taxes, access to venture capital and so forth. This micro-oriented approach has gained much popularity in recent years.

The theoretical contributions to the study of the location of new activities mentioned above are fairly general in nature and do not offer any specific insight into the urban-regional dimensions of new activities (either inside or outside the high- technology sector). Therefore, it is understandable that in recent years a search for more satisfactory explanatory paradigms has started. One of the main paradigms investigated has been the incubation hypothesis. In order to provide a proper context for discussion of this hypothesis, we will briefly discuss the locational dimensions of innovation production and innovation diffusion.

2.3 Spatial Aspects of Innovation Production

In this paper we define as *local* determinants of the production of innovations those factors which are *external* to the firm, which are expected to have an important influence on its *production* decisions and which are characterized (among other things) by a specific *spatial* orientation. It is evident that such firm-specific factors have a more or less heterogeneous pattern in space. We will ignore here factors which are internal to the firm (e.g. if a new permanent communication structure or interaction between various functional divisions in a firm has an important influence on the production of innovations, we consider this as a factor internal to the firm). So the question to be treated here is: what are the major locational determinants of the production process of innovations? Or otherwise: why are certain regions particularly fertile concerning the production of innovations? As empirical evidence is scarce, our discussion on this issue will mainly be theoretical in nature.

Now the most intriguing question is which external factors can be considered locational determinants of the innovations produced?

In our key factor analysis we will identify four clusters or major driving forces that - according to an extensive literature survey - may be assumed to exert a significant impact on the production or innovations in a certain place, i.e. (l) the composition and spatial size distribution of sectors, (2) the demography and population structure or an area, (3) the information infrastructure, and (4) the physical and institutional infrastructure. These four clusters will now briefly be discussed.

1 Composition and spatial size distribution of economic sectors

In the first place, the most straightforward factor in this context is of course the *number and type of firms* belonging to sector s in region r. The regional sectoral composition is clearly of decisive importance, so in testing the fertility of certain regions concerning the production of innovations we have to be aware of the spatial distribution of the various sectors. The mere occurrence of a large number of innovations produced in a region may to a large extent be explained by a relatively high concentration of innovative sectors within this region (cf. Andersson and Johansson, 1984).

Next, the spatial concentration of *similar* firms may play a role, as this may result in lower costs (a joint sharing of certain overhead costs, for example), but also in a higher awareness of the actions of the competitors or a greater need to innovate in order to attain a share of the market (cf. Mouwen 1984; Hansen 1980). As a result of these incentives to innovate, more innovations may be produced in regions with a higher concentration of similar firms.

On the other hand, the spatial clustering of *dissimilar* firms may also have a positive influence on the production of innovations because of a diversity of buyers and suppliers close at hand (cf. Thwaites 1981; Carlino 1978; Hoover and Vernon 1959; Vernon 1960). As innovations often have to be modified during the 'introduction phase', the 'producer' of the innovation frequently has to change his product more or less drastically. A higher diversity of suppliers may reduce the risks involved in this process of change and consequently increase the number of innovations being produced. In this context one can also consider the spatial clustering of R&D departments of *different* firms and public R&D institutions (e.g. Silicon Valley) and the resulting positive effects (cross-fertilization of ideas, good opportunities to start new innovative firms because of subcontracting, spin-off effects and so on; cf. Aydalot 1985 and Stohr 1985). It should be added however, that according to some authors there may also be certain limits with regard to spatial scale advantages (cf. Mouwen, 1984, Nijkamp and Schubert 1983; Camagni and Diappi 1985). The implication of this argument is that certain *diseconomies* may come to the fore when the spatial concentration of firms (and population) becomes too large with respect to the existing capacity of the area. With regard to the important centres of R&D just mentioned, Malecki (1979c) has indeed found some evidence concerning these hypotheses (implying deconcentration of R&D divisions of firms away from the largest urban areas).

Furthermore, the *spatial distribution of the various size categories of firms* may be relevant for the spatial distribution of innovations produced. It is important to note in this context that large firms seem to spend more on R&D (as a percentage of sales) than small firms (cf. Cappellin 1983; Hoogteijling 1984; Malecki 1979a; Dasgupta 1982). Large firms also produce the highest number of innovations (cf. Kok et al. 1985; Thomas 1981). Small firms, however, seem to have a higher productivity concerning the production of innovations (cf. Cappellin 1983; Hoogteijling 1984 and Malecki 1979a). Cappellin (1983, p. 464) states: "It seems, moreover, that the productivity of R&D expenditure is highest for small firms in terms of patent or output growth". It is interesting to observe in this context that Rothwell and Zegveld (1982) claim that small and large firms are complementary. This is best illustrated by means of the following citation: "Thus we see a certain complementary interaction between the large and the small, the nature of the relationship being based on their relative strengths (e.g. the small firm's entrepreneurship; the large firm's access to resources)" (p.247). In their opinion there exists an 'optimal mix' between the number of small and large firms. At first sight, we are tempted to infer that large firms may be more favourable to the production of innovations than small firms. But, if the argument of Rothwell and Zegveld holds true, the 'mix effect' may be more important. In that case, we may state that the spatial size distribution is of considerable importance.

In the current literature on regions a considerable amount of attention is being paid to multi-locational firms (cf. Pred 1977). The spread of *multi-plant firms* is expected to influence the spatial production of innovations. So the location of the various types of R&D may be determined by the spatial distribution of the various 'components' of these firms. Basic research could be attracted to the head offices, and development and 'applied' research to the of branch plants (cf. Howells 1984; Thwaites 1981).

2 Demography and population structure

An important factor often mentioned in the literature is the total number of people living in a region (or city). Malecki (1979, p. 226), for example, remarks: 'There is an increasing relationship between R&D activity and city size, suggesting increasing economies of urban size for R&D". In this context one should be aware of the fact that R&D can be considered as an input to the production process of innovations. The relationship between R&D input and output (innovations) is certainly not independent (cf. Mansfield 1968).

Often a particular subgroup of the total population is considered to be especially important: such as the number of technical, managerial or R&D personnel (cf. Andersson and Johansson 1984; Mouwen and Nijkamp 1985; De Jong and Lambooy 1985; Johansson and Nijkamp 1986; Bushwell et al. 1983; Malecki 1979b; Oakey 1983; Thwaites 1981). As R&D personnel can be considered a labour input in the production process of innovations, it is logical to include this variable in group (2).

From a *demographic* viewpoint, the existence of minority groups in a region is sometimes assumed to influence the production of innovations positively (cf. De Ruijter 1983; Pred 1977). This can be 'explained' by the fact that such groups have nothing to lose, so that they have a lower risk aversion and a higher propensity to try something new. It may also be the case that those people start to introduce products or services which are familiar in their native country but not in their new home country. So the influence of a segmented population composition in a region on the production of innovations is expected to be positive.

Another important key factor is the influence of *agglomeration size* on the production of innovations. Malecki and Varaiya (1986, p. 7) for example remark: "Agglomeration economies play a multiple role: they promote technical progress and higher productivity." It is convenient to distinguish these economies into *localization* and *urbanization economies* (cf. Carlino 1978; Hansen 1980). By localization economies we mean advantages in a system that result from the spatial concentration of firms in the same sector, while urbanization economies refer to advantages which result from the spatial concentration of firms in different sectors. To a certain extent these advantages are related to the 'locational determinants' discussed above (for example, information exchange). In Biehl et al. (1986) it is stated that agglomeration economies refer to 'external economies' which are related to *size* and *concentration*.

3 Information infrastructure

Information availability is generally regarded as a major determinant of innovation. The term itself incorporates several diverse components of knowledge transfer. Various components may be distinguished by discriminating between the different 'senders' of the information (the 'receivers' being in each case the innovation producing firms). We will distinguish three different sources of information.

Inter-firm contact patterns deal with a mutual private information exchange between firms. As Pred (1977) has already noted, every exchange of goods and services between firms is accompanied by information exchange. Thus the flow of goods and services between regions can be seen as a 'proxy' for this kind of information exchange. In this way Pred shows that these kinds of information flows are highly spatially biased (i.e. urban based) (see also Norton 1979; Andersson and Johansson 1984). The production of innovations may be stimulated by means of this kind of information exchange, mainly because firms in the centre of these information flows are aware of market opportunities, market imperfections, and so on (cf. Nijkamp 1982; Pred 1977). So we may expect that a higher intensity of goods and services exchange results in a more intensive information exchange and, consequently, in more innovations being produced.

Public research institutes, universities, institutes of technology and knowledge transfer centres are expected to influence the innovation potential of the regions in which they are located in a positive way (cf. Malecki 1979b; Gibbs and Thwaites 1985; De Jong and Lambooy 1985; Feldman 1984; Mouwen 1985). Firms may consult these organizations if they have technical and marketing problems. Also the spin-off implications need to be mentioned in this context (cf. Van Tilburg and Van der Meer 1983; Rothwell and Zegveld 1982). Oakey (1983) even remarks: "However, the importance of local American universities may be in their role of providers of 'spin-off' entrepreneurs and skilled workers rather than in terms of interactive collaboration with existing firms' (see p. 244). People working in research institutions may also begin the production of a new commodity by establishing a new innovative firm. Often this firm will be located near the parent institution because of subcontracting with this institution, or because a founder locates his new firm near his place of residence (cf. Gudgin 1978; Aydalot 1985). It is thus expected that a higher concentration of such institutions will influence the innovation production potential of a region in a positive way.

Demographic and spatial interaction patterns exert in general a positive influence on the innovation potential of a region (cf. Vernon 1960; Pred 1977). For example, an intensive flow of customers to a certain region (or city) means that firms located in this region have the advantage of a permanent and intensive information exchange with their customers. In this way they can keep abreast of certain developments in consumer tastes and may adjust their innovation strategy accordingly. So a higher intensity of personal interaction is likely to have a positive influence on the total number of innovations produced. In this context we may also mention the importance of 'face to face' contacts for innovation production (cf. Lambooy 1984; Pred 1977; Batten 1981 Kok et al. 1985; Moss 1985; Nijkamp and Schubert 1983). Clearly, with regard to this communication factor, metropolitan regions are normally in a favourable position.

4 Physical and institutional infrastructure

Cultural and educational *amenities* (theatres, cinemas, libraries, art galleries and so on) may influence the production of innovation in a positive way. This can be explained as follows. Highly educated R&D personnel can be considered a scarce input in relation to innovation (cf. Malecki and Varaiya 1986), so R&D institutions of private firms cannot be considered as footloose. In order to attract adequate research personnel they often have to locate where their personnel wants to live in order to attract adequate research personnel (cf. Feldman 1984, Bushwell et al. 1983). Those people seem to be attracted to cities with many cultural and educational amenities (cf. Aydalot 1984; Cappellin 1983; Malecki and Varaiya 1986; Howells 1984). Malecki and Varaiya (1986) remark: "These workers favour attractive urban regions where cultural, educational and alternative employment opportunities are abundant". The locational preferences of R&D personnel may have a significant influence on the location of R&D activity and in this way on the production of innovations in amenity-rich regions.

Physical climate and environmental qualities may also influence the production of innovations in a positive way by means of the mechanism outlined above: the attraction of highly skilled R&D personnel. Malecki and Varaiya (1986) explain the rise of certain Sunbelt states in the U.S. by means of such a mechanism. Also in France some evidence concerning this hypothesis can be found (cf. Aydalot 1984).

Moreover, it is also worth noting that the *availability of public (physical) infrastructure* is sometimes considered a necessity to the production of innovations. "In conclusion, the availability or a satisfactory infrastructure capital stock (in its broadest sense) shapes the necessary conditions for innovative capacities in an area" (Nijkamp 1982, p. 6). In this context Feldman (1984) points to the need of many firms in bio-technology to be located in the vicinity of an airport.

The financing of innovative products or services is often problematic (cf. Feldman 1984). Therefore, the spatial pattern of institutions offering *venture capital* can be considered an important 'explanatory' variable to the regional distribution of the innovations produced (cf. Bushwell et al. 1983; Thwaites 1981; Lambooy 1978; Stöhr 1985; Oakey 1983; Mouwen 1984; Rothwell and Zegveld

1982). Stöhr for example emphasizes the role of venture capital in his discussion on the Mondragon project in Spain.

A last factor in this context concerns *institutional arrangements* (cf. Aydalot 1984; Rothwell and Zegveld 1982; Brown 1981). It is possible to identify a number of regulations which may stimulate or demotivate the production of innovations. There has been, however, a lack of serious investigation into the precise effects of such institutional measures.

In conclusion, one can derive from the literature many factors which many influence the total number of innovations produced. The empirical evidence concerning the individual relevance of each factor, however, is rather scarce. The overall impacts are summarized in Figure 2.1. Clearly, the explanatory variables may be interdependent. In this scheme we have not drawn all these links, but in the following section more attention will be paid to this issue in relation to the diffusion of innovation.

In Figure 2.1 the dependent variable is the total number of innovations produced in a region. So in comparing and explaining the innovativeness of different regions we should compare the different intensities ('values') of each of these explanatory (locational) factors. Thus, region A may produce more innovations than region B simply because region A has ceteris paribus a larger number of innovative firms within its boundaries. If one is not interested in this scale effect, one has to compare the total number of innovations produced per firm (in a certain sector) in each region.

2.4 Diffusion of Innovations in a Spatial Context

Both the theoretical and empirical literature on innovation diffusion is quite extensive. In this section we will concentrate in particular on the diffusion of technical-economic innovations. As stated before, the diffusion process may be more important than the pure effects of production of innovations. Consequently, a region or country may even attain a strong position in the world market just by importing innovations that have been produced elsewhere (cf. Rothwell and Zegveld 1985). This does not imply, however, that the production of innovations is irrelevant. In the first place, firms located in the producing region may have easier access to the innovations produced. Secondly, exporting innovations may be quite profitable.

The question to be raised in this section is: What are the locational determinants of the innovation diffusion process? Or stated in another way: Why are certain regions so attractive regarding the adoption or innovations? In the following we will again focus attention on the external determinants of the diffusion of innovations. In some cases a few remarks in relation to the explanatory factors will be sufficient, because they have already been discussed in the previous section. It will be useful to distinguish between supply and demand side factors (cf. Brown 1981).

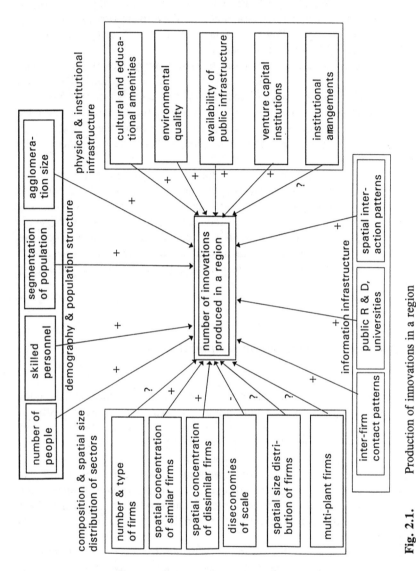

Fig. 2.1. Production of innovations in a region

We will start with some *supply side factors*. Brown (1981) stated that although innovation diffusion research has tended to concentrate on demand side factors (cf. the adoption perspective of Hagerstrand 1967), supply side factors may be (equally) important. The essence of the supply side approach is that the strategy pursued by a supplier of an innovation may have important locational implications. This strategy may of course in part be determined by factors internal to the firm (profit rate, market perception and so on), but these factors will not be discussed here, as they are to a large extent be firm-specific. We will only concentrate on those locational factors which may influence the choices made by firms concerning the distribution of produced innovations. The literature concerning the supply side factors is rather scarce, as only recently has attention been paid to this issue.

Three main classes of determinants will be discussed here, viz. (1) sectoral composition, (2) information network, and (3) agglomeration economies.

1 Sectoral composition

To a certain extent we can expect that the *total number of (innovative) firms in a region* will influence the supply of the various innovations positively. *The role of multi-locational firms* should also be stressed in this context (cf. Pred 1977). In fact, innovations are sometimes only diffused *within* a (multi-plant) firm. So in this case the spatial organization of the multi-locational firm will determine the spatial supply pattern of innovations.

2 Information network

The *spatial concentration of information flows* is a second factor which needs to be mentioned in this context. It is noteworthy that information between (dis-)similar firms may result in imitative behaviour. When a firm recognizes that a competitor offers a certain innovation, it may 'follow the leader' (cf. Brown 1981; Nijkamp and Schubert 1983). An adequate communication infrastructure may reduce the costs of supplying the innovation or it may even occur that the availability of certain infrastructural components is a necessity for offering the innovation (cf. Brown 1981).

3 Agglomeration economies

Agglomeration economies may stimulate the number of innovations supplied since the supplier can expect a higher demand potential and fewer risks in improving the innovation (because of the diversity of input supplies and an intensive contact with potential customers) (cf. Andersson and Johansson 1984; Mouwen 1984; Heinemeyer 1978). The introduction of a new product or service often requires a minimum 'threshold' concerning the number of firms and/or the size of population (cf. Koerhuis and Cnossen 1982; Andersson and Johansson 1984; Lambooy 1978).

Next, we will discuss *demand side factors*. Four different categories will be considered here: (1) sectoral structure, (2) communication network, (3) agglomeration advantages, and (4) physical infrastructure.

1 Sectoral structure

The *number and size of firms* will in general exert a positive influence upon the number of innovations demanded in a region (cf. Cappellin 1983; Howells 1984). As the influence of this variable is again straightforward, it will not be discussed any further in this section.

The *spatial distribution of the various size categories of firms* is also expected to influence the innovation demand potential of a region (cf. Davies 1979). In Davies' model, for example, the most significant explanatory variable of innovation adoption is the size of the firm. Large firms may be in a favourable position concerning the adoption of innovations because of larger (internal) resources.

In regional research there is a tendency to stress the role of *multi-locational organizations* in explaining regional economic phenomena (cf. Holland 1979). This phenomenon can also be observed in the study of innovation diffusion. Pred (1977); Malecki (1979a) for example stress the role of multi-plant firms, although it is very difficult to identify this factor. One can often observe a reasonably integrated production and diffusion process. By this we mean that the R&D departments of these firms 'produce' certain innovations which are then diffused to the various branch plants of the firm (there is only limited information, goods and service exchange between establishments of a multi-locational firm). Although it is difficult to hypothesize the expected influence, the role of these firms in relation to innovation diffusion may be quite surprising and even cause peripheral regions to have a relatively high degree of innovation adoption (the branch plants of these firms may be located in peripheral regions simply because of relatively low wages and low congestion; cf. Thwaites 1981; Oakey 1983).

2 Communication network

Not everybody seems to be equally receptive to the *adoption* of innovations. In this context the term 'opinion leaders' is sometimes used for first adopters of an innovation (cf. Brown 1981; Malecki 1982). These opinion leaders are essential for the further diffusion of innovations, because informal communication between adopters and potential adopters seems to be essential to the adoption of the innovation among the great mass of people (cf. Rogers 1983; Pred 1977; Batten 1981; Kok et al. 1985). So the spatial distribution (and number) of 'opinion leaders' may be an essential component of the demand for innovations.

Information availability will influence the demand for innovations positively (cf. Brown 1981; Hagerstrand 1967; Nijkamp 1982; Malecki and Varaiya 1985; Batten 1981; Pred 1977). In the first place one can expect a greater awareness among potential adopters when the information availability increases. Secondly, greater information availability may increase imitative behaviour of the potential adopters (cf. Pred 1977; Hagerstrand 1967; Brown 1981). Brown (1981) even states that in accordance with the 'adoption perspective' (cf. Hagerstrand 1967) information flows are fundamental.

3 Agglomeration advantages

Some authors are of the opinion that *agglomeration economies* will stimulate the number of innovations adopted in a region (by households and firms). To a certain extent these agglomeration economies are related to the other variables mentioned in this section. These mutual relations will be studied later on in this paper. Now some more attention will be paid to the reason why agglomeration economies will influence the innovations adoption positively. In the first place one can point to the positive influence of *spatial clustering on the information exchange between firms and people* (cf. Mouwen 1984; Heinemeyer 1978; Lambooy 1978). Secondly, it may be less costly to supply the innovations because of *lower transportation and communication costs* (cf. Brown 1981). Thirdly, the number of *face to face contacts* will increase in large agglomerations. These contacts can of course be considered as a special subgroup of information exchange, but according to many authors this element is of such importance that it needs to be mentioned separately (cf. Lambooy 1973; Heinemeyer 1978; Pred 1977; Nijkamp and Schubert 1983; Norton 1979; Batten 1981). Batten (1981), for example, states that the social dimension of innovation adoption is of paramount importance.

4 Physical infrastructure

A last factor to be mentioned in this section is the *influence of physical infrastructure* (cf. Brown 1981; Mouwen 1984; Nijkamp and Schubert 1983). Brown for example states: "Thus the characteristics of the relevant public and private infrastructures - such as service, delivery, information, transportation - also have an important influence upon the rate and spatial patterning of diffusion" (p. 9). Brown also shows that some innovations (computers) can only be adopted when physical infrastructure is close at hand. The availability of physical infrastructure will of course also increase the agglomeration economies to be gained in a certain region and, consequently, the intensity of information exchange between firms and people.

In this section we have discussed several locational factors influencing the diffusion of innovations, and indicated how and why the innovation diffusion process may be spatially biased. Schematically the above arguments can be summarized by means of a key factor analysis (see Figure 2.2). In this figure, it is shown how the perception of innovation supply vis-à-vis potential demand in the various regions influences the probability that certain innovation distribution centres will be located in the region. The more such centres a region has, the more innovations will be supplied in this region. It is also shown how several demand side factors may influence the total numbers of innovations demanded in this region. The actual number of innovations adopted (over a certain period of time) will depend on the interaction between total demand and supply.

In the present paper, we are especially interested in the effects of *agglomeration economies* and *social overhead capital* on the innovation potential of firms in a region (i.e. the potential of firms to produce and adopt innovations), in the framework of the *urban incubation hypothesis*. These issues will be taken up in the next section.

30

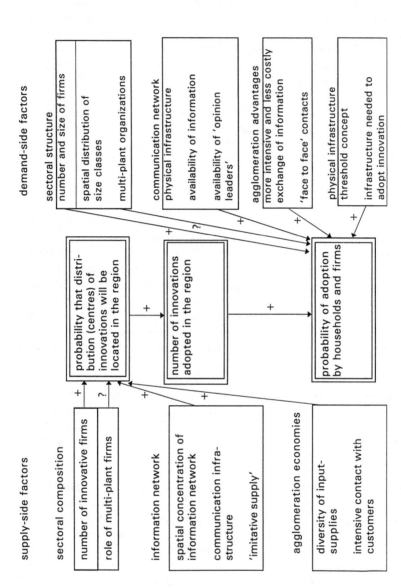

Fig. 2.2. Spatial diffusion of innovations

31

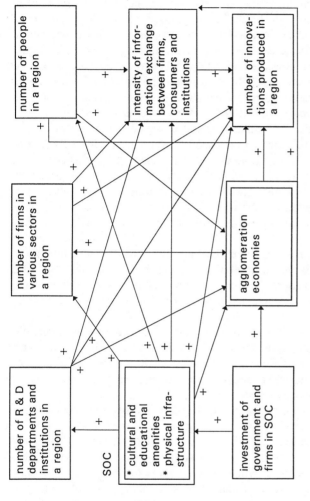

Fig. 2.3. Determinants of agglomeration economies and social overhead capital and their influence on the production of innovations

2.5 The Concepts of Agglomeration Economies and Social Overhead Capital Reconsidered

In this section we concentrate on the influence of two major determinants of the innovation potential of a region. From the foregoing it will be clear that we will again distinguish between the *production* and *diffusion* process of innovations. Figures 2.1 and 2.2 depict these processes. In the following we will discuss how the elements of agglomeration economies and social overhead capital are expected to influence (in a theoretical sense) the *production* and *diffusion* of innovations. Before turning to this issue however, we have to define more precisely the concepts of *social overhead capital* (SOC) and *agglomeration economies*.

Agglomeration economies mean that a spatial concentration of people, firms, institutions and so on, is favourable to economic development. Hirschman (1958) defines social overhead capital as follows: "SOC is usually defined as comprising those basic services without which primary, secondary, and tertiary productive activities cannot function" (p. 83). The term social overhead capital does not only refer to physical infrastructure in the sense of roads, railways ('band' infrastructure according to Biehl et al. 1986), but also includes educational and cultural amenities.

Agglomeration economies on the other hand are determined by two dimensions, i.e. the *number* of people and firms in a region and the *distance* between them. Thus, various locational factors sketched in Figure 2.1 are related to this concept (for example, the number of people and firms in a region, physical infrastructure, and so on).

In the following figures we will indicate how an increase in agglomeration economies or social overhead capital may affect the total number of innovations produced in a region. In these figures it will be assumed, for ease of presentation, that there is no limit (bottleneck) to agglomeration economies (thus assuming away for the moment the existence of diseconomies).

In Figure 2.3 we have supposed that the agglomeration economies in a region are independent of the characteristics of other regions. The important point in this context is that agglomeration economies do not refer to the mere *existence* of firms, but to the positive effects to be gained from *spatial clustering* of these firms. Although it is difficult to define the term *spatial cluster*, we will assume that the distances between industrial concentrations for each pair of regions are large enough to allow each region to possess agglomeration economies (the only exception being the case in which a certain industrial concentration crosses several boundaries). In the following we will assume, however, that the size and shape of the regions concerned (no agglomerations crossing regional boundaries) have been composed in such a way that any increase in the number of firms (or people) - or decrease in the distances between them - can be classified as agglomeration economies in only one region. Figure 2.3 shows that the term agglomeration economies comprises several factors concerning the location of innovation production. Up to a certain point (beyond which diseconomies might start), we expect the influence of agglomeration economies on the production of innovations to be positive. The same can be said concerning the influence of social overhead capital, which is to a large extent an instrument of government policy. In the first, place the effect of physical infrastructure on agglomeration economies will be positive (by reducing the average distance between producers and consumers). Secondly, the effect of cultural and educational amenities (both being part of social overhead capital) on the production innovations in a region is expected to be positive, because this will increase the attractiveness of the region for highly educated R&D personnel (and as a consequence attract R&D departments and institutions).

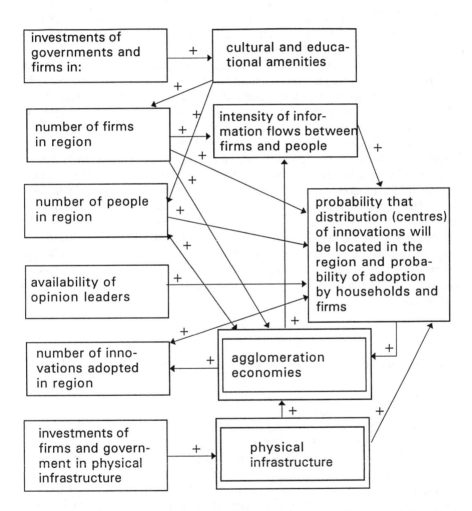

Fig. 2.4. The effects (and determinants of) agglomeration economies and social overhead capital on innovation diffusion

We will now turn to the expected influence of agglomeration economies and social overhead capital on the adoption (diffusion) process of innovations in a region. Figure 2.4 shows which factors of Figure 2.2 determine the extent of agglomeration economies and social overhead capital. This figure indicates how agglomeration economies and social overhead capital may influence the diffusion of innovations. As in Figure 2.3 we can see that the two concepts are not independent. An increase in physical infrastructure will diminish the relative distances between firms and consumers and consequently result in increasing agglomeration economies.

A certain part of these agglomeration economies will consist of intensified information flows in the specific region. It may also be the case that these flows will increase between regions because of physical infrastructure investments. As we have already seen before in Figure 2.2, both concepts are (in a theoretical sense) expected to influence both supply and demand of innovation positively, so that more innovations will be adopted in the region.

In conclusion, the influence of both agglomeration economies and social overhead capital on both production and diffusion of innovations is expected to be positive. It is however quite a difficult task to test such hypotheses empirically. A major problem is of course to collect relevant data, since both concepts are difficult to quantify. With regard to *agglomeration economies* one possibility is to use relevant proxies (cf. Kawashima 1974; Carlino 1978; Hansen 1983). With regard to *social overhead capital* it may be possible to use a compound score as suggested in the Biehl report (1986). Another possibility would of course be to use the data of inquiries concerning the innovative behaviour of firms (cf. Kleinknecht 1984; Hoogteijling 1984) and to assess the influence of these stimuli. An important problem, however, is that firms/entrepreneurs are not explicitly aware of the influence of these factors. The mere *awareness* of an innovation by a certain firm may be due to the fact that it is located in a region (city) with very intensive information flows. Yet it may occur that the entrepreneur will not unambiguously attribute his adoption of the innovation to his location in an information-rich environment.

2.6 Assessment of the Production and Diffusion of Innovations

1 Production of innovations

In most countries data concerning the production of innovations are inadequate. Usually, there is no registration or monitoring of the number and quality of the innovations produced. This implies that one has to resort to indirect indicators such as the following:

- **Output data**. These data may refer to the rise in sales and/or productivity of a certain firm. The main problem here is of course to identify whether such changes in economic performance can be ascribed to innovations or to general labour or capital productivity rises, market changes, etc.
- **Licence or patent data**. The use of these indicators has the disadvantage that not all firms will apply for licences or patents because this often does not prevent imitative behaviour by rivals. Nor does licence or patent data usually

give any information about the quality of the innovation. In addition, patent statistics are biased as larger companies are usually over-represented. One can make a distinction between important and less important innovations, the important innovations being those for which the producer will apply for licences or patents more than once.

- **The volume of inputs** being used in the production process of innovations. In this case one may think of the percentage of non-manual workers in the total workforce of a region or firm (cf. Oakey 1983), R&D as a percentage of total sales of a firm, the number of R&D personnel working in a firm or region, or related measures. The idea behind this approach is essentially based on a production function of innovations which relates output (number and quality of innovations) to certain inputs (R&D personnel, R&D costs, and so on). Although the existence of a relationship between R&D inputs and outputs cannot be denied (cf. Mansfield 1968), it is not at all clear how such a production function should be specified. One may even question the production function approach in this context (cf. Nelson and Winter 1982).
- **The number of awards** regarding innovations that have been produced. The fact that not all innovations will be rewarded is, however, is an important disadvantage of this approach.
- Design of firm-oriented **inquiries** (cf. Hoogteijling 1984; Kleinknecht 1984). One can try to establish whether a certain innovation in a firm has been produced within a given firm or has been adopted from outside. An important weakness of this approach, however, is the non-uniformity of the interpretation of the innovation concept from the side of the entrepreneurs.

Given the above complications it is clear that the measurement of the production of innovations can be considered a difficult task. One can only try to use proxies which map out the production of innovations as reliably as possible.

2 Diffusion of innovations

In regard to the diffusion process one should be aware of the fact that diffusion is a dynamic phenomenon. In general however, it is only possible to measure the degree Oof diffusion (of a certain innovation) at *a certain moment in time*.

When the objective is to try to study the whole diffusion path through time, this will require a high standard of reliability and completeness in the innovation diffusion data (cf. Hagerstrand 1967). This will be possible for a limited number of innovations however: for example, when the 'seller' of the innovation has complete records concerning the buyers of the innovation (cf. Pred 1977; Hagerstrand 1967).

As already noted, complete data will only be available for a very limited number of innovations, and as a consequence one has to rely on an 'approximation' method, for example, by means of inquiries (cf. Hoogteijling 1984; Kleinknecht 1984) or licence data. These approximations have the same disadvantages already discussed under the heading of the production process of innovations.

2.7 Influence of Agglomeration Economies and Social Overhead Capital on the Innovation Potential of a Region: Retrospect

In this concluding section we will summarize our findings in the form of a series of concise propositions.

- Innovations are often considered to be of paramount importance for the economic well-being of a country, region or firm. Due to this (supposed) importance, innovation studies are at present enjoying much popularity. Further *behavioural* analyses of the production and adoption of innovation are no doubt warranted.
- The innovation potential of a region should be divided into the potential of a region to *produce* innovations and the potential to *adopt* innovations. To a certain extent, both processes are complementary.
- Innovations are produced and adopted in a certain spatial context, so that a central issue is the *locational determinant(s)* of innovation production and adoption in these regions.
- According to these determinants some regions may be more fertile concerning the innovation potential. In this context, one may use the term *'incubation milieu'* analogous to the incubation hypothesis of cities with regard to the generation of new firms (cf. Hoover and Vernon 1959; Jacobs 1961).
- *Agglomeration economies* and *social overhead capital* are two elements of the incubation milieu of a region. They are generally expected to have a positive influence.
- Given the importance of these factors for the innovation potential of a region, a *key factor analysis* which studies the (expected) influences of these concepts on the production and diffusion of innovations is needed.
- In general, it is plausible to assume that agglomeration economies are not transferable across regional boundaries, because the central element of this concept is the synergy to be gained from *spatial* clustering.
- With regard to social overhead capital it is meaningful to follow the *multidimensional profile* approach, as *inter alia* adopted in the Biehl report (1986). In this case, however, we should stress those elements of social overhead capital which are expected to be of special relevance to innovation production and diffusion.
- The *subdivision of agglomeration economies* into different constituents, proposed among others by Hansen (1980); Carlino (1978); Kawashima (1974), may be useful for measuring the extent of agglomeration economies to be gained in the various regions.

- It is of paramount importance to test whether the influence of agglomeration economies and social overhead capital on innovation production and diffusion is *significantly different* in the various regions of a given spatial system.

References

Andersson A E and B Johansson (1994) Knowledge Intensity and Product Cycles in Metropolitan Regions, Contributions to the Metropolitan Study: 8. IIASA. Laxenburg

Aydalot P (1984) Reversals of Spatial Trends In French Industry since 1974. In: Lambooy J G (Ed) New Spatial Dynamics and Economic Crisis. Tampere 41-61

Aydalot P (1985) Some Comments on the Location of New Firm Creation. Paper presented at the Sixth Italian Conference of the RSA Genova (mimeographed)

Batten D F (1981) On the Dynamics of Industrial Evolution. Umea Economic Studies 97 Umea

Bianchi G, Bruckmann G, Delbeke J and T Vasko (Eds) (1985) Long Waves, Depression, and Innovation. International Institute for Applied Systems Analysis Laxenburg

Biehl D et al (Ed) (1986) The Contribution of Infrastructure to Regional Development. Directorate-General Regional Policy. Commission of European Communities Brussels

Boer N A de and W F Heinemeyer (Eds) (1978) Het Grootstedelijk Milieu. Van Gorcum Assen

Brown L A (1981) Innovation Diffusion, A New Perspective. Methuen London

Bushwell R J (1983) Research and Development and Regional Development: A Review. In: Gillespie A (Ed) Technological Change and Regional Development. London 9-22

Camagni R and L Diappi (1985) Urban Growth and Decline in a Hierarchical System: A Supply-Side Dynamic Approach. Paper presented at the IIASA Workshop on Dynamics Or Metropolitan Areas Rotterdam

Capello R (1994) Spatial Economic Analysis of Telecommunication Network Externalities. Avebury Aldershot UK

Cappellin R (1983) Productivity Growth and Technological Change in a Regional Perspective. Giornale degli Economisti e Analisi Economia 459-482

Carlino G A (1978) Economies of Scale in Manufacturing Location. Martinus Nijhoff Leiden

Cuadrado-Roura J R, Nijkamp P and P Salva (Eds) (1994) Moving Frontiers. Avebury Aldershot UK

Dasgupta P (1982) The Theory of Technological Competition. London School of Economics London

Davelaar E J (1992) Regional Economic Analysis of Innovation and Incubation. Avebury Aldershot UK

Davelaar E J and P Nijkamp (1986) De Stad als Broedplaats van Nieuwe Activiteiten: Theorie en Onderzoek. Stedebouw en Volkshuisvesting 2:61-66

Gudgin G (1978) Industrial Location Processes and Regional Employment Growth. Saxon House Farnborough

Gunning J W, Hoogteijling E and P Nijkamp (1986) Spatial Dimensions of Innovation and Employment: Some Dutch Results. In: Nijkamp P (Ed) Technological Change, Employment and Spatial Dynamics. Springer Verlag Berlin

38

Hagerstrand T (1967) Innovation Diffusion as a Spatial Process. University of Chicago Press Chicago

Hansen E R (1983) Why Do Firms Locate Where They Do?, World Bank Discussion Paper Wudd 25. Washington DC

Hirschman A O (1958) The Strategy of Economic Development. Yale University Press Connecticut

Holland S (1979) Capital Versus the Regions. The MacMillan Press London

Hoogteijling E M J (1984) Innovatie en Arbeidsmarkt (ESI VU) Amsterdam

Hoover E M and R Vernon (1959) Anatomy of a Metropolis. Harvard University Press Cambridge

Howells J R L (1984) The Location of Research and Development: Some Observations and Evidence from Britain. Regional Studies 18:13-29

Jacobs J (1961) The Life and Death of Great American Cities. Vintage Books London

Jansen A C M (1981) "Inkubatie-milieu": Analyse van een Geografisch Begrip. KNAC Geografisch Tijdschrift XY 1981. 4:306-314

Johannson B and P Nijkamp (1984) Analysis of Episodes in Urban Event Histories. In Burns L and L K Klaassen (Eds) Spatial Cycles. Gower Aldershot 133-151

Jong M W de and J G Lambooy (1984) De Informatica-sector Centraal. Perspectieven voor de Amsterdamse Binnenstad. EGI University of Amsterdam

Kawashima T (1975) Urban Agglomeration Economies in Manufacturing Industries. Papers of the RSA 157-175

Kleinknecht A (1986) Crisis and Prosperity in Schumpeterian Innovation Patterns. MacMillan London

Koerhuis H and W Gnossen (1982) De Software- en Computerservice-bedrijven. Sociaal Geografische Reeks 23 Groningen

Kok J A A M, Offerman G J D and P H Pellenbarg (1985) Innovatie in het Midden- en Kleinbedrijf. In Molle WTM (Ed) Innovatie en Regio. The Hague 143-165

Lambooy J G (1973) Economic and Geonomic Space: Some Theoretical Considerations in the Case of Urban Core Symbiosis. Papers of the RSA 145-158

Lambooy J G (1978) Het Grootstedelijk Milieu. In: Boer N A de and W F Heinemeyer (1978) (op cit)

Lambooy J G (Ed) (1984) New Spatial Dynamics and Economic Crisis. Jyvaakyla

Leone R A and R Struyk (1976) The Incubator Hypothesis: Evidence from Five SMSAs. Urban Studies 13:325-331

Malecki E J (1979a) Corporate Organization of R&D and the Location of Technological Activities. Regional Studies 14:219-234

Malecki E J (1979b) Agglomeration and Intra-firm Linkage in R&D Locations in the United States. Tijdschrift voor Economische en Sociale Geografie 70 6

Malecki E J (1979c) Locational Trends in R&D by Large U.S. Corporations. Economic Geography 5:309-323

Malecki E J (1979d) Corporate Organizations of R and D and the Location of Technological Activities. Regional Studies 14:219-234

Malecki E J (1982) Technology and Regional Development: A Survey. International Regional Science Review 122-148

Malecki E J and P Varaiya (1985) Innovation and Changes in Regional Structure. In Nijkamp P O (Ed) Handbook Regional Economics. North-Holland Publishing Company Amsterdam 484-515

Mansfield E (1968) Industrial Research and Technological Innovation. Longman New York

Meer J D van der and J J Tilburg van (1983) Spin-offs uit de Nederlandse Kenniscentra. Projekt Technologiebeleid/Ministerie van Economische Zaken

Mensch G (1979) The Stalemate Technology. Ballinger Cambridge Massachusetts

Molle W T M (Ed) (1985) Innovatie en Regio. Staatsuitgevertij Den Haag

Moss M L 1985) Telecommunications and the Future of Cities. Paper presented at the Conference of Landtronics London

Mouwen A (1984) Theorie en Praktijk van Lange-termijn Stedelijke Ontwikkelingen. Discussion Paper 1984-7. Free University Amsterdam

Mouwen A (1985) Transfercentra en Stedelijke Herstructurering. Discussion Paper. Free University Amsterdam

Mouwen A and P Nijkamp (1985) Knowledge Centres as Strategic Tools in Regional Policy. Research Memorandum. Dept of Economics. Free University Amsterdam

Nelson R R and S G Winter (1982) An Evolutionary Theory of Economic Change. Harvard University Press Cambridge

Norton R D (1979) City Life Cycles and American Urban Policy. Academic Press New York

Nijkamp P (1982) Technological Change, Policy Response and Spatial Dynamics. In: Griffith D A (Ed) Evolving Geographical Structure. The Hague 269-291

Nijkamp P and U Schubert (1983) Structural Change in Urban Systems. Contributions to the Metropolitan study: 5. IIASA Laxenburg

Nijkamp P et al (1985) Innovatie en Regionale Omgeving: een Test van enkele Hypotheses. In: Molle W T M (Ed) Innovatie en Regio. Den Haag 221-223

Nijkamp P (Ed) (1986) Technological Change, Employment and Spatial Dynamics, Springer-Verlag Berlin

Oakey R P (1981) High Technology Industry and Industrial Location. Gower Aldershot

Oakey R P (1983) Innovation and Regional Growth in Small High Technology Firms: Evidence from Britain and the USA. Regional Studies. 18.3:237-251

OECD (1984) Changing Market Structures in Telecommunications. Ergers H and J Okayama (Eds) North-Holland Amsterdam

Office of Technology Assessment (1984) Technology, Innovation and Regional Economic Development. US Congress OTA-SII-238. Washington DC

Pred A (1977) City Systems in Advanced Economies. Hutchinson London

Rogers E M (1962) Diffusion of Innovations. The Free Press of Glencoe New York

Rothwell R and W Zegveld (1982) Innovation and the Small and Medium Sized Firm. Pinter London

Rothwell R and W Zegveld (1985) Reindustrialization and Technology. Longman Somerset

Ruijter P A (1983) De Bruikbaarheid van het Begrip "Incubatiemilieu". KNAC Geografisch Tijdschrift XVII 2:106-110

Steed G P F (1976) Standardizations Scale, Incubation and Inertia: Montreal and Toronto Clothing Industries. Canadian Geographer 20:298-309

Stohr W B (1985) Industrial Structural Change and Regional Development Strategies, Towards a Conceptual Framework. IIR Discussion 21. Vienna

Stoneman P (1983) The Economic Analysis of Technological Change. Oxford University Press Oxford

Suarez-Villa L and J R Cuadrado-Roura (1994) Regional Economic Integration and the Evolution of Disparities. Papers in Regional Science 4:369-387

Thomas M D (1981) Growth and Change In Innovative Manufacturing Industries and Firms. Discussion Paper. IIASA Laxenburg

Thwaites A T (1982) Some Evidence on Regional Variations in the Introduction and Differentiation of Industrial Products within British Industry. Regional Studies, vol. 16 5:357-81

Vernon R (1960) Metropolis 1985. Harvard University Press Cambridge

3 Technological Innovation and Diffusion Models: A Review

Amnon Frenkel and Daniel Shefer

O31
O32
R12
R32

3.1 Introduction

There has been growing interest over the years in the role of technological innovation and diffusion processes in regional development and growth. The objective has been to identify new approaches able to foster, in particular, the growth of peripheral regions and to reduce regional disparities in general.

Many countries are interested in developing their peripheral regions. This is especially true of developed countries where there are socio-economic gaps between central areas and peripheral and rural regions - gaps that very often result in social and political agitation. In order to stimulate economic growth in peripheral areas, it is necessary, among other things, to create employment opportunities that will attract populations to migrate and settle in these regions.

Regional disparity, inherent in the capitalistic system, is based on spatial and regional inequality as a means of survival. The centre-periphery theory explains the concept of spatial dichotomy existing in many countries between the centre and the periphery. Most industrial development and capital projects are concentrated in central areas, which are characterized by intensive economic activity and population concentration. They are more developed than are other regions of the country and usually coincide with metropolitan regions. Peripheral regions, by contrast, dependent on the central regions, are usually sparsely populated and characterized by lower income and social welfare levels (Soja 1990, Friedmann 1973).

The difference between central and peripheral regions may be measured in three ways:

- level of involvement in decision-making;
- level of economic activity;
- level of individual social welfare.

Several writers hypothesize that as the economic development level of a country increases, inter-regional differences gradually decrease (Williamson 1965, Lipshitz 1992). Justification for governmental intervention and the implementation of public policies in national spatial organization is related to the desire to hasten the process of closing these differences (Friedmann 1973).

Many studies have been conducted to identify the principal factors affecting industrial plant location. Location considerations of high-tech industries differ significantly from those of traditional industries; they are also related to the product and plant life cycles. In a large number of industries, growth in production brings about an evolution in manufacturing technology (Scott et al. 1988). Initial product development and innovation result in the same firm types locating close to scientific research activities and to places where the chances are greatest for the product to succeed on the market (Shefer and Bar-El 1993, Shefer and Bar-El 1989).

Geographical preference - centre versus periphery - is related to the type of industry or activity (R&D versus production) of firms belonging to the same type of industry (Shefer and Frenkel 1986, Malecki 1979a, b, 1981). Close correlation also exists between plant location and type of ownership (public or private) (Razin 1988). Local plants in peripheral regions normally seek out locations having a large unskilled labour force that can be employed in the production process. Plants located in metropolitan areas place greater importance on highly-skilled labour.

There is general agreement that during the period of structural adjustment, economic growth is based on existing industries and on existing markets. Future economic growth will need to be based on the expansion and development of new markets (Freeman et al. 1982, Rothwell and Zegveld 1985). Schumpeter's theory that innovative entrepreneurship turns the wheels of capitalistic economic growth has become more and more relevant (Schumpeter 1934).

The contribution of innovation to regional economic change is often cited when dealing with economic growth and development (see, among others, Freeman et al. 1982, Jorgenson et al. 1988, Schmookler 1966, Rosenberg 1972). Regional development in a location where discoveries, inventions and technological changes are prevalent is generally accompanied by the sprouting of new economic activities, new markets, and new technological applications. Regions with a sound infrastructure of inventive capability become preferred locations for highly-skilled labour, as well as a target for reinforcing the educational and cultural infrastructures through attracting higher-level populations from other regions (Suarez-Villa 1993).

The first part of this article will review the nature of technological innovation and its influence on economic growth. The second part will give a critical presentation of the life cycle approach in industry and technology and discuss the interrelationship of the different stages of economic development. The third part of the article will focus on the definition of regional innovative ability, and the fourth section will present the structure of several analytical diffusion models and their development. The fifth and concluding section will provide a thorough review of the development of research in the field of diffusion of technological innovation.

3.2 Technological Change and Economic Growth

3.2.1 Technological Opportunities, Invention and Innovation

One of the definitions of the term 'technological change' reads as follows:

"Technical change refers to all the changes in technology and techniques which lead to new products, new processes and new methods in industrial and distributional organization and covers all the activities related to the innovation process, but also those related to the transfer and diffusion of knowledge." (Fischer 1989:47)

Over the past few years, the concept of 'technological opportunities' has become much better understood. Improvement in production over time is related to the extent of technological opportunities presented to the firm. The greater these opportunities, the greater the chances that the firm will learn to produce more efficiently compared to its competitors. In other words, firms having a higher level of innovative ability also increase their chances to compete better (Dosi 1988).

Many researchers have come to the conclusion that in most industries, successful technological development processes are those with strong internal logic, induced by existing and forecast market demand. The choice of production technology depends, among other things, on decision-making relating to required investment. In the past, technological opportunities were perceived as an external variable influencing the firm; that is, a source of knowledge fed by scientific development. Today, there is a tendency to view opportunities, as internal variables with considerable influence on decision-making processes (Sharp 1990).

The term 'innovation' refers to the first appearance of a new product on the market or the first application of a new process (Fischer 1989, Davelaar 1991, Rogers 1983). Innovative development is thought to be a process that includes a series of logical operations, but not necessarily a continuous linear process.

The development process consists of three separate stages: (a) the recognition and conceptualization stage; (b) the R&D stage; and (c) the innovative stage. The innovative solution, therefore, includes the discovery component as well as the creation component (Dosi 1988). A new and/or improved product will lead the competition, particularly if there is great demand for this product in comparison with alternative products or substitutes. Competition based on a new product is, therefore, effective only to the extent that it increases the firm's sales and profits. In addition, innovative processes based on technological changes in the production process of existing products and services play an important role in improving production efficiency and competitiveness.

A firm's opportunity to innovate depends on the flow of new knowledge and reliance on internal and external information sources. Communication channels connect not only between functionally separate departments within the firm, but also between the firm and external information and technology sources from outside firms and consumers. The existence of opportunity important, but so is the firm's ability to innovate, the latter depending on the flow of information as well as on the access to sources of required capital (Fischer 1989).

The difference between a discovery and an innovation is that a discovery emanates from basic research and technological know-how, whereas innovation includes the application and development of the same discovery in such a way as to bring about economic change (Suarez-Villa 1990, 1993). R&D generally leads to both discovery and innovation. The risk component and uncertainty of an innovation are less than those of a discovery and the efforts involved in its development are less pronounced. Skill levels required for the development of innovations are more prevalent in the market than are the unique skills required for a discovery. The latter are generally difficult to define and identify in advance.

The most important long-term goal is to bring about constant growth in the innovativeness of the region. A stock of discoveries combined with a high level of innovativeness will ensure the generation of additional innovations and, consequently, rapid recovery during recession periods. In the latter case, rapid production of technological substitutes can increase efficiency and productivity and open up new markets (see, for example, Kamann and Nijkamp 1990, Nijkamp 1986, Berry 1991, Adams 1990, Ayres 1990).

One of the first models to deal with technological changes and innovation processes was that of Schumpeter (1950) and its subsequent extension, as reviewed by Nelson and Winter (1982). According to this model, ideas and technological innovations compete for limited resources within the firm's environment, and the most suitable technology is the one that survives. The process may be lengthy and is often inefficient: it is not an equilibrium model in a competitive and dynamic market of entrepreneurs (Freeman et al. 1982). Schumpeter related the technological innovation process with scientific research, the incentive of entrepreneurs to invent new technologies being based on the possibility of enjoying temporary monopolistic power. Other entrepreneurs, who have observed the success of their predecessor, then enter the market as 'free riders' adopting these technological changes and improving on them. Economic growth occurs when an increasing number of entrepreneurs adopt technological innovations. As the number of entrepreneurs increases, however, profits decrease. Schumpeter's model emphasizes that technology acts as an accelerator in the economic growth process.

3.2.2   Innovation and Economic Growth

A significant portion of technological innovation is incremental, resulting from a combination of expanding demand and the firm's increased costs of production and from 'learning by doing'. From this time radical innovation or 'frog leaps' appear. Nelson and Winter describe the series of new opportunities created by a radical change in technology as a 'technology regime' (Nelson and Winter 1977, 1982), the introduction of radical innovation marking a new kind of 'technology paradigm' (Dosi 1988).

The increased interest in the innovation process emanates from the existing relationship between innovation, market competition and economic growth (see, for example, Pavitt and Walker 1976, Freeman 1982, 1990, Freeman et al. 1982, Dosi 1984, 1989). The net effect of innovation may be expressed in the short-term by labour savings and, in the long-term, by the expansion of the labour market. As a result, producers begin operating more efficiently, the relative competitiveness of the economic unit is improved, and product sales and marketing ability in local, national, and international markets are improved.

Technological innovation appears to follow a series of distinct stages, manifested in the technological and economic characteristics of products and processes. Related to this is the concept of 'technological trajectories' (Dosi 1982, 1984), which considers the process of technological change to be a factor responsible for the appearance of new products or services. It is not seen as a certain or random event, but as a dynamic process extending over specific transition stages.

A technological trajectory may be defined as a process characterized by trade-offs between the economic and technological dimensions (Dosi 1988). It may reasonably be assumed that economic growth and the division of labour in production activities will increase the range of demand on the macro level. Conversely, a change in the technology paradigm generally implies a change in the technological trajectory. New technological trajectories are thus transitions in innovation activities, resulting from the introduction of new products and services (Davelaar 1991).

Technological trajectories describe the course over which technology develops with time. They do not consist of radical changes, but the incremental innovation resulting from continuous marginal improvements in the product. The trajectories result, on the one hand, from internal scientific/technological logic and, on the other hand, from market forces. In other words, improvement reveals itself in product design and controlled processes derived from production and marketing experience. Nelson summarizes the process as follows:

"Where these easily pursued directions correspond to user needs, and where innovations have some mechanisms for assuring that they receive a non-trivial fraction of the use value of innovation, technological change tends to proceed along these tracks." (Nelson 1988:220)

The concept embedded in the basic principle of incremental innovation is termed the 'natural' trajectory, i.e. the natural course of development resulting from the *basic knowledge of the firm*. Most firms limit the search process for new ideas, new products, and new processes to familiar areas, tending to look for ways of improving their technology by seeking out regions that will enable them to use existing technologies and relying on the markets for their products. The first is related to the fact that there are two reasonable explanations for this assumption. The second is that since these innovation-searching procedures are expensive, firms will tend to rely on existing expertise. At the same time, the character of the innovative process will encourage them to enhance existing expertise. Most incremental innovation, therefore, results from learning by doing. This is something which is not easily transferable (Sharp 1990) and means that the search for new technologies is a process with a cumulative effect. In most cases, the firm's future technological development will be directed along narrow paths, utilizing its past implementational ability (Teece 1982, 1986).

One serious consequence is, of course, that new firms with no experience will encounter great difficulty in breaking into the market place once this trajectory has been created. Incremental innovation and the 'learning by doing' process may be very useful for existing firms working in the market place, but not for new firms trying to break into the market. Consequently, the competitive advantage of veteran firms over a new firm's potential is strengthened. This helps to explain the relative stability of the oligopoly market structure that develops in mature industries (Sharp 1990). New technological trajectories will generally invoke new opportunities to produce profits; therefore many new firms are attracted. This stage is characterized by accelerated economic growth, expressed by an increase in investments in the new industries, in the number of employees, and in the quantity of sales.

The two models described above represent two contrasting approaches regarding the performance of technological change. The first perceives these changes as the result of what may be called 'technological push', the second approach as a result of 'market pull' or the influence of demand. Growth in demand is seen as a stimulus to technological innovation and the search for new knowledge through basic economic activities (Schmookler 1966, Mowery and Rosenberg 1979). These two separate influences may be observed in the different stages of the life cycle of new industries and technologies (Davelaar 1991).

The firm's ability to innovate derives, to a great extent, from the existence of a sound infrastructure of institutions and information:

"...Innovation is dependent upon a broad technological infrastructure or social structure of innovation to mobilize resources, knowledge, and other innovative inputs which are essential to the innovation process."
(Feldman and Florida 1992:15-16)

The spatial aspect of a firm's innovativeness, is influenced by two primary components: (a) the firm's structural component and (b) the production milieu.

The structural component includes the age of the firm, its affiliation to a large industrial concern or investment company, its area of activity, the character of its management method, its integration of research activities and marketing, the level of expenditures on R&D and the internal organization. The production milieu includes the economic environment and the essential infrastructures needed for the transfer of information. The business milieu provides the psychological, cultural and social background necessary and will determine the educational level, willingness to take risks, organizational ability and access to technology (Camagni 1985)

3.2.3   The Life Cycle Model, Competition and Economic Growth

The product life-cycle model is the dominant model through which the dependency relationships between industries, technologies, and economic growth are explained (Vernon 1966). The present point of view regarding product life cycle and the strategy of moving from the initial S-shaped learning curve to the new innovation curve was originally put forward by Schumpeter:

"...innovation breaks of any such 'curve' and replaces it by another, which ...
displays higher increments or product thought."
(Schumpeter 1939:88)

According to this view, innovation brings about a shift in the curve, resulting
from the growth in the number of products. For every point in time, therefore,
some firms will be found on the new higher 'S' curve and others on the lower
curve. As technology becomes progressively newer, it may be expected that the
measure of homogeneity between the firms exploiting it increases.

Three main stages characterize the life cycle of a new product or innovation and
influence the firm's location considerations: the innovation and development stage,
the maturing or growing stage, and the mature product stage (Vernon 1966). The
life-cycle approach has been extensively adopted in various studies (Norton and
Rees 1979, Norton 1979, Taylor 1987, Barkley 1988, Karlsson 1988) represents
a valuable tool for understanding the influence of industrial change on the work
place and on employment (Ford and Ryan 1981, Shanklin and Ryans 1984).
Understanding the successive stages of the technology life-cycle can contribute to
an early prognosis of changes on the verge of taking place in products and
production processes.

Just as changes occur in production processes during the product life cycle,
changes also occur in the skills and training needs of labour during the technology
life cycle. The early stage of technology innovation generally requires highly
skilled labour, such as scientists and engineers, whereas most jobs in the following
stages require skilled management and marketing personnel specialized in
production. Location incentives influence the diverse production activities in
different ways. In the early stages of the product life-cycle, competition among
firms primarily focuses on aspects of product innovation. Salary subsidization and
tax incentives may be very valuable for firms operating in the later stages of the
product life-cycle, but these strategies are not as effective for firms engaged in
R&D and entrepreneurial activities. Similarly, whereas short-term training
programs may be suitable for firms dealing in mass production, they are not as
effective for firms engaged in complex, non-standard activities requiring training
of highly skilled labour (Flynn 1994).

The conventional view of the relationship between the product life-cycle model
and technological changes ignores space and time dimensions. The model does not
consider the contribution of new technological systems which represent basic
technological innovation for the creation of a new life-cycle for industries and
technologies (Davelaar 1991).

The product life-cycle model is related to activities carried out in both the central
and peripheral regions during different stages of the product life-cycle. By
contrast, technological change appears in the model only as an external factor and
does not influence the spatial transfer of production processes or employment. As
a result, the model is unable to explain the empirical findings of studies showing
that firms located outside the metropolitan regions also engage in the creation of
innovations (Davelaar and Nijkamp 1992).

3.2.4   The Spatial Diffusion Process

Diffusion of innovation is defined as a process which, with time and through various means, causes the transfer of innovation among groups and individuals belonging to a certain social system (Rogers 1983). It may be compared to a specific type of communication concerning in the transfer of information and new ideas. Diffusion also concerns a certain type of social change, since the adoption of new ideas may result in socio-economic change. It requires four main components: innovation; means of innovation transfer; time and a socio-economic system.

Technology diffusion is a complex process, involving changes in the behaviour of economic agents. Many studies emphasize the great importance of the technology-diffusion process for market development. Nevertheless, it is surprising to find that only a few policies are designed to foster this process. The expected societal return on new technology without the diffusion process will be insignificant (Metcalfe 1990).

The diffusion process may be understood by integrating three basic elements: companies, environment and technology (Camagni 1985). Their integration creates the necessary conditions to adopt innovation, and each pair plays a central role in one of the decision-making three stages, which are as follows:

1   Awareness stage - the integration of technology and the environment. The strategic element at this stage is the availability of information, representing a necessary condition to the adoption of an innovation.

2   Consideration stage - the integration of technology and the firm. This stage is strategically characterized by problems relating to the measure of adaptability and the relative advantage of the new technology or innovation over the old. This stage is related to a second essential necessary condition for adopting innovation, the relative profitability of the new technology or innovation.

3   Adoption stage - the integration of the firm and the environment. This is the third and final stage, during which the firm evaluates the cost of replacing the old technology with the new (Scherer 1980). It is not enough that new technology is available to the firm; it is also necessary to ascertain that the present value of the expected income will be greater than the cost of adapting the firm (Camagni and Cappellin 1984). This cost is a function of the firm's internal factors, such as expenditures on R&D and sales ability, as well as of external factors, such as the ability to tap loan opportunities, remove political and union-dependent obstacles, etc.

A common distinction made in diffusion studies in the regional context relates to the division between product innovation and process innovation (Davelaar 1991). Development regions are able to adopt technologies associated with production processes, although they may face severe difficulties in adopting an advanced product innovation. Process innovation can usually be bought 'off the shelf' on the open market. Product innovation, on the other hand, is not as readily available. Since innovation is the means by which the firm can maintain a competitive advantage over its rivals, product innovation is less transferable in terms of diffusion (Okey et al. 1982, Alderman 1990).

Innovation transfer involves a component of risk or uncertainty. Hence the importance of information which lies, among other things, in its ability to reduce uncertainty. Greater importance should be placed on the uncertainty component in innovation activity than is presently ascribed to it by popular economic models. Uncertainty is concerned not only with the lack of information regarding the exact income and expenditure associated with the various alternatives, but most often with the limited knowledge concerning the nature of the alternatives (Freeman 1982, Nelson and Winter 1982).

Dosi (1988) believes that a distinction should be made between uncertainty resulting from the partial availability of information concerning known events and what is termed 'strong uncertainty'. The latter exists when a set of possible events is not just unknown but unknowable; it is therefore impossible to determine the results of a given occurrence. Innovation is characterized in most cases by strong uncertainty.

The diffusion model is described by using different distribution functions, such as the normal cumulative function, the Gompertz function, and the logistic distribution function. All of these functions can be expressed graphically with an 'S' curve (Griliches 1957, Davies 1979, Metcalfe 1981, Andersson and Johansson 1984, McArthur 1987). The behavioural logistic model, derived from the model describing the process of disease spreading in a homogeneous population over time (Camagni 1985) is the most common one used to describe the diffusion process. Diffusion models were primarily developed in order to describe, explain and forecast the spread of innovation by a given number of potential adopters. The increase in the number of adopters over time may be illustrated with simple mathematical functions (Mahajan and Peterson 1985).

The diffusion curve can have a normal cumulative structure or a normal cumulative log structure, depending on the particular innovation properties. A log-normal cumulative diffusion-curve structure is attained when the diffusion concerns a simple technological innovation, of relatively low cost and with a rapid spatial-transfer process. This type of innovation is normally replaced very rapidly by an even more advanced technology. In contrast to this model, the normal cumulative diffusion curve is attained when the innovation diffusion is concerned with complex technology which is generally expensive and has a long learning curve, even though it apparently yields higher long-term profits (Alderman and Davies 1990).

3.3 Innovation Diffusion Models

3.3.1 The Basic Diffusion Model

The basic diffusion model may be expressed by the following differential equation:

$$\frac{dN(t)}{dt} = g(t) \, [\bar{N} - N(t)] \qquad (3.1)$$

$$N(t=t_0) = N_0 \qquad (3.1)$$

where:

$N(t)$	= cumulative number of adopters during period t
$\bar{N}$	= total number of potential adopters in the system during period t
$\dfrac{dN(t)}{dt}$	= rate of diffusion during period t
$g(t)$	= diffusion coefficient
N_0	= cumulative number of adopters during period t_0

Equation (3.1) depicts the rate of diffusion of innovation during the time period t, which is a function of the difference between the total number of possible adopters during this period and the number of prior adopters $[\bar{N} - N(t)]$ during this period.

The reciprocal relationships between the rate of diffusion and the number of potential adopters during the period $[\bar{N} - N(t)]$ are shown by the diffusion coefficient $g(t)$. This coefficient depends on factors such as properties of the innovation, means of information transfer and properties of the social system in which the process takes place. $g(t)$ can even express the probability of the adoption during period t, so $g(t)[\bar{N} - N(t)]$ expresses the number of expected adopters during period t. $N(t)$ expresses the number of individuals/firms associated with the socio-economic system that moved from the stage of potential adopters to that of actual adopters during period t, where $g(t)$ is the conversion coefficient or the transfer mechanism. In the more accepted approach, coefficient $g(t)$ represents a function of the number of previous adopters, whereas in the less popular approach it represents the time function.

$g(t)$ is expressed as a linear function of $N(t)$ by $g(t) = (a+b)N(t)$. In this case, the mathematical expression of the basic diffusion model will be as follows:

$$\frac{dN(t)}{dt} = a+bN(t)[\bar{N} - N(t)] \qquad (3.3)$$

where the coefficient 'a' expresses the system's external influence and the coefficient 'b' expresses the internal influence.

The typical model describing the innovation diffusion process assumes a full flow of information between potential and former adopters. This assumption is based on the free flow of information concerning the innovation between regions. If this basic assumption does not hold true, a significant diversion could occur in the results obtained with the basic model (Coleman 1964). In such circumstances, Lavaraj and Gore (1992) suggested using the two following differential equations:

$$\frac{dN_{1t}}{dt} = bN_{1t}(k_1 - N_{1t})$$ (3.4)

$$\frac{dN_{2t}}{dt} = (b_1N_{1t} + b_2N_{2t} + b_3N_{1t}N_{2t})(k_2 - N_{2t})...$$ (3.5)

where N_{1t} defines the innovation adopters until time t in the central region; N_{2t} defines the innovation adopters until time t in the peripheral regions; k_1 and k_2 define the maximum potential adopters in the central and peripheral regions, respectively; and b, b_1, b_2, b_3 represent the model's parameters. Equation (3.4) is equivalent to the basic logistic model. In Equation (3.5), the term $b_1N_{1t}(k_2-N_{2t})$ depicts the increase in the number of adopters owing to the interaction between the centre and the periphery; the term $b_2N_{2t}(k_2-N_{2t})$ depicts the influence of internal regional communication on the periphery; and the term $b_3N_{1t}N_{2t}(k_2-N_{2t})$ represents the impact of all potential contacts between region 1 and region 2 in relation to the periphery. Clearly, this model may incorporate chaotic properties.

Another assumption of the basic model is the setting of N̄ as the number of potential adopters of the system. This may be either a pre-determined or an estimated number. In fact this approach simply states that the size of the socio-economic system is fixed and final and, therefore, the diffusion model is static. The socio-economic system included in the model is not permitted either to increase or to decrease during the diffusion process (Mahajan and Peterson 1979, Shrif and Ramanathan 1981). A topic to be studied further is the basic structure of technology and its properties.

"Clearly, in the study of spatial diffusion of new technologies, the changing nature and expanding range of applications of these technologies 'moves the goal posts' in that the population of potential adopters can change if major new developments occur, while saturation levels can also be affected." (Alderman 1990:295)

The dynamics of innovation processes is reflected in the variability of diffusion rates. These rates change over time as a result of the introduction of new techniques deriving from the accumulation of information by previous adopters. The diffusion rate might also be a reaction to the external development of the innovation itself, concerning its technological and marketing characteristics

(Davies 1981, Gold 1979). Thus, technology adoption also depends on relative profitability. This could vary due to a reduction in the cost of acquisition and the resulting increase in potential applications (Metcalfe 1981, 1982). Innovation, therefore, may not be statistically defined at the beginning of the diffusion process, since its characteristics depend on the time variable, which changes the number of possible adopters. In analytical model terms, it is not possible to express this with a single curve; what is involved is an envelope of continuous logistic curves, each referring to a separate set of environmental and technological characteristics (Metcalfe 1982).

3.3.2   Limitations of the Model

Use of the logistic model entails several assumptions, discussed in detail by Davies (1979). He assumed that the potential adopter will decide to innovate when the reward meets his expectations. Davies maintains that this time is a function of firm size, among other things, and that adoption will take place once the critical size is reached. The influence of new technology on every relevant sector, after the superiority of the old technology has already been proven, may occur, only in the long-term however (Camagni 1985). A kind of technological pluralism may exist for many years if the relative profitability of the new technology varies significantly among regional markets. The appearance of a new technology does not automatically guarantee that it will replace the old one in the short-term.

Another criticism of the basic assumption of the logistic model is that the diffusion process is binary, suggesting that potential adopters choose between adopting or not adopting innovation. In other words, the process does not represent continuous events, but disjointed events. The basic model, therefore, does not take into account the existence of stages in the adoption process, such as awareness, knowledge, etc. (Rogers 1983). The solution to this is a dual-stage diffusion model (Camagni 1985). The first stage refers to diffusion among firms belonging to a specific manufacturing sector; in the second stage the diffusion takes place among firms belonging to different sectors. The reason for this is that the innovation itself undergoes an incremental change over time; as the information on advantages resulting from its adoption is widely known. The logistic model is capable of describing the behaviour of inter-sectorial diffusion.

An additional assumption is that the innovation itself does not vary during the diffusion process, i.e. no adjustments to innovations and their exposure to new technology occur during the adoption period. Moreover, the innovation under consideration is assumed to be independent of other innovations. Therefore, the adoption of any innovation represents neither a complementary technology nor a substitute, neither a reduction nor an inducement with respect to the adoption of another innovation.

3.3.3   Applications of the Basic Model

Despite the criticisms and the binding constraints discussed above, the basic model was extensively applied when some of the later extensions attempted to resolve its drawbacks. One of the extensions deals with the dynamic nature of the basic model, but assumes that the number of potential adopters during the process period is fixed (Mahajan and Peterson 1979). Implementation of the basic model presented for the dynamic diffusion process will yield an incorrect estimate for the parameters or for the number of the adopters $\bar{N}$. Therefore, we suggest a correction that converts the basic static model into a dynamic model, in which $\bar{N}$ may vary with time and is defined: $\bar{N}(t) = f\underline{S}(t))$, where $\underline{S}(t)$ is the vector (potential) of relevant exogenous and endogenous factors, controlled or not controlled, that influence $\bar{N}(t)$. Thus when the expression $(\underline{S}(t))$ replaces $\bar{N}$ in the basic model equation, the following dynamic diffusion model is obtained:

$$\frac{dN(t)}{dt} = [a+bN(t)][f(\underline{S}(t)) - N(t)] \qquad (3.6)$$

Examples of relevant variables include the socio-economic conditions of the system in which the diffusion process takes place; an increase or decrease in the size of the population of the potential adopters; government-initiated programs and programs designed to induce the diffusion process such as advertising, etc. The desired number of variables to be included in $\underline{S}(t)$ is a function of a number of conceptual and practical factors. The character of innovation under examination is a limiting factor, as is the extent of data availability. The level of precision of the dynamic diffusion model depends, in part, on the definition of the variables for $\underline{S}(t)$, which really influences $\bar{N}(t)$.

A more comprehensive development of the basic diffusion model is related to a simultaneous combination of time and space dimensions. In the basic model, the fundamental assumption is that the process is static, i.e. that no changes occur in the time and space dimensions. Most diffusion studies cannot cope with the complexity involving a dual-dimension combination and resort to dealing with one dimension only. While in various disciplines the time dimension is dealt with extensively in diverse disciplines, the space dimension is investigated primarily by geographers, one of the most outstanding of whom is Hagerstrand (1967). He perceives diffusion to be a process causing change in the population, from a region having a small number of adopters to one with a large number. This change is observed through information transfer by means of communication and inter-personal interactions. Hagerstrand defined three empirical laws in his studies: (a) the process is characterized by an S-shaped curve; (b) the process is characterized by a hierarchical effect, whereby diffusion is expected to spread from large to small centres; and (c) the process is characterized by the 'neighbourhood effect', whereby diffusion is expected to spread in a wavy or oscillatory fashion to outside city centres, initially mostly to locations close to these centres and less to more distant locations.

One attempt to develop a model that deals simultaneously with the time and space dimensions suggests that the diffusion process be examined in the context not only of predicting the cumulative number of adopters over time, but also of evaluating the way innovation spreads over different geographic areas. This model can supply information to allow for a comparison to be made between the level and quantity of adopters and different geographical regions and to assess the degree of innovation dissemination to different regions. In this way, the model helps determine a dissemination policy suitable for attaining the process (Mahajan and Peterson 1979).

Two of the three empirical laws documented in the spatial diffusion literature are incorporated into this model: (a) the S-shaped curve and (b) the 'neighbourhood effect'. Firstly, it is assumed that the growth in the number of adopters in every region can be expressed with the diffusion model, consisting of unique parameters, a, b, and N. Secondly, it is also assumed that, relatively speaking, the number of adopters will be greater in areas closer to regions where innovation has been created.[1] The model simplifies the complexity of the problem by determining that the innovations initially appear in one region only, and that the distance parameter (x) between the innovation region and the other regions is measured as the distance between their centres of gravity. The cumulated number of adopters, N, is therefore a function of two variables: time and distance - $N = f(x,t)$. The combined model is expressed by the following equation:

$$\frac{dN(x,t)}{dt} = [a(x) + b(x)N(x,t)] \times [N(x) - \bar{N}(x,t)] \qquad (3.7)$$

Inter-regional gaps may be expressed with this model, not only by physical distance measures, but also by economic measures, thereby showing the economic differences among regions (Camagni 1985).

One of the more interesting recent developments of the spatial diffusion model is the dynamic incubator model (Davelaar 1991), which is a conceptual model, not tested empirically, that examines several basic principles relating to the time and space dimensions. The dynamic incubator model is based on two fundamental assumptions: The first states that a radical change in technology and in the framework of the socio-economic system, manifested by the appearance of new technological systems, also brings about structural change in the spatial economics. The second assumption is that a change occurs as a result of the spatial diffusion process and is itself naturally expressed in the resulting technological changes in different types of products and services. The model focuses on spatial applications regarding both the launching of new technological trajectories and their diffusion process. It defines three stages (similar to the stages of the product life-cycle model): (a) the incubation stage, which is related to the first stage of agglomeration processes (it primarily applies to metropolitan areas); (b) the competition stage, related to the adoption-process stage and the migration of firms to intermediate and peripheral regions; and (c) the decline stage, in which the new products and processes created during the technological trajectories reach market

---

[1]If innovation is created in more than one region and, as frequently happens, this occurs at different times, it may be assumed that every region will be influenced simultaneously by different waves of diffusion (see, also, Corvers and Giaoutzi 1995).

saturation in all regions. Market demand places limitations on the expansion possibilities of new industries, and the cost is dominant in determining competitiveness.

3.3.4 Spatial Diffusion Research of Technological Innovation

Interest in the multi-disciplinary subject of innovation research grew significantly in the 1980s. Economists showed particular interest in innovation inputs (such as expenditure on R&D), technological opportunities and innovation yields such as patents (Cohen 1992), while geographers dealt with the location considerations of R&D activities (Malecki 1986) or hi-tech industries (Markusen, Hall and Glasmeier 1986); and regional planners examined the organizational and institutional structures necessary for the development of innovation activities (Feldman and Florida 1992).

Technological innovation research up to the 1960s was primarily illustrative, bibliographical, or purely technical. There was very little empirical research, despite the awareness by economists of the importance of innovation-fostering productivity and competitiveness. In the 1950s most innovation research was carried out by geographers and sociologists, particularly in the field of diffusion. Only in the 1960s did economists begin to carry out more systematic research, including empirical studies, in their innovation research. Until the early 1970s, most studies focused on specific individual innovations, whose goal was to identify those characteristics of innovation that would ensure technical and commercial success (Freeman 1991).

One of the important early empirical studies concerning innovation was the SAPPHO project, in which attempts were made to identify the success factors of innovation by making a comparison of pairs of innovations that succeeded or failed (Rothwell et al. 1974). The project compared about 100 characteristics of 40 pairs of innovations in two industrial sectors: the chemical industry and the precision instruments industry. Later studies dealing with the mechanical and electronics industries arrived at similar findings (Lundval 1985, Maidique and Zirger 1984). One of the main contributing factors identified in this latter study was an early understanding of the special need by potential users of the new products (or of the new processes) under development (see also Lundvall 1985, 1988). The success of an innovation depends, to a great extent, on the generation of joint techniques of research and development, production and marketing in the early stages of the innovation. This finding was later confirmed in other studies (see Lami et al. 1985, Takeuchi and Nonaka 1986).

Although many pioneering efforts in diffusion theory have been made by geographers (see, for example, Hagerstrand 1967, Brown 1981, Clark 1984), the success of the tools developed by geographers has been very limited in research into the ways in which new technologies penetrate individual firms (Alderman 1990). Geographers dealing with innovation and research relating to regional development have emphasized the location of R&D activities and basic technology industries, and indicated that spatial distribution of high-tech industries and employment are interrelated (Malecki 1980b, 1985, 1986, 1990). Some studies pointed out the influence of wage levels in the region and of the strength of the labour unions on the location considerations of high-tech industries and on employment (Markusen et al 1986). Other studies focused on the influence of organizational and institutional factors on the creation of conditions for innovation development, and usually included a case study such as Route 128 near Boston (Dorfman 1983, Roberts 1991), Silicon Valley (Saxanian 1985), or Orange County (Scott and Stroper 1988).

Other studies dealing with innovation processes focused on the spatial factor and its effect on the process and its intensity, and the importance of regional characteristics for industrial innovation processes (Bramanti and Senn 1991). The influence of the central city, the metropolitan region, and the periphery was investigated in studies dealing with the creation and transfer of innovation (Davelaar and Nijkamp 1989). (Concerning this important point, see also Aydalot and Keeble 1988, Nijkamp 1986, Oakey, Thwaites and Nash 1980, Malecki and Nijkamp 1988, Aydalot 1984, Camagni 1984, Ewers 1986, Perrin 1986, Malecki and Variya 1986.)

Stohr (1986) emphasized the combined effect that emanated from the interaction between innovation and regional characteristics, such as the capacity to acquire skills, obtain information and business advice on decision-making. Perrin (1988) identified new organizational types with reciprocal relations between large, medium-sized and small firms and public institutions. One of the important preconditions for creating innovation potential is proximity between the 'players'. The importance of spatial proximity stems, among other things, from the fact that intra-regional mobility of human capital is greater than inter-regional mobility (Camagni, 1991). Camagni suggests that the relationship between 'players' is often based on personal contact and therefore common cultural aspects, psychological factors, and shared political backgrounds all help create the cumulative effect.

In the area of industrial innovation, most theoretical and empirical studies have been carried out by economists (Alderman and Davies 1990) who have focused on the reciprocal relationship between the inputs and outputs of innovation and generally ignored the spatial dimension. Nevertheless, much importance must be placed on defining the milieu in which the diffusion process takes place and the attributes of the competing technologies. It is not enough to describe the adoption of a new technology in terms of its properties. New technology is not transferred into a vacuum, but into a known environment which determines the advantages and disadvantages of existing technologies. Mature existing technology can sometimes make an improvement under the pressure of competition which is equivalent to that of a new technology. The diffusion process depends on the competitive advantage of the innovation. In certain cases, existing technology reaches its optimum level only after being subjected to fierce competition with a newly developed technology (Metcalfe 1990).

3.4 Summary

Interest in the contribution of technological change to development and economic growth has recently played an important role in regional science research, largely because the new technologies have constituted a primary factor in bringing about structural change in the spatial economy. Understanding the relationship that exists between economic efficiency and innovation activities has brought about a change in regional development policies: firms are encouraged to develop (or adopt) new technologies. In parallel, regions are pursuing a policy of creating a favourable milieu to attract innovative firms and thereby influence economic growth . This policy concentrates on the influence of the production-milieu component on the firm's ability to innovative, a component that includes environmental characteristics representing the necessary economic and infrastructure conditions for information transfer. The environment provides a psychological, cultural and social basis for defining the willingness to take risks for high organizational ability and for access to technology.

The life-cycle model helps determine the role played by economic development policies. Based on this model, the attractiveness of a region and its economic development will change in accordance with the labour skills required for producing at different development stages. Changes by industries in the time-space dimension, which generally occur as a result of the variety of technologies employed, affect new products at different development stages and influence requirements for skilled labour during these changes.

Much importance may be placed on the economic characteristics of new technology and its application potential with respect to the diffusion process. The early users of newly developed technologies may not fully enjoy them, and the technologies themselves may have limited application opportunities. In many instances, therefore, support of embryonic technologies must be provided by the government. Technology diffusion requires a selective competitive environment open to changes, and one in which more profits are accrued than with the competing technologies. The profit element is of utmost importance from the point of view of manufacturers and users of new technologies. Profits allow investment in R&D activities and in training manpower. Despite this, a shortage of skilled labour may limit the diffusion process. Here, the government has a key role to play in ensuring national support for R&D of new technology and in providing the required level of skilled labour.

Many studies emphasize the importance of the technology-diffusion process on economic development; nonetheless, it is surprising how few policies are designed to enhance this process. If the process does not occur and new technologies are not transferred, their influence on the economy will be limited. The creative ability of firms and national research institutions are insufficient conditions. There is a need to continue to invest in new technologies that will completely or partially replace existing ones. Without the diffusion process, the expected return on the creation of new technology will be insignificant. Numerous empirical studies from diverse disciplines have pointed out that many new technologies have not been transferred or duplicated immediately to all parts of the economy, and that these differences affect the growth of the national economy.

58

References

Adams J D (1990) Fundamental Stocks of Knowledge and Productivity Growth. Journal of Political Economy 98:673-702

Alderman N (1990) New Patterns of Technological Change in British Manufacturing Industry Sistemi Urbani 3:287-299

Alderman N and S Davies (1990) Modeling Regional Patterns and Innovation Diffusion in the UK Metalworking Industries. Regional Studies 24 6:513-528

Anderson A and B Johansson (1984) Knowledge Intensity and Product Cycles in Metropolitan Regions, Contributions to the Metropolitan Study 8. IIASA Luxembourg (Unpublished paper)

Aydalot Ph (1984) Reversals on Spatial Trends in French Industry Since 1974. In: Lambooy J G (Ed) New Spatial Dynamics and Economic Crisis. Finnpublishers Tampere Finland:41-63

Aydalot Ph and D Keeble (Eds) (1988) High Technology Industry and Innovative Environments: The European Experience. Routledge London

Ayres R U (1990) Technological Transformations and Long Waves, Part I. Technological Forecasting and Social Change 37:1-37

Bar-El E L (1989) The Suitability of High Technology Industries for Israel's Peripheral Regions. Final Paper, Submited in Partial Fulfillment of the Requirement for the Degree of Master of Science in Urban and Regional Planning. Technion Haifa Israel

Barkley D L (1988) The Decentralization of High-Technology Manufacturing to Non Metropolitan Areas. Growth and Change 19 1:13-30

Baumol W J, Blackman S A B and E N Wolf (1989). Productivity and American Leadership: The Long View. MIT Press Cambridge

Berry B J L (1991) Long-Wave Rhythms in Economic Development and Political Behavior. Johns Hopkins University Press Baltimore

Bramanti A and L Senn (1991) Innovation, Firms and Milieu: a Dynamic and Cyclic Approach. In: Camagni R (Ed) Innovation Networks: Spatial Perspectives. Belhaven Press London:89-104

Brown L A (1981) Innovation Diffusion. A New Perspective. Menthuen London

Camagni R (1984) Spatial Diffusion of Pervasive Process Innovation. Paper presented at the 24th European Congress of the Regional Science Association. Milan:28-31 August

Camagni R (1985) Spatial Diffusion of Pervasive Process Innovation. Papers of the Regional Science Association 58:83-95

Camagni R (Ed) (1991) Innovation Networks: Spatial Perspectives. Belhaven Press London

Camagni R and R Cappellin (1984) Sectoral Productivity and Regional Policy. Research Report for the Directorate General for Regional Policy of the Commission of the European Communities. Final Report October

Clark G (1984) Innovation Diffusion: Contemporary Geographical Approaches. Concepts and Techniques in Modern Geography 40. Geo Books Norwich

Cohen W (1992) Empirical Studies of Innovation and Performance. (22 June). Unpublished paper

Coleman J S (1964) Introduction to Mathematical Sociology. Free Press New York

Corvers F and M Giaoutzi (1995) Borders and Barriers in Europe. Paper presented at the NECTAR conference on European Transport and Communication Networks: Policies on European Network. Espinho Portugal April 17-23

Davelaar E J (1991) Regional Economic Analysis of Innovation and Incubation. Billings & Sons Worcester UK

Davelaar E J and P Nijkamp (1989) Spatial Dispersion of Technological Innovation: A Case Study for the Netherlands by Means of Partial Least Squares. Journal of Regional Science 29 3:325-346

Davelaar E J and P Nijkamp (1992) Operational Models on Industrial Innovation and Spatial Development: A Case Study for the Netherlands. Journal of Scientific & Industrial Research 51:273-284

Davies S (1979) The Diffusion of Process Innovation. Cambridge University Press Cambridge

Dosi G (1982) Technological Paradigms and Technological Trajectories. Research Policy 11:142-162

Dosi G (1984) Technical Change and Industrial Transformation. MacMillan Hong Kong

Dosi G (1988) Sources, Procedures, and Microeconomic Effects of Innovation. Journal of Economic Literature XXVI:1120-1171

Dosi G (1989) The Nature of the Innovation Process. In: Dosi G G, Freeman C, Nelson R, Silverberg G and L Soete (Eds). Technical Change and Economic Theory. Pinter London:221-238

Ewers H J (1986) Spatial Dynamics of Technological Development and Employment Effects. In: Nijkamp P (Ed) Technological Change, Employment and Spatial Dynamics. Springer-Verlag Berlin:157-176

Feldman P M and R Florida (1992) The Geography of Innovation. (Unpublished paper)

Fischer M M (1989) Innovation, Diffusion and Regions. Chapter 5. In: Andersson A E, Batten D F and C Karlsson (Eds) Knowledge and Industrial Organization. Springer Verlag Berlin, Heidelberg and New York:47-61

Flynn P M (1994) Technology Life Cycles and State Economic Development Strategies. New England Economic Review:17-30

Ford D and C Ryan (1981) Taking Technology to Market. Harvard Business Review 59 2 2 (March/April):117-126

Freeman C (1982) The Economics of Industrial Innovation. MIT Press Cambridge MA

Freeman C (1990) The Economics of Innovation. Aldershot Hants

Freeman C (1991) Network of Innovation: A Synthesis of Research Issue. Research Policy 20:499-514

Freeman C, Clark J and L Soete (1982) Unemployment and Technical Innovation. A Study of a Long Waves and Economic Development. Frances Pinter London

Friedmann J (1973) Urbanization Planning and Development. Sage Publications Beverly Hills California

Gold B (1981) Technological Diffusion in Industry: Research Needs and Shortcomings. The Journal of Industrial Economics 3:247-269

Griliches Z (1957) Hybrid Corn: An Exploration in the Economics of Technological Change. Econometrica 25:502-522

Hagerstrand T (1967) Innovation Diffusion as a Spatial Process. University of Chicago Press Chicago

Jorgenson D, Gollop F and B Fraumeni (1988) Productivity and U.S. Economic Growth. Harvard University Press Camgridge

Kamann D J F and P Nijkamp (1990) Technogenesis: Incubation and Diffusion. In: Cappellin R and P Nijkamp (Eds) The Spatial Context of Technological Development. Avebury Aldershot:257-302

Karlsson C (1988) Innovation Adoption The Product Life Cycle. Umea Economic Studies 185. University of Umea Sweden

Lami K et al. (1985) Managing the New Product Development Process: How Japanese Companies Learn and Unlearn. In: Clark K B et al. (Eds) The Uneasy Alliance: Managing the Productivity-Technology Dilemma. Harvard Business School Press Boston

Lavaraj U A and A P Gore (1992) Modelling Innovation Diffusion - Some Methodological Issues. Journal of Scientific & Industrial Research 51:291-195

Lipshitz G (1992) Divergence Versus Convergence in Regional Development. Journal of Planning Literature 7 2:123-138

Lundval B A (1985) Product Innovation and User-Producer Interaction. Alborg University Press Alborg

Lundvall B A (1988) Innovation as an Interactive Process: From User Product Interaction to the National System of Innovation. In: Dosi G G, Freeman C, Nelson R, Silverberg G and L Soete (Eds) Technical Change and Economic Theory. Pinter London:349-369

Mahajan V and R A Peterson (1985) Models for Innovation Diffusion. Sage Publications Beverly Hills California

Mahajan V and R A Peterson (1979) Integrating Time and Space in Technological Substitution Models. Technological Forecasting and Social Change 14:127-146

Maidique M A and B I Zirger (1984) The New Product Learning Cycle. Research Policy 14:299-313

Malecki E J (1979a) Agglomeration and Intra-Firm Linkage in R&D Location in the United States. TESyG 70:322-331

Malecki E J (1979b) Locational Trends in R&D by Large U.S Corporations 1965-1977. Economic Geography 55:309-323

Malecki E J (1980b) Firm Size, Location and Industrial R&D: A Disagregated Analysis. Review of Business and Economic Research 16:29-42

Malecki E J (1985) Industrial Location and Corporate Organization High Technology Industries. Economic Geography 75:345-369

Malecki E J (1986) Research and Development and the Geography of High-Technology Complexes. In: Rees J (Ed) Technology, Regions and Policy. Rowman and Littlefield Totowa NW:51-74

Malecki E J (1990) Technological Innovation and Paths to Regional Economic Growth. In: Schmandt J and R Wilson (Eds) Growth Policy in the Age of High Technology: The Role of Regions and States. Unwin Hyman Boston:97-119

Malecki E J and P Nijkamp (1988) Technology and Regional Development: Some Thoughts on Policy. Environment and Planning C. Government and Policy 6:383-399

Malecki E J and P Varaiya (1986) Innovation and Changes in Regional Structure. In: Nijkamp P (Ed) Handbook of Regional and Urban Economics. North-Holland Amsterdam 7:629-645

Markusen A, Hall P and A Glasmeier (1986) High Tech America: The What, How, Where and Why of Sunrise Industries. Allen & Unwin London

McArthur R (1987) Innovation, Diffusion and Technical Change: A Case Study. In: Chapman and G., Humphrys K (Ed) Technical Change and Industrial Policy. Basic Blackwell Oxford:26-50

Metcalfe J (1981) Impulse and Diffusion in the Study of Technical Change. Futures 13 5:347-359 (special issue)

Metcalfe J (1982) On the Diffusion of Innovation and the Evolution of Technology. Discussion Paper. University of Manchester January

Metcalfe S (1990) On Diffusion, Investment and the Process of Technological Change. In: Deiaco E, Hornell E and G Vickery (Eds) Technology and Investment Crucial Issues for the 1990s. Pinter London

Mowery D C and N Rosenberg (1979) The Influence of Market Demand Upon Innovation: A Critical Review of Some Recent Empirical Studies. Research Policy 8:102-153

Nelson R R (1988) Innovation and the Evolution of Firms. In: Dosi G G, Freeman C, Nelson R, Silverberg G and L Soete (Eds) Technical Change and Economic Theory. Preface to part IV. Pinter London:219-220

Nelson R R and S G Winter (1977) In Search of a Useful Theory of Innovation. Research Policy 6:36-77

Nelson R R and S G Winter (1982) An Evolutionary Theory of Economic Change. Belknap Cambridge Massachusetts

Nijkamp P (Ed) (1986) Technological Change, Employment and Spatial Dynamics. Springer Verlag Berlin

Norton R D (1979) City Life Cycle and American Urban Policy. Academic Press New York

Norton R D and J Rees (1979) The Product Life Cycle and the Spatial Decentralization of American Manufacturing. Regional Studies 13:141-151

Oakey R P (1984) Innovation and Regional Growth in Small High Technology Firms: Evidence from Britain and the USA. Regional Studies 18:237-251

Oakey R P, Thwaites A T and P A Nash (1980) The Regional Distribution of Innovative Manufacturing Establishments in Britain. Regional Studies 14:235-253

Pavitt K and W Walker (1976) Government Policies Towards Industrial Innovation: A Review. Research Policy 5:11-97

Perrin J C (1986) Les Synergies Locales - Elements de Theorie et d'Analyse. In: Aydalot Ph (Ed), Milieux Innovateurs en Europe. GREMI Paris

Perrin J C (1988) New Technologies, Local Synergies and Regional Policies in Europe. In: Aydalot & Keeble (Eds) High-Technology Industry and Innovative Environments: The European Experience. Routledge London:139-162

Razin E (1988) The Role of Ownership Characteristics in the Industrial Development of Israel's Peripheral Towns. Environment and Planning A 20:1235-1252

Rogers E M (1983) Diffusion of Innovation. Free Press New York

Rosenberg N (1972) Technology and American Economic Growth. Harper and Row New York

Rothwell R, Freeman C, Horlsey A, Jervis V T P, and A B Robertson (1974) SAPPHO Update Project SAPPHO Phase II. Research Policy 3 3:258-291

Rothwell R and W Zegveld (1985) Reindustrialization and Technology. Longman Essex

Saxenian A (1985) Silicon Valley and Route 128: Regional Prototype or Historical Exceptions? In: Castells M (Ed) High Technology, Space and Society. Sage Publications Beverly Hills California:81-115

Scherer F M (1980) Industrial Market Structure and Economic Performance. Rand McNally Chicago

Schmookler J (1966) Invention and Economic Growth. Harvard University Press Cambridge Massachusetts

Schumpeter J (1939) Business Cycles I, II. McGraw-Hill New York

Schumpeter J (1950) Capitalism, Socialism and Democracy. Harper Scott New York

Scott A J and M Stroper (1988) High Technology Industry and Regional Development: A Theoretical Critique and Reconstruction. International Social Science Journal 112:215-232

Shanklin W L and J K Ryans (1984) Marketing High Technology. Lexington Books Lexington Massachusetts

Sharp M (1990) Technological Trajectories and Corporate Strategies in the Diffusion of Biotechnology. In: Deiaco E, Hornell E and G Vickery (Eds) Technology and Investment Crucial Issues for the 1990s. Printer publisher London

Shefer D and A Frenkel (1986) The Impact of Communications on High-Tech Industrial Development in Israel: An Empirical Study. Haifa, Israel. The Samuel Neaman Institute for Advanced Study. Technion - Israel Institute of Technology

Shefer D and E Bar-El (1993) High-Technology Industries as a Vehicle for Growth in Israel's Peripheral Regions. Environment and Planning C, Government and Policy 11:245-261

Soja E W (1990) Postmodern Geographies. Verso London

Stöhr W (1986) Regional Innovation Complexes. Papers of the Regional Science Association 59:29-44

Suarez-Villa L (1993) The Dynamics of Regional Invention and Innovation: Innovative Capacity and Regional Changes in the Twentieth Century, Geographical Analysis 25 2:147-164

Suarez-Villa L (1990) Invention, Inventive Learning, and Innovative Capacity. Behavioral Science 35:290-310

Takeuchi H and I Nonaka (1986) The New Product Development Game. Harvard Business Review. Jan-Feb 137-145

Taylor M (1987) Enterprise and the Product Life Cycle Model: Conceptual Ambiguities. In: Van der Knaap G A and E Wever (Eds) New Technology and Regional Development. Croom Helm London:75-93

Teece D J (1982) Toward an Economic Theory of the Multiproduct Firms. Journal of Economic Behavior 3 1:39-63

Teece D J (1986) Profiting From Technological Innovation: Implication, for Integration, Collaboration, Licensing and Public Policy. Research Policy 15 6:285-306

Vernon R (1966) International Investment and Institutional Trade in the Product Cycle. Quarterly Journal of Economics 8:190-207

4 Modelling Innovation Diffusion

Karmeshu and V.P.Jain

4.1 Introduction

Diffusion of innovation occupies a central place as an instrument of social change. Numerous attempts have been made to model this phenomenon in disciplines such as economics, sociology, marketing, geography etc. Considerable research effort has been devoted to the mathematical modelling in the field, focusing mainly on temporal aspects of diffusion of innovations in the form of growth curves in a given social system. These models have been constructed within both deterministic and stochastic frameworks. Some of the recent modelling efforts have analyzed the process as an evolutionary phenomenon in the framework of self-organizing systems. Such models offer considerable insight into the dynamics of the success of an innovation in the form of niche acquisition and the self-reinforcing 'lock-in' mechanism. The framework employed in these models makes it possible to enlarge the scope of enquiry by their ability to examine the critical phenomenon in the form of phase transformation characterizing take off and sustainability of an innovation. In this chapter we present a brief review of the literature in the field of mathematical modelling of temporal aspects of innovation diffusion.

4.2 Deterministic and Stochastic Models

Most of the early studies on diffusion of innovations were essentially empirical in nature and were sought to be explained with the help of various distribution functions (Banks 1994). These studies did not offer much of an insight into the process of diffucion and, therefore, served only a limited purpose. Consequently, attempts have been made to develop theory-based diffusion models evolving in time (see Mahajan and Peterson 1985), for a detailed review (Jain and Rai 1988). Bartholomew (1976) extended the framework by incorporating loss of interest factor to capture the phenomenon of discontinuance of an innovation (Rogers 1983). Another refinement pertains to introduction of time lage between knowledge and actual adoption (Lal, Karmeshu and Kaicker 1988). Several authors have also emphasized the factors influencing the supply side determining the diffusion process (see for example, Stoneman 1976; Davies 1979; Brown 1981; Metcalfe 1987; Vij, Sushil and Vrat 1992). However, the supply side finds its full expression in the system's aspect of diffusion where it is studies as an interactive process between supply and demand side factors (Freeman 1988). Another aspect which has generated considerable interest deals with interactive media like electronic messaging systems. Such systems exhibit a distinctive quality termed as the critical mass. The viability potential in which the critical mass acts as a catalyst to change the cost-benefit perception of an adopter from negative to positive (see Rogers 1990). A framework for the definition and measurement of externality on corporate and regional performance has been offered by Capello and Nijkamp (1993).

One of the early stochastic innovation diffusion models is also attributable to Rapoport (1953a, 1953b). The stochastic approach also found wide applicability in the fields of diffucion in social systems. Subsequently, several attempts have been made by scholars to incorporate intrinsic stochasticity. It soon became apparent that the mnodel had similar structural characteristics as those obtained in stochastic epidemic models (Bailey 1975). This parallel inspired researchers to draw heavily on epidemic models to study innovation diffusion processes. Subsequently, the scope of the models has been enlarged to include the features of imperfect mixing of the adopter's population and the distribution of adopters by gereration to account for possible distortions in the dissemination of information.

Recently, Jain and Raman (1992) have developed a stochastic version of the Bass model (1969). The resulting model reduces to a nonlinear stochastic differential equation with additive noise. It was found that the deterministic model over-predicts adoption levels; a finding which is in agreement with the studies by Bartholomew (1976) and Karmeshu and Pathria (1980a).

4.3 Critical Behaviour in Diffusion Models

Bartholomew (1976) extended the mixed influence model by including the additional feature of 'loss of interest' in the diffusion process. In the absence of external sources this model depicts critical behaviour which can be viewed as a phase transition in the form of switch over from microscopic to macroscopic propagation of information when the control parameter attains a critical value. In this model, the transition from microscopic to macroscopic propagation of information involves a change in the form of the probability distribution from geometric to quasi-Poissonian. One finds that the relative fluctuations in the number of adopters is greatly enhanced at the threshold point, thus elucidating the inherent feature that below the threshold the system is in a state of relative disorder, whereas above the threshold it attains a state of relative order (Karmeshu and Pathria 1979; Sharma, Pathria and Karmeshu 1982). This phenomenon is reminiscent of phase transition in physico-chemical systems (Haken 1977; Nicolis and Prigogine 1977). The above mentioned models are characterized by transition rates which are quadratic in the state variable. In a recent study, Karmeshu, Sharma and Jain (1992) have extended the non-uniform influence (NUI) model by incorporating the discontinuance phenomenon and positing it in a stochastic framework. The model is based on the transition probabilities assuming the form of a third degree polynomial or a rational function yielding three steady states. Based on stochastic catastrophe theory, the model brings out a rich variety of features like bimodality, sudden jumps and hysteresis which become manifest in the diffusion process. (Haken 1977; Nicolis and Prigogine (1977). This phenomenon has several important policy implications. The promoters of an innovation must ensure that the 'loss of interest' parameter does not cross the critical value at which the innovation adoption level slumps to the lower branch of the growth curve. Once this is allowed to happen, retrieval of the situation would require disproportional efforts. The relevance of this feature has its guidelines in marketing, political or social group membership and ideological following. The qualitative features of hysteresis and threshold phenomena have important policy implications in the sense that it may not always be possible to generalize from the effects of minor policy actions to large ones (Silverberg 1988).

The relevance of the framework of synergetics to the quantitative description of a broad class of collective dynamical phenomena in a social system has been underlined by Weidlich (1991) whose model of collective opinion formation exhibits the transition of unimodal distribution to bimodal distribution, implying transition from consensus to polarization. In another study, de Palma and Lefevre (1988) have used compartmental models to explain similar phenomena with the help of a random utility function. The model has also been applied to the distribution of two competing products to explain the relative prevalence of one type of product over the other.

In another class of models, the influence of a random environment has been studied by regarding the parameters of the process as stochastic in nature (Karmeshu and Pathria 1980b). An interesting finding relates to the phenomenon of bistability which emerges on account of randomness in internal influence parameter. Noise induced bistability has been studied extensively for a variety of systems by Horsthemke and Lefever (1984). The significance of this approach in the context of innovation diffusion lies in the fact that the phenomenon of bistability could occur even in the absence of the 'loss of interest' parameter (Karmeshu 1995).

4.4 Micro-level Models

A key aspect of the models considered so far is that they deal with the diffusion process at the aggregate level. A considerable amount of research has, however, been conducted at the microlevel, particularly in the context of new technology (new products and services) corresponding to marginal changes in the behaviour of firms and individuals. The basic thrust of these models is to provide rationale for the adoption of an innovation by an individual in terms of enhancing some objective function such as expected utility or returns. The need for the development of micro-level models arises from the heterogeneity of the adopters' population, which accounts for the systematic difference in adoption times among individuals.

The maximization of such an objective function relates to the perception of uncertainty regarding the product or technology. Hiebert (1974) examines the attitude towards risk and learning in the context of uncertainty associated with adoption at the level of the individual actor. During the unfolding of the process of innovation diffusion, more and more information regarding the innovation accumulates as data is updated. Employing the Bayesian framework for updating of uncertain perception, Stoneman (1981) has studied intra-firm diffusion of new technology. Chatterjee and Eliashberg (1990) have offered a hypothesis which explicitly considers the behavioural determinants of adoption of an innovation at the individual level in a decision analytic framework.

In these models, heterogeneity of the potential adopter's population is usually accounted for as a feature governed by chance. However, in reality it is well-known that potential adopters differ with regard to their propensity to adopt an innovation. Underlying the need to incorporate this feature in a deterministic framework Granovetter and Soong (1983) have modelled the phenomenon in terms of an adoption propensity threshold. The model is shown to find wide applicability in the sphere of socio-political diffusion. Recently Braun (1995) has extended this framework in relation to social diffusion.

Diffusion models have also been studied within the hazard function modelling framework. The hazard function is the conditional probability that an adoption will occur in the time interval $(t, t + \delta t)$ given that it has not yet occurred by time t. Lavaraj and Gore (1992) have used this framework to investigate the diffusion of crossbred goats among the heterogeneous rural population in Pune, a district in western India. In a totally different context, Diekmann (1989, 1992) has demonstrated the relevance of this framework to the study of social diffusion in relation to the demographic process of entry into marriage. The contention of the author is that the hazard function framework may well explain different diffusion processes of social and technical innovations.

4.5 Replacement Dynamics - An Innovation Diffusion Paradigm

The models of innovation diffusion considered in the previous sections treat innovation as an independent entity in isolation. In reality, however, an innovation is perceived as a new alternative by individuals, organizations or society. In this sense, an innovation can be considered as a replacement of one technique, idea or product by another (Montroll 1982). Fisher and Pry (1971) have offered a replacement dynamics model in the context of technology replacement which provides significant insight into the pattern of diffusion of an innovation. Employing the logistic growth model, they have analyzed a large number of industrial replacements and have observed that the fits are remarkably good.

Montroll (1978) has extended this idea to a variety of situations. He observes that technological and social evolution is a consequence of a sequence of replacements of one technique, idea, tradition or artifact by another. Marchetti and Nakicenovic (1979) have investigated world energy use and its substitution on the basis of the logistic model. "The immensely complex phenomenon of using energy in various forms with all the interfacing of economics, technologies and politics over a period of more than hundred years, showed up as a crystalline substitution i.e. a multiple diffusion process" (Marchetti 1991). Another interesting application of the replacement hypothesis figures in a study in the context of urbanization as a process of replacement of rural population by urban population (Karmeshu,1988; Rao, Karmeshu, Jain 1989).

Sahal (1979) has drawn an analogy between technology substitution and morphological changes in organisms. Employing the model of Batten (1982) in the context of industrial evolution, Karmeshu, Bhargava and Jain (1985) have provided a rationale for technology substitution obtained in the Fisher-Pry law.

Fisher and Pry (1971) have demonstrated that if a new technology succeeds in attaining a critical size, capturing market share of say 10 to 15 percent, it is highly probable that it will ultimately succeed in replacing the old technology. Based on the stochastic evolutionary model of technological change, Montano and Ebeling (1980) substantiate this by observing that the new technology must exceed the old one by a certain amount of quality to have real chance of survival.

An attempt has also been made to integrate the diffusion and substitution features of innovation in the form of a demand growth model (Norton and Bass 1987). The model is relevant for manufacturers of high-technology products, especially in the field of electronics where the time interval of replacement of successive technologies gets smaller and smaller. An interesting feature captured by the model pertains to the fact that the innovation, in addition to creating its own demand, also cannibalizes the diffusion of its predecessors. The model has been found to be promising in supporting the generalization that the "parameters stabilize considerably once a single generation of technology attains a turning point" (Norton and Bass 1987).

Mahajan and Muller (1990) have employed this framework in a recent paper to model the adoption and substitution pattern for each successive generation of durable goods with regard to the IBM mainframe computer. The model brings out the implications of optimal timing strategy, which calls for the choice of a new generation of product as soon as it is available or delay introduction until its predecessor reaches maturity. The feature of vertically differentiated durable goods has been analyzed to show that different quality goods are not targeted to different economic strata of the adopter's population and the market is eventually saturated by competing products of high and low quality (Deneckere and de Palma 1993).

4.6 The Evolutionary Paradigm

In the previous sections, models of innovations based on diffusion and substitution processes have been discussed. Attention is being increasingly drawn to the idea of approaching the issue from the point of view of systemic analysis. Most of these studies form part of the family of models which can be broadly classified as evolution models due to their twin features of variation and selection. These studies have been inspired by the classic works of Kuznets and Schumpeter, which relate diffusion of technical innovation to changes in the structure of the industry and the development of the whole economy.

The evolutionary paradigm derives from the earlier works on self-organization models of complex systems (Nicolis and Prigogine 1977; Haken 1977). Subsequently, a rapidly growing number of authors have proposed models of innovation diffusion in the form of induced fluctuations as an evolutionary process (Allen 1976; Batten 1982; Silverberg 1988; Bruckner et al. 1994). In a similar vein a number of studies have been made pertaining to technology trajectories (Nelson and Winter 1977), the technological paradigm (Dosi 1982), technology systems (Freeman 1988), self-reinforcing mechanisms (Arthur 1988), collective choice - "homo socialis" (Sonis 1992) and the like.

The evolutionary framework has been postulated in the form of self-organization models capable of depicting series of successive innovations replacements. Such systems are essentially nonlinear and stochastic, giving rise to the possibility of multiple equilibria, bifurcation, chaos and pattern formation. As observed by Bruckner et al (1994) "the whole history of self-organization and synergetics is focused on a surprising new understanding of the relationship between micro and macro level descriptions".

According to Allen (1976) "Biological evolution due to selective advantage can be described by just such a mechanism, whereby the equations describing the populations of interacting genotypes in an ecosystem must be unstable to the appearance of a genotype or species if the evolution is to occur. The mean value of each population is described by dynamics involving such concepts as birth rate and death rate, which will be characteristic of competition for vital factors, the reproductive mechanism and the form of the trophic network describing the ecosystem. An individual event, however, such as the spontaneous mutation of a single individual corresponds to a different level of description from this average density dynamics". Thus a more appropriate approach to study the dynamics of technical change is the biophysical perspective which treats each technology as an interacting species in population dynamics. An innovation can thus be taken as a structural fluctuation in a given social system. Various conditions under which these fluctuations are amplified and generate instability in the system will eventually determine whether the new innovation will succeed in replacing the old one or not. A variety of scenarios, characterized by complete replacement, coexistence of competing innovations or technologies may emerge. (Batten 1982).

The evolutionary paradigm assumes particular significance in the context of modern industrial society where one often witnesses the phenomenon of increasing returns. The prevalence of increasing returns in an economy operates in the form of a positive feedback self-reinforcing mechanism. This emerges as a non-conventional economic theory, giving rise to multiple equilibria which correspond to different outcomes in the form of adoption shares (Arthur 1988). "The cumulation of small 'random' events drives the adoption process into the domain of one of these outcomes, not necessarily the most desirable one. And the increasing-returns advantage that accrues to the technology that achieves dominance keeps it locked-in to its dominant position" (Arthur 1988). The contextual relevance of such a framework is to examine the pattern of ultimate diffusion of the innovation. The eventual outcome of the innovation acquiring monopoly position or coexistence in the form of a shared market would depend on certain conditions (Hanson 1985). The framework offers important policy options in the form of guiding the path for competing technologies and tilting the market in favour of superior technology. However, as observed by David (1986), there are only 'narrow windows' in which such a policy would be effective.

The successful use of the evolutionary paradigm has given further impetus to researchers who have tended to utilize the developments in nonlinear system theory to understand the complexities of the diffusion process. It has been suggested by Mort (1992) that innovation diffusion may be examined as an analogy to mathematical percolation theory. The author stresses that the timing of the introduction of an innovation is crucial and must be guided by the percolation threshold. Bhargava, Kumar and Mukherjee (1993) have analyzed, with the help of stochastic cellular automata, diffusion processes in the case of two competing technologies.

A multi-layer approach based on nested time niche capacities has been proposed by Reggiani and Nijkamp (1994) to examine the technology diffusion process. They have developed a simulation model to explain transport technology replacement within the overall bounds of environmental constraints. It has been observed by several authors that diffusion models may exhibit chaotic behaviour for certain range of values of parameters (Gordon and Greenspan 1988). Dixon (1994) has argued, however, that chaotic behaviour is unlikely to emerge in discrete time diffusion models characterized by logistic type growth patterns. It would be worthwhile examining this issue with the help of available data sets on diffusion patterns.

Silverberg and Verspagen (1994) have proposed a model in the form of collective learning, innovation and growth in an evolutionary framework where bounded rationality plays a crucial role in the diffusion process. The authors have emphasized that innovations appear on the scene as a corollary of R & D efforts and are stochastic in nature. They have highlighted the emergence of bifurcation phenomenon in the long run growth patterns when the innovation function is logistic.

4.7 Epilogue

In conclusion one may observe the "the real message of the new concepts in science is that change and disequilibria are probably more natural than equilibrium and stasis. Those who can adapt and learn will survive and this will depend on their creativity" (Allen 1988).

References

Allen P M (1976) Evolution, Population Dynamics and Stability. Proceedings National Academy Sciences 73:665-668

Allen P M (1988) Evolution, Innovation and Economics. In: Dosi G, Freeman C, Nelson R, Silverberg G and L Soete (Eds) Technical Change and Economic Theory. Pinter Publishers London

Arthur W B (1988) Competing Technologies: an Overview in Technical Change and Economic Theory. Dosi G, Freeman C, Nelson R, Silverberg G and L Soete (Eds) Pinter Publishers London

Bailey N T J (1975) The Mathematical Theory of Infectious Diseases and its Applications. (2nd Edn) Griffin London

Banks R B (1994) Growth and Diffusion Phenomena (Mathematical frameworks and Applications). Springer-Verlag Berlin

Bartholomew D J (1976) Continuous Time Diffusion Models with Random Duration of Interest. Journal of Mathematical Sociology 4:187-199

Bartholomew D J (1982) Stochastic Models for Social Processes 3rd Edition John Wiley

Bartholomew DJ (1984) Recent Developments in Nonlinear Stochastic Modelling Social Processes. Canadian Journal of Statistics 12:39-52

Bass F M (1969) A New Product Growth Model for Consumer Durables. Management Science 15:215-227

Batten D (1982) On the Dynamics of Industrial Evolution. Regional and Urban Economics 12:449-462

Bhargava S C, Kumar A and A Mukherjee (1993) Stochastic Cellular Automata Model of Innovation Diffusion. Technological Forecasting and Social Change 44:87-97

Braun N (1995) Individual Thresholds and Social Diffusion. Rationality and Society 7 (forthcoming)

Brown L A (1981) Innovation Diffusion. A New Perspective. Methuen London

Bruckner E, Ebeling W, Montano M A J and A Scharnhorst (1994) Hyperselection and Innovation Described by a Stochastic Model of Technological Change in Evolutionary Economics and Chaos theory. Leydesdorff L and P V den Besselaar (Eds) Pinter Publishers London

Capello R and P Nijkamp (1993) Measuring Network Externalities Their Role on Corporate and Regional Performance. Research Memorandum. Vrije Universiteit Amsterdam The Netherlands

Chatterjee R and J Eliashberg (1990) The Innovation Diffusion Process in a Heterogeneous Population: A Micromodelling Approach. Management Science 9:1057-1079

David P (1986) Some New Standards for the Economics of Standardization in the Information Age. Paper 79. Center for Economic Policy Research Stanford

Davies S (1979) The Diffusion Process of Innovations. Cambridge University Press Cambridge

Deneckere R J and A de Palma (1993) The Diffusion of Consumer Durables in a Vertically Differentiated Oligopoly. Technical Reports. Université de Geneve Switzerland

de Palma A and C Lefevre (1988) The Theory of Deterministic and Stochastic Compartmental Models and its Applications in Urban Systems. Bertuglia C S, Leonard G, Leonardi G, Occelli S, Rabino G A, Tadei R and A G Wilson (Eds). Croom Helm London

Diekmann A (1989) Diffusion and Survival Models for the Process of Entry into Marriage. Journal of Mathematical Sociology 14:31-44

Diekmann A (1992) The Log-logistic Distribution as a Model for Social Diffusion Processes. Journal of Scientific and Industrial Research 51:285-290

Dixon R (1994) The Logistic Family of Discrete Dynamic Models in Chaos and Nonlinear Models in Economics. Creedy J and V L Martin (Eds). Edward Elgar

Dosi G (1982) Technological Paradigms and Technological trajectories. Research Policy 11:147-62

Eliashberg J and R Chatterjee (1986) Stochastic Issues on Innovation Diffusion Models in Innovation Diffusion Models of New Product Acceptance. Mahajan V and Y Wind (Eds). Ballinger Publishing Company Cambridge

Fisher J C and R H Pry (1971) A Simple Substitution Model of Technological Change. Technological Forecasting and Social Change 3:75-88

Freeman C (1988) Diffusion: The Spread of New Technology to Firms, Sectors and Nations in Innovation Technology and Finance. Heertje A (Ed) Basil Blackwell

Gordon T J and D Greenspan (1988) Chaos and Fractals: New Tools for Technological and Social Forecasting. Technological Forecasting and Social Change 34:1-26

Granovetter M and R Soong (1983) Threshold Models of Diffusion and Collective Behaviour. Journal of Mathematical Sociology 9:165-179

Haken H (1977) Synergetics. (also 3rd edition (1983) Springer-Verlag

Hanson W (1985) Band Wagons and Orphans: Dynamic Pricing of Competing Systems Subject to Decreasing Costs. PhD Dissertation. Stanford USA

Hiebert L D (1974) Risk, Learning and the Adoption of Fertilizer Responsive Seed Varieties. American Journal of Agricultural Economics 764-768

Horsthemke W and R Lefever (1984) Noise-induced Transitions. Springer-Verlag Berlin

Jain A and L P Rai (1988) Diffusion Models for Technology Forecasting. Journal of Scientific and Industrial Research 47:419-429

Jain D and K Raman (1992) A Stochastic Generalization of the Bass Diffusion Model. Journal of Scientific Industrial Research 51:216-228

Karmeshu and R K Pathria (1979) Co-operative Behaviour in a Non-linear Model of Diffusion of Information. Canadian Journal of Physics 57:1572 - 1578

Karmeshu and R K Pathria (1980a) Stochastic Evolution of a Nonlinear Model of Diffusion of Information. Journal of Mathematical Sociology 7:59-71

Karmeshu and Pathria R K (1980b) Diffusion of Information in a Random Environment. Journal of Mathematical Sociology 7:215-227

Karmeshu, Bhargava S C and V P Jain (1985) A Rationale for Law of Technological Substitution. Regional Science and Urban Economics 15:137-141

Karmeshu (1988) Demographic Models of Urbanization. Environment and Planning B 15:47-54

Karmeshu, Sharma C L and V P Jain (1992) Nonlinear Stochastic Models of Innovation Diffusion with Multiple Adoption Levels. Journal of Scientific and Industrial Research 51:229-241

Karmeshu (1995) Noise Induced Bistability in Innovation Diffusion Model (submitted)

Lal V B, Karmeshu and S Kaicker (1988) Modelling Innovation Diffusion with Distributed Time Lag. Technological Forecasting and Social Change 34:103-113

Lavaraj U A and A P Gore (1992) Modelling Innovation Diffusion Some Methodological Issues. Journal of Scientific and Industrial Research 51:291-295

Mahajan V and R A Peterson (1985) Models for Innovation Diffusion. Sage University Paper Series on Quantitative Applications in Social Sciences 07-001. Sage Publications Beverly Hills and London

Mahajan V and E Muller (1990) Timing, Diffusion and Substitution of Successive Generations of Desirable technological Innovations. The IBM Mainframe Case. Working paper No.71/90. Israeli Institute of Business Research Tel Aviv University Israel

Marchetti C (1991) Branching Out into the Universe in Diffusion of Technologies and Social Behaviour. Nakicenovic N and A Grubler (Eds). Springer-Verlag Berlin

Marchetti C and N Nakicenovic (1979) The Dynamics of Energy Systems and the Logistic Substitution Model. International Institute for Applied Systems Analysis (IIASA). Laxenburg Austria

May R M (1974) Stability and Complexity in Model Ecosystems. Princeton University Press Princeton

Metcalfe J S (1987) The Diffusion of Innovations: An Interpretive Survey, Paper presented for IFIAS Workshop on Technical Change and Economic Theory: The Global Process of Development. Maastricht

Montano M A J and W Ebeling (1980) A Stochastic Evolutionary Model of Technological Change. Collective Phenomena 3:107- 114

Montroll E W (1978) Social Dynamics and the Quantifying of Social Forces. Proceedings National Academy Sciences USA 75:4633-37

Montroll E W (1982) On the Dynamics of Technological Evolutions: Phase Transitions in Self-Organization and Dissipative Structures. Schieve W C and P M Allen (Eds). University of Texas Press Austin

Mort J (1992) Innovation as a Percolation Phenomenon, Presented at Third International Conference on Management of Technology. University of Miami USA

Nelson R R and S G Winter (1977) In Search of a Useful Theory of Innovation. Research Policy 36-76

Nicolis G and I Prigogine (1977) Self-organization in Nonequilibrium Systems. John Wiley New York

Norton J A and F M Bass (1987) A Diffusion Theory Model of Adoption and Substitution for successive Generations of High Technology Products. Management Science 33:1069-1086

Reggiani A and P Nijkamp (1994) Evolutionary Dynamics in Technological Systems: a Multi-layer Niche Approach in Evolutionary Economics and Chaos Theory. Leydesdorff L and P V den Besselaar (Eds). Pinter Publishers London

Rao D N, Karmeshu and V P Jain (1989) Dynamics of Urbanization: The Empirical Validation of the Rreplacement Hypothesis. Environment and Planning B 16:289-295

Rapoport A (1953a) Spread of Information Through a Population with Socio-cultural Bias I. Assumption of Transitivity. Bulletin of Mathematical Biophysics 15:523-533

Rapoport A (1953b) Spread of Information Through a Population with Socio-structural bias II. Various Models with Partial Transitivity. Bulletin of Mathematical Biophysics 15:536-546

Rogers E M (1983) Diffusion of Innovations 3rd Edition. The Free Press New York

Rogers E M (1990) The "Critical Mass" in the Diffusion of Interactive Technologies in Modelling the Innovation. Cornevale M, Lucertini M and S Nicosia (Eds). North Holland Amsterdam

Sahal D (1979) The Temporal and spatial Aspects of Diffusion of Technology. IEEE Transactions on Systems. Man and Cybernetics 9:829-839

Sharma C L, Pathria R K and Karmeshu (1982) Critical Behaviour of a Class of Nonlinear Stochastic Models of Diffusion of Information. Physical Review 26:3567-3574

Silverberg G (1988) Modelling Economic Dynamics and Technical Change: Mathematical Approaches to Self-organization and Evolution in Technical Change and Economic Theory. Dosi G, Freeman C, Nelson R, Silverberg G and L Soete (Eds). Pinter Publishers London

Silverberg G and B Verspagen (1994) Collective Learning, Innovation and Growth in a Boundedly Rational, Evolutionary Model. Journal of Evolutionary Economics 4:207-26

Sonis M (1992) Innovation Diffusion, Schumpertrian Competition and Dynamic Choice: A New Synthesis. Journal of Scientific and Industrial Research 51:172-186

Stoneman P (1976) Technological Diffusion and the Computer Revolution: The UK Experience. Cambridge University Press Cambridge

Stoneman P (1981) Intrafirm Diffusion, Bayesian Learning and Profitability. Economic Journal 91:375-388

Vij A K, Vrat P and Sushil (1992) Technological Change in Multisector Modelling. Journal of Scientific and Industrial Research 51:266-272

Weidlich W (1991) Physics and Social Science - The Approach of Synergetics. Physics Reports 204:1-163

PART B

Theories and Models

5 Firms R&D Investments, Innovation and Market Shares

Domenico Campisi, Agostino La Bella,
Paolo Mancuso and Alberto Nastasi

Õ3|

Õ3ユ

G3|

6 3ᵭ

5.1 Introduction

Technological innovation is one of the major factors in determining the long term performance of firms and economies. Here, this subject is analyzed from the point of view of the relationship between the innovative behaviour of a given firm within an oligopolistic industry and its market share. As a matter of fact, the firm capacity to change is closely connected to the size of its R&D activity which generate directly product and/or process innovation, and also develops the firm's ability to identify, assimilate, and exploit knowledge from the external environment, taking advantage of the public research effort and even of the rivals' R&D expenditure (Tilton 1971; Allen 1977; Mowery 1983; Cohen and Levinthal 1989; Campisi and Nastasi 1993). The traditional approach to industrial innovative activity devotes little attention to this absorptive capacity of R&D: following Arrow (1962) and Nelson (1959), economists have assumed that any innovation created by one firm provides usable information to other firms at little or no cost, so according to an unpatented innovation the status of a public good. In other terms, the intra-industry technological knowledge transfer requires negligible R&D efforts (Spence 1984; Tirole 1988; Quirmbach 1993).

An alternative analysis of R&D competition has been based on the decision-theoretic approach which assumes either that there is only one firm doing R&D or, more generally, that a firm's environment, including the rivals R&D expenditures, is taken as exogenous, so that each firm assumes that its R&D spending has no influence on its rivals spending (Kamien and Schwartz 1972, 1982; Grossman and Shapiro 1986). This formulation has been characterized as a dynamic analysis in the sense that each firm chooses a path of spending on R&D over time by solving an optimal control problem. Reinganum (1982) presented an interesting combination of game and decision theoretic approaches by modelling firms as strategic agents engaged in a race for a particular technological breakthrough where the decisions of each firm evolve over time. The resulting dynamic game model synthetizes the analytic methods utilized by the two alternative formulations described above: static game theory and optimal control.

The present chapter can be set within this analytical framework in the sense that firms R&D investments plans take the shape of a dynamic multi-stage (multi-period) decision process which can be properly analyzed through the dynamic game theory. In particular, we characterize the R&D competition as a dynamic noncooperative feedback game (Baser and Oldser 1982) where firms compete over time in order to maximize their market shares. However, we depart from models of race for innovation (with or without patent protection), where only the successful firms receive a payoff, since we simply assume a deterministic relation between firms R&D expenditures and their payoffs expressed in terms of market shares. In our analysis, we refer to the concept of firm's stock of technological knowledge provided by Cohen and Levinthal (1989), according to which the external R&D (conducted outside the firm) contributes to a firm's stock only if R&D activity is systematically carried out of by the firm itself. So, a firm R&D expenditure plays two roles: it enhances directly its stock of technological knowledge by increasing the accumulated investment in R&D, and indirectly by improving its capability of exploiting externally available knowledge. We set the concept of firm's stock of technological knowledge within a dynamic model of oligopoly where at each period firms decide simultaneously and non-cooperatively their R&D efforts. Thus, each firm chooses its optimal R&D strategy over time taking into account that an increase in a firm stock of technological knowledge allows to introduce a new production process. Consequently, it allows a firm to move downwards its cost function and improve its ability of exerting the price competition to gain an increasing market share to the detriment of its competitors. This approach permits to analyze the intertemporal impacts of firms R&D expenditure on market shares in the industry and the conditions for the existence of the optimal feedback Nash equilibrium level of R&D investments for all the firms in the industry. Furthermore, the traditional result that increasing spillovers reduces the incentive to invest in R&D is discussed. As a matter of fact, the possibility of exploiting the knowledge generated by rival firms provides a positive incentive to invest in R&D as the degree of spillovers increases; and the disincentive associated with rival firms' assimilation of firm i R&D is dampened because rival firms' absorptive capacity requires a considerable resource allocation in their R&D activity.

The chapter is organized as follows. The next section analyses the relations among firms R&D investments, the extra-industry R&D activity and firms stock of technological knowledge. In Section 5.3, the features of an industry in which R&D activity is the crucial competitive instrument for the firms are discussed, and the relations between production cost, R&D cost, firms stock of technological knowledge and market shares examined. On this basis, in Section 5.4, the intertemporal impacts of firms R&D investments on their market shares are evaluated, and the influence of intra-industry spillovers and extra-industry R&D activity on the optimal amount of firms R&D expenditure discussed, providing the conditions for the existence of a Nash equilibrium R&D strategy for all the firms in the industry. Section 5.5 presents some numerical simulation concerning a duopolistic industry and Section 5.6 concludes our discussion summarizing the main conclusions.

5.2 The Formation of the Stock of Technological Knowledge

Consider an industry with n firms ($2 \leq n < \infty$) where time elapses in a succession of discrete time periods indexed by h (h=1, 2,...). Let m_h^i be the R&D expenditure of firm i in period h and

$$M_h^i = \sum_{k=h-T}^{h-1} m_k^i \quad i=1,...,n; \; h=1, 2,...; \; T>1 \qquad (5.1)$$

the stock of R&D accumulated investments available to firm i for improving the technological performance of its production process at time h. Notice that firm i available stock at time h, M_h^i, is equal to the R&D capital accumulation from h-T up to the previous period h-1, where T-1 is the relevant number of periods with respect to the state of technology. Analogously, we denote with v_h the extra-industry R&D expenditure (R&D expenditures of government, university laboratories and the firms in other industries) in period h and

$$V_h = \sum_{k=h-T}^{h-1} v_k \quad h=1, 2,...; \; T>1 \qquad (5.2)$$

the extra-industry R&D stock at time h. Firm i's stock of technological knowledge at date h, Z_h^i is defined by

$$z_h^i = M_h^i + g^i(M_h^i) \; [\theta \sum_{j \neq i} M_{h-1}^j + V_{h-1}] \quad i,j=1,...,n; \; h=1,2,... \qquad (5.3)$$

where $g^i(M_h^i)$ is the absorptive capacity of firm i in period h, that is, firm i's ability to assimilate and exploit external (conducted outside firm i) R&D; $\sum_{j \neq i} M_{h-1}^j$ is the total amount of R&D stock at h-1 of the other firms in the industry; $0 \leq \theta \leq 1$ is the degree of intra-industry spillovers, that is, the degree to which the research effort of one firm may spill over to a pool of knowledge potentially available to all other firms. If $\theta=0$, there is no benefit for firm i arising from its rival firms' R&D, whereas if $\theta=1$ the R&D stock of rival firms at h-1 are potentially and completely available to firm i at time h. The degree of spillovers is independent of the firms strategies and only depends on the patent system. We also assume that θ is constant over time. Observe that relation (5.3) implies that the external R&D activities, conducted by the other n-1 firms in the industry and outside the industry, also contribute to firm i's stock of technological knowledge. Furthermore, we assume that the external stock of R&D can be incorporated in firm i's stock of technological knowledge with a one period lag. The contribution of the other n-1 firms in the industry is filtered by θ, whereas, the extra-industry R&D is postulated unprotected by patents. Actually, while R&D expenditures of government and university laboratories are unprotected by definition, the results of other industries R&D are not. However, these results have the same importance for all the firms in the industry. So, without loss of generality, the spillover parameter applicable to extraindustry R&D can be uniformly settled equal to one. Thus, the total stock of external accumulated investments in R&D, potentially available to firm i in period h is represented by $[\theta\Sigma_{j\neq i}\, M^j_{h-1} + V_{h-1}]$. The portion of this external stock which is effectively assimilated and exploited by firm i depends on its absorptive capacity $g^i(M^i_h)$. In this connection, we assume the following.

Assumption 1: The absorptive capacity function $g^i(M_h^i)$ is twice continuously differentiable and satisfies $0 \leq g^i(M_h^i) < 1$ for

all $M_h^i \geq 0$, $g^i(M_h^i) = 0$ for $M_h^i = 0$, $g^{i\prime}(M_h^i) < 0$ and $g^{i\prime\prime}(M_h^i) > 0$ for

all $M_h^i \geq 0$ (i=1,2,...,n; h=1,2,...)

Assumption 1 incorporates some fundamental properties of the firm's absorptive capacity. It is an increasing function of its accumulated investments in R&D, though at a decreasing rate. Moreover, if M^i_h is equal to zero, then firm i does not have any absorptive capacity: a firm is able to assimilate the external technological knowledge only if it carries out systematically R&D activity (Cohen and Levinthal 1989). This is different from the traditional approach where the external R&D provides usable information to firm i at little or no cost, so that $g^i(M^i_h)$ would be equal to one independently of firm i R&D investments and the absorptive capacity of R&D is completely ignored. On the contrary, according to relation (5.3) and assumption 1, firms invest in R&D not only to pursue directly innovations but also to enhance and maintain their capabilities to assimilate and exploit externally available information. These two faces of R&D can be highlighted by differentiating relation (5.3) and taking into account relations (5.1) and (5.2). In fact, from these three relations, it is easy to verify that the marginal impact of firm i R&D investments on its own stock of technological knowledge is measured by

$$\frac{\partial z^i_{h+t+1}}{\partial m^i_h} = \frac{\partial z^i_{h+t+1}}{\partial M^i_{h+t+1}} = 1 + g_i^{\,\prime}(M^i_{h+t+1})[\theta \sum_{j\neq i} M^j_{h+t} + V_{h+t}] \qquad (5.4)$$
$$i,j=1,...,n; \quad h=1,2,...; \quad t\geq 0$$

while the marginal impacts of rival firms and extra-industry investments in R&D on firm i stock are measured respectively by:

$$\frac{\partial z^i_{h+t+1}}{\partial \sum_{j\neq i} m^j_h} = \frac{\partial z^i_{h+t+1}}{\partial \sum_{j\neq i} M^j_{h+1}} = \theta \; g^i(M^i_{h+t+1}) \qquad (5.5)$$
$$i,j=1,...,n; \quad h=1,2,...; \quad t>0$$

and

$$\frac{\partial z_{h+t+1}^i}{\partial v_h} = \frac{\partial z_{h+t+1}^i}{\partial V_{h+t}} = g^i(M_{h+t+1}^i) \quad i=1,\ldots,n; \quad h=1,2,\ldots; \quad t>0 \qquad (5.6)$$

By virtue of Assumption 1, we observe that

$$\frac{\partial z_{h+t+1}^i}{\partial m_h^i} > 0; \quad \frac{\partial^2 z_{h+t+1}^i}{\partial m_h^{i^2}} < 0 \quad i=1,\ldots,n; \quad h=1,2,\ldots; \quad t\geq 0 \qquad (5.7)$$

$$\frac{\partial z_{h+t+1}^i}{\partial \sum_{j\neq i} m_h^j} \geq 0 \quad i,j=1,\ldots,n; \quad h=1,2,\ldots; \quad t>0 \qquad (5.8)$$

$$\frac{\partial z_{h+t+1}^i}{\partial v_h} \geq 0 \quad i=1,\ldots n; \quad h=1,2,\ldots; \quad t>0 \qquad (5.9)$$

Relations (5.4) and (5.7) show that firm i R&D expenditure at time h, m_h^i, always produce a positive impact on its own stock of technological knowledge as from one period lag onwards, z_{h+t+1}^i for all $t\geq 0$, though at a decreasing rate. On the contrary, relations (5.5) and (5.8) indicate that firm i competitors R&D expenditures at time h, $\Sigma_{j\neq i} m_h^i$, can benefit firm i stock of technological knowledge as from two periods lag onwards, z_{h+t+1}^i for all $t>0$. Moreover, the marginal impact produced by rival firms R&D investments can be equal to zero. In particular, relations (5.5) shows that firm i exploitation of its competitors R&D expenditures is realized through the interaction of intra-industry spillovers and firm i absorptive capacity as from two periods lag onwards. So, if $\theta = 0$ firm i cannot assimilate what is not spilled out; on the other hand, if firm i does not invest in R&D, then it is not able to assimilate any external R&D even if $\theta > 0$ (see Assumption l). In accordance, relations (5.6) and (5.9) indicate that extraindustry R&D, unprotected by definition, cannot be passively assimilated either.

5.3 Firms R&D Investments and Market Shares

In the previous section we have specified the relations between (industry and extraindustry) R&D investments and firms' stocks of technological knowledge. Now we will focus the features of an oligopolistic industry in which R&D activity is the crucial competitive instrument for the firms.

Let us assume that the n firms ($2\leq n< \infty$) produce non-homogeneous goods, so that the cross-elasticity of demand is not infinite at equal prices. Each firm produces a single good and behaves as a price-taker on the market of its inputs. Moreover, each firm acts so as to maximize its own market share which is assumed to be a decreasing function of its own product price and an increasing function of the prices charged by rival firms. Now assume that, within each period h, the n firms "meet only once" in the market, and simultaneously and non-cooperatively charge their price. In the short run, i.e., in a context of rigid cost structures and product characteristics, firms can handle their prices consistently with their cost functions. So, within each period h, firms exert their competitive pressure by reducing prices to the level of their unit costs, and consequently the market share function of firm i at time h, s_h^i, can be written as

$$s_h^i = s^i(c_h^i, c_h^{-i}) \qquad i=1,\dots,n; \ h=1,2,\dots \qquad (5.10)$$

where c_h^i is firm i unit cost at time h and

$$c_h^{-i} = (c_h^1,\dots, c_h^{i-1}, c_h^{i+1},\dots,c_h^n).$$

Assumption 2: The market share function $s_h^i = s^i(c_h^i, c_h^{-1})$ is twice continuously differentiable and satisfies $0 \le s_h^i \le 1$,

with $\Sigma_i s_h^i = 1$ $(i=1,2,\dots,n; \ h=1,2,\dots)$, $\partial s_h^i / \partial c_h^i < 0$ for

all $c_h^i > 0$ $(i=1,2,\dots,n; \ h=1,2,\dots)$, and $\partial s_h^i / \partial c_h^j > 0$ for

all $c_h^j > 0$ $(i \ne j; \ i,j,=1,2,\dots,n; \ h=1,2,\dots)$. Furthermore, $\partial^2 s_h^i / \partial c_h^{i^2} < 0$ for all

$c_h^i > 0$ $(i=1,2,\dots,n; \ h=1,2,\dots)$, and $\partial^2 s_h^i / \partial c_h^{j^2} > 0$ for

all $c_h^j > 0$ $(i \ne j; \ i,j=1,2,\dots,n; \ h=1,2,\dots)$.

We split firm i unit cost at time h, c_h^{-i} , into two components: the unit production cost, q_h^i , which depends on firm i stock of technological knowledge at time h, $q_h^i = q^i(z_h^i)$, and the unit R&D cost, d_h^i , which depends on firm i R&D accumulated investments up to h-1, $d_h^i = d^i(M_h^i)$

$$c_h^i = q^i(z_h^i) + d^i(M_h^i) \qquad i=1,\dots,n; \ h=1,2,\dots \qquad (5.11)$$

Assumption 3: The unit production cost function $q_h^i = q^i(z_h^i)$ is twice continuously differentiable and satisfies $0 < q_h^i \le \omega^i$ for all $z_h^i \ge 0$, $q_h^i = \omega_i$ for $z_h^i = 0$, $q^{i'}(z_h^i) < 0$ and $q^{i''}(z_h^i) > 0$ for all $z_h^i \ge 0$ $(i=1,2,\dots,n; \ h=1,2,\dots)$.

Assumption 4: The unit R&D cost function $d_h^i = d^i(M_h^i)$ is twice continuously differentiable and satisfies $d_h^i \ge 0$ for all $M_h^i \geq 0$, $d_h^i = 0$ for $M_h^i = 0$, $d^{i'}(M_h^i) > 0$ and $d^{i''}(M_h^i) \geq 0$ for

all $M_h^i \geq 0$ $(i=1,2,\ldots,n;\ h=1,2,\ldots,)$.

Relations (5.1), (5.2), (5.3), (5.10) and (5.11), together with assumptions 1, 2, 3 and 4 establish the intertemporal links between firms R&D efforts and market sharing. In particular, from relations (5.1) and (5.3) we note that the R&D expenditure of firm i at time h, m_h^i, produces an increase in M_{h+t}^i, and consequently in z_{h+t}^i and z_{h+t+1}^i for all $t \geq 1$. In addition, taking also into account relations (5.10) and (5.11) we observe that

$$
\partial \frac{s_{h+t+1}^i}{\partial m_h^i} =
\begin{cases}
\dfrac{\partial s_{h+t+1}^i}{\partial s_{h+t+1}^i} \left[q^{i'}(z_{h+t+1}^i)\, \dfrac{\partial z_{h+t+1}^i}{\partial m_h^i} + d^{i'}(m_h^i) \right] & \text{for } t=0 \quad (5.12a) \\[3ex]
\dfrac{\partial s_{h+t+1}^i}{\partial c_{h+t+1}^i} \left[q^{i'}(z_{h+t+1}^i)\, \dfrac{\partial z_{h+t+1}^i}{\partial m_h^i} + d^{i'}(m_h^i) \right] + \\[3ex]
\quad + \displaystyle\sum_{j \neq 1} \dfrac{\partial s_{h+t+1}^i}{\partial c_{h+t+1}^j}\, q^{i'}(z_{h+t+1}^j)\, \dfrac{\partial z_{h+t+1}^j}{\partial m_h^i} & \text{for } t>0
\end{cases}
$$

$$ i \neq j; i,j = 1,2,\ldots,n;\ h=1,2,\ldots \qquad (5.12b) $$

Relations (5.12a) and (5.12b) measure the total impact of a marginal increase in firm i R&D expenditure at h on its market share in the following periods. In particular, from assumptions 1, 2, 3 and 4, and relation (5.3) one obtains that the first two addenda of the right-hand side of (5.12a) and (5.12b) satisfy

$$ \frac{\partial s_{h+t+1}^i}{\partial c_{h+t+1}^i}\, q^{i'}(z_{h+t+1}^i)\, \frac{\partial z_{h+t+1}^i}{\partial m_h^i} > 0 \quad i=1,2,\ldots,n;\ h=1,2,\ldots;\ t \geq 0 \quad (5.13) $$

$$ \frac{\partial s_{h+t+1}^i}{\partial c_{h+t+1}^i}\, d^{i'}(m_h^i) < 0 \quad i=1,2,\ldots,n;\ h=1,2,\ldots;\ t \geq 0 \qquad (5.14) $$

Relation (5.14) indicates that an increase in m_h^i causes an increase in d_{h+t+1}^i, which, in its turn, causes a negative impact on s_{h+t+1}^i, for all $t \geq 0$. On the contrary, relation (5.13) indicates that an increase in m_h^i causes an increase in z_{h+t+1}^i and consequently a decrease in q_{h+t+1}^i which, in its turn, produces a positive impact on s_{h+t+1}^i, for all $t \geq 0$. Furthermore, from the definition of the stock of technological knowledge given by (5.3), we observe that the last term of (5.13) satisfies

$$\frac{\partial z^i_{h+t+1}}{\partial m^i_h} = 1 + g^{i'}(m^i_h)\, [\theta\, \Sigma_{j\neq i}\, M^j_{h+t} + V_{h+t}] \geq 0 \quad i,j=1,\dots,n;\ h=1,2,\dots;\ t\geq0$$

where $g'(m^i_h) > 0$ (see assumption 1) and $0 \leq \theta \leq 1$. So, we can state

Proposition 1: The impact of a marginal increase in firm i R&D expenditure at h, m^i_h, on its own market share in the following period h+1, s^i_{h+t+1} for t =0, is positively correlated to the degree of intra-industry spillovers, θ, the total amount of R&D stock available to the other firms in the industry at h, $\Sigma_{j\neq1}\, M^j_{h+1}$, and the level of the extra-industry R&D stock at h, V_{h+t}, for t=0. Furthermore, if $\theta=0$ the impact of an increase in m^i_h on s^i_{h+t+1} is independent of $\Sigma_{j\neq i}\, M^j_{h+1}$, for t=0.

Relation (5.3) and assumptions 1, 2 and 3, assures that the third addend of the right hand side of (5.12b) satisfy

$$\sum_{j\neq i} \frac{\partial s^i_{h+t+1}}{\partial c^i_{h+t+j}}\, q^{j'}(z^j_{h+t+1})\, \frac{\partial z^j_{h+t+1}}{\partial m^i_h} \leq 0 \quad i,j=1,2,\dots,n;\ h=1,2,\dots;\ t>0 \quad (5.15)$$

Relation (5.15) indicates that an increase in m^i_h can benefit rival firms' stock of technological knowledge, z^j_{h+t+1}, and consequently can reduce their unit production cost, q^i_{h+t+1}, for any $j\neq i$ and for all $t>0$, so as to produce a negative impact on firm i market share s^i_{h+t+1}, for all $t>0$. Furthermore, from the definition of the stock of technological knowledge given by (5.3), we observe that the last term of (5.15) satisfies

$$\frac{\partial z^j_{h+t+1}}{\partial m^i_h} = \theta\, g^j(M^j_{h+t+1}) \geq 0 \quad i\neq j;\ i,j=1,2,\dots,n;\ h=1,2,\dots;\ t>0$$

So, we can state

Proposition 2: If $\theta=0$ or $M^j_{h+t+1}=0$ for all $j\neq i$ and all $t>0$, a marginal increase in firm i R&D expenditure at h, m^i_h does not produce any change in its own market share from period h+1 onwards. If $\theta>0$ and $M^i_{h+t+1}>0$ for at least one $j\neq i$ and $t>0$, the impact of a marginal increase in m^i_h on firm i market share from period h+2 onwards, s^i_{h+t+1} for $t>0$, depends on θ and M^j_{h+t+1}.

5.4 Conditions for the Existence of Optimal R&D Strategies

Propositions 1 and 2 imply that the effect a firm decision to invest in R&D at a given period is related to the choices of its competitors and has to be evaluated over a multi-period horizon. In addition, the effects of a firm's decision depend on its past choices, which are summarized by its accumulated stock of R&D investments (see relation (5.1)), and influence the future ones. This dynamic multi-stage decision process can be properly described through the dynamic game theory. In this framework, we define

$$M_{h+1} = M_h + m_h \quad h=1,2,\ldots$$

as the state equation of the system, where $M_h = (M_h^1, M_h^2,\ldots, M_h^n)$ and $m_h = (m_h^1, m_h^2,\ldots, m_h^n)$ are the state and the decision vector at time h respectively. According to Section 5.3, the objective of firm i is to maximize its market shares over the future T-1 periods:

$$\sum_{h=K-T}^{K-1} s_{h+1}^i = \sum_{h=K-T}^{K-1} s^i (M_h, m_h) = S^i(m^1, m^2,\ldots,m^n) \quad i=1,\ldots,n$$

where M_{K-T}, the initial state vector, is specified a priori and $m^i=(m_{K-T}^i,\ldots, m_{K-1}^i)$. Moreover, we denote with Q_h^i the set of all the possible choices available to firm i at period h with regard to R&D investments, $m_h^i \in Q_h^i$, and assume the following rules.

Rule A: Firms are not able to make binding agreements.

Rule B: At every stage h each firm knows: the set of possible choices available to the n firms Q_h^i, i=1,\ldots,n, all payoff functions s^i, i=1,\ldots,n, and the state vector of the system M_h^i, i=1,\ldots,n.

Rule A implies that at each stage, firms decide simultaneously and non-cooperatively their R&D efforts whereas rule B assures that at every stage each firm knows the set of possible R&D investments, the market share functions si and the available stock of R&D investments of all the n firms in the industry. The above rules characterize a dynamic noncooperative feedback game. This class of multi-stage game can admit a feedback Nash equilibrium which is defined as follows. Given the set of inequalities

$$
\begin{bmatrix}
S^{1*} \equiv S^1(m^{1*}, m^{2*}, \ldots, m^{n*}) \geq S^1(m^1, m^{2*}, \ldots, m^{n*}) \\
\cdots\cdots\cdots\cdots\cdots\cdots\cdots\cdots\cdots\cdots\cdots\cdots\cdots\cdots \\
\cdots\cdots\cdots\cdots\cdots\cdots\cdots\cdots\cdots\cdots\cdots\cdots\cdots\cdots \\
S^{n*} \equiv S^n(m^{1*}, m^{2*}, \ldots, m^{n*}) \geq S^n(m^1, m^{2*}, \ldots, m^n)
\end{bmatrix}
$$

and the following T-1 n-tuple of inequalities

$$
\text{level } K-1 \begin{bmatrix}
S^1(m^1_{K-T}, \ldots, m^1_{K-2}, m^{1*}_{K-1}; m^2_{K-T}, \ldots, m^2_{K-2}, m^{2*}_{K-1}; m^n_{K-T}, \ldots, m^n_{K-2}, m^{n*}_{K-1}) \\
\geq S^1(m^1_{K-T}, \ldots, m^1_{K-2}, m^1_{K-1}; m^2_{K-T}, \ldots, m^2_{K-1}, m^{2*}_{K-1}; m^n_{K-T}, \ldots, m^n_{K-2}, m^{n*}_{K-1}) \\
\cdots\cdots\cdots\cdots\cdots\cdots\cdots\cdots\cdots\cdots\cdots\cdots\cdots \\
\cdots\cdots\cdots\cdots\cdots\cdots\cdots\cdots\cdots\cdots\cdots\cdots\cdots \\
S^n(m^1_{K-T}, \ldots, m^1_{K-2}, m^{1*}_{K-1}; m^2_{K-T}, \ldots, m^2_{K-2}, m^2_{K-1}; m^n_{K-T}, \ldots, m^n_{K-2}, m^{n*}_{K-1}) \\
\geq S^n(m^1_{K-T}, \ldots, m^1_{K-2}, m^{1*}_{K-1}; m^2_{K-T}, \ldots, m^2_{K-2}, m^{2*}_{K-1}; m^n_{K-T}, \ldots, m^n_{K-2}, m^n_{K-1})
\end{bmatrix}
$$

$$
\text{level } K-2 \begin{bmatrix}
S^1(m^1_{K-T}, \ldots, m^{1*}_{K-2}, m^{1*}_{K-1}; m^2_{K-T}, \ldots, m^{2*}_{K-2}, m^{2*}_{K-1}; m^n_{K-T}, \ldots, m^{n*}_{K-2}, m^{n*}_{K-1}) \\
\geq S^1(m^1_{K-T}, \ldots, m^1_{K-2}, m^{1*}_{K-1}; m^2_{K-T}, \ldots, m^2_{K-1}, m^{2*}_{K-1}; m^n_{K-T}, \ldots, m^{n*}_{K-2}, m^{n*}_{K-1}) \\
\cdots\cdots\cdots\cdots\cdots\cdots\cdots\cdots\cdots\cdots\cdots\cdots\cdots \\
\cdots\cdots\cdots\cdots\cdots\cdots\cdots\cdots\cdots\cdots\cdots\cdots\cdots \\
S^n(m^1_{K-T}, \ldots, m^{1*}_{K-2}, m^{1*}_{K-1}; m^2_{K-T}, \ldots, m^{2*}_{K-2}, m^{n*}_{K-1}; m^n_{K-T}, \ldots, m^{n*}_{K-2}, m^{n*}_{K-1}) \\
\geq S^n(m^1_{K-T}, \ldots, m^{1*}_{K-2}, m^{1*}_{K-1}; m^2_{K-T}, \ldots, m^{2*}_{K-2}, m^{2*}_{K-1}; m^n_{K-T}, \ldots, m^n_{K-2}, m^{n*}_{K-1})
\end{bmatrix}
$$

$$
\cdots\cdots\cdots\cdots\cdots\cdots\cdots\cdots\cdots\cdots\cdots\cdots\cdots
$$
$$
\cdots\cdots\cdots\cdots\cdots\cdots\cdots\cdots\cdots\cdots\cdots\cdots\cdots \tag{5.17}
$$
$$
\cdots\cdots\cdots\cdots\cdots\cdots\cdots\cdots\cdots\cdots\cdots\cdots\cdots
$$

$$
\text{level } K-T \begin{bmatrix}
S^1(m^{1*}_{K-T}, \ldots, m^{1*}_{K-2}, m^{1*}_{K-1}; m^{2*}_{K-T}, \ldots, m^{2*}_{K-2}, m^{2*}_{K-1}; m^{n*}_{K-T}, \ldots, m^{n*}_{K-2}, m^{n*}_{K-1}) \\
\geq S^1(m^1_{K-T}, \ldots, m^{1*}_{K-2}, m^{1*}_{K-1}; m^{2*}_{K-T}, \ldots, m^{2*}_{K-2}, m^{2*}_{K-1}; m^{n*}_{K-T}, \ldots, m^{n*}_{K-2}, m^{n*}_{K-1}) \\
\cdots\cdots\cdots\cdots\cdots\cdots\cdots\cdots\cdots\cdots\cdots\cdots\cdots \\
\cdots\cdots\cdots\cdots\cdots\cdots\cdots\cdots\cdots\cdots\cdots\cdots\cdots \\
S^n(m^{1*}_{K-T}, \ldots, m^{1*}_{K-2}, m^{1*}_{K-1}; m^{2*}_{K-T}, \ldots, m^{2*}_{K-2}, m^{n*}_{K-1}; m^{n*}_{K-T}, \ldots, m^{n*}_{K-2}, m^{n*}_{K-1}) \\
\geq S^n(m^{1*}_{K-T}, \ldots, m^{1*}_{K-2}, m^{1*}_{K-1}; m^{2*}_{K-T}, \ldots, m^{2*}_{K-2}, m^{2*}_{K-1}; m^n_{K-T}, \ldots, m^{n*}_{K-2}, m^{n*}_{K-1})
\end{bmatrix}
$$

which are to be satisfied for all m^i_h, $i=1,\ldots,n$; $h=K-T,\ldots,K-1$, then

Definition: Every n-tuple $(m^{1*}, m^{2*},\ldots, m^{n*})$ that satisfies the set of inequalities (5.16) and (5.17) constitutes a (pure) feedback Nash equilibrium solution for the feedback game.

In this connection, we have the following:

Proposition 3. Every n-tuple $(m^{1*}, m^{2*},\ldots, m^{n*})$ that satisfies the set of inequalities (5.17) also satisfies the set of inequalities (5.16).

Proof. Follows directly from Basar and Oldser (1982, p. 117).#

In view of proposition 3, the inequalities (5.16) become redundant for determining the feedback Nash equilibrium, so that we can only consider the set of inequalities (5.17) without any loss of generality. The first set of inequalities of (5.17) are fulfilled for all m_h^i, $i=1,\ldots,n$; $h=K-T,\ldots,K-2$. Then, according to the state equation above defined, they have to hold for all possible values of state m_{K-1}.which are reachable by utilization of some combination of m_h^i. Therefore, the first set of inequalities of (5.17) is connected to the problem of seeking a Nash equilibrium of a static noncooperative norms game with the objective function

$$\sum_{h=K-T}^{K-1} s^i(M_h, m_h) \quad i=1,\ldots,n.$$

If this game admits a Nash equilibrium m_{K-1}^*, it will only depend on M_{K-1}. By substituting m_{K-1}^*, into the objective function we obtain

$$s^i(M_{K-1}, m_{K-1}^*) + \sum_{h=K-T}^{K-2} s^i(M_h, m_h) \quad i=1,\ldots,n$$

which represents the objective function of the static game related to the second set of inequalities (5.17). Here again, if this game admits a Nash equilibrium m_{K-2}^*, it will only be function of M_{K-2}. Then, using recursively this procedure backward to K-T, i.e., solving sequentially the T-1 static games, we obtain the feedback Nash solution for the whole game. The essential requirement of this procedure is the existence of a Nash solution for all the T-1 static games. In this connection, we need the following lemmas.

Lemma 1. If Q_h^i is closed and bounded, then it is compact and convex for all $i=1,..,n$ and $h=K-T,..,K-1$.

Proof. Since Q_h^i is a subset of $\Re$, it is convex. Moreover, by the Heine-Borel theorem if Q_h^i is closed and bounded, it is compact.#

Lemma 2. Let Q_h the Cartesian product of the sets of all the possible choices available to the n firms at time h,

$$Q_h = \overset{n}{\underset{i=1}{\otimes}} Q_h^i.$$

Then, at every stage h, h=K-T,..., K-1, the market share function of each firm i is concave with respect to $m_h^i \in Q_h^i$ for all $m_h \in Q_h$ and all i=1,..., n.

Proof. From assumption 1, 2, 3 and 4, and relations (5.1), (5.3), (5.10) and (5.11), one obtains that $\partial^2 s^i / \partial m_h^{i^2} \leq 0$ for all $m_h \in Q_h$; h=K-T,..., K-1; i=1,...,n. So, the lemma follows.#

Now, we can state the following

Theorem. Let G_h be the static non-cooperative n-firms game at period h, where Q_h^i is closed and bounded for all i=1,...,n and h=K-T,..., K-1. Then, every G_h, h=K-T,..., K-1, admits at least one Nash equilibrium solution.

Proof. From assumptions 2, 3 and 4 it follows that the market share function s^i of each firm is continuous and bounded for all $m_h \in Q_h$. Taking also into account lemmas 1 and 2, we note that all the sufficient conditions required by Friedman (1990, Th 3.1, p. 72) for the existence of at least one Nash equilibrium solution for a static noncooperative n-players game are fulfilled.#

5.5   A Numerical Simulation

In this section the effect of inter-industrial and extra-industrial R&D expenditures on the optimal R&D investments at firm level are explored by means of numerical simulations. For simplicity sake we consider an industry with two firms, α and β, that compete over a three periods horizon (h=0,1,2). Therefore, according to Section 5.4, firms choose the levels of investments at time 0 and 1. Two cases are examined. In the first one we suppose perfect symmetry, so that both firms have the same budget constraint on the R&D expenditure, i.e. $m_h^i \leq 3$ (i=α, β; h=0, 1). In the second case we suppose that firm β presents a lower budget constraint on R&D expenditure, i.e. $m_h^\beta \leq 1.5$ (h=0, 1). In both cases we compare two situations characterized by opposite assumptions on the absorptive capacity function. The first assumption is closely related to the model previously discussed, so that the firm's absorptive capacity depends on the level of the R&D investments and it always less than one:

$$g^i(M_h^i) = \frac{M_h^i}{R} \le 1 \quad i=\alpha, \beta; \ h=0,1,2$$

where $M_h^i = M_{h-1}^i + m_{h-1}^i$ and $R = 100$ (observe that the firms' R&D expenditure constraints assure that R exceeds the maximum level of R&D investments that each firm is able to accumulate over the three periods horizon). In the second assumption, the absorptive capacity equal to 1 whatever the level of firms' R&D investment is:

$$g^i(M_h^i) = 1, \ \forall \ M_h^i \ge 0 \quad i=\alpha, \beta; \ h=0,1,2$$

Furthermore, the extra-industry R&D stock of accumulated investments, V, is set constant over the whole time horizon, while the remaining functions and parameters of the model have been specified as follows:

$$z_h^i = M_h^i + g^i(M_h^i) \ [\theta M_{h-1}^j + V] \quad i \ne j; \ i,j=\alpha, \beta; \ h=0,1,2$$

$$q_h^i = \frac{\omega^i}{1+z_h^i} \quad i=\alpha, \beta; \ h=0,1,2$$

$$\omega^i = 20 \quad i=\alpha, \beta$$

$$d^{h^i} = \frac{M_h^i}{2} \quad i=\alpha, \beta; \ h=0,1,2$$

$$c_h^i = q_h^i + d_h^i \quad i=\alpha, \beta; \ h=0,1,2$$

$$s_h^\alpha = \frac{1}{2}\left[1+\frac{c_h^\beta}{\omega^\alpha}-\frac{c_h^\alpha}{\omega^\alpha}\right]; \ s_h^\beta = 1 - s_h^\alpha = \frac{1}{2}\left[1+\frac{c_h^\alpha}{\omega^\alpha}-\frac{c_h^\beta}{\omega^\beta}\right] \quad h=0,1,2$$

Each firm maximizes the following objective function:

$$\max_{m_1^i}\left[\max_{m_2^i}\left[s_1^i + s_2^i\right]\right] \quad i=\alpha, \beta$$

Moreover, we suppose that the firms do not invest in R&D up to period 0, so that the initial state vector is:

$$M_0 = (M_0^1, M_0^2) = (0, 0)$$

The above choices, while reducing computational complexity, cause no loss of generality.

Tables 5.1, 5.2 and 5.3 refer to the symmetric case (both firms in each period have the same R&D expenditure constraint, i.e., $m_h^i \leq 3$; $i = \alpha$, β; $h = 0,1$). In particular Table 5.1 displays firms' optimal R&D choices when their absorptive capacity depends on the level of their accumulated R&D investments. It can be seen that for each level of extra-industry R&D stock, the optimal level of firms' accumulated (up to period 1) R&D investments decreases as the degree of intra-industry spillovers increases. In fact, a rise in the degree of intra-industry spillovers allows a greater amount of rival firms' R&D stock to be absorbed, so that each firm's optimal level of R&D expenditure decreases. This result confirms the traditional approach according to which an increase in spillovers discourages firms' R&D investments. However, Tables 5.2 and 5.3 reveal that the spillover effect is extremely mitigated (but not ambiguous as hypothesized by Cohen and Levinthal (1989)) when the firm's absorptive capacity depends on the level of firms' R&D investments. In particular, from Table 5.3 we can see that when the absorptive capacity is set equal to 1, independent of the level of firms' R&D investments, an increase in the degree of spillovers from 0 to 1 causes a reduction of 56.34% in the optimal level of firms' R&D accumulated investments if the amount of extra-industry R&D stock is equal to 0, and a reduction of 90.24% if the amount of extra-industry R&D stock is equal to 2. When the absorptive capacity depends on the level of firms' investments, the reduction rates are 1.20% and 1.18% respectively. This strong attenuation of spillover effect is due to the fact that in the first case a firm is able to assimilate (without its own R&D effort) a much greater amount of R&D from its rival than in the second case.

As far as the effects of changes in the level of extra-industry R&D stock are concerned, from Table 5.1 it can be seen that for each level of inter-industrial spillovers, the optimal levels of firms' accumulated R&D investment decreases when as the level of extra-industrial R&D stock increases. As a matter of fact, an increase in the amount of the exploitable external R&D allows the firms to reduce their optimal level of R&D investments. Also in this case, the magnitude of these effects is connected to the assumptions on the firms' absorptive capacity function (see Table 5.2). When the absorptive capacity is set equal to 1 independent of the level of firms' R&D investments, a variation in the extra-industry R&D stock produces a variation of the same amount and opposite sign in the optimal level of firms' R&D accumulated investments. Therefore, in such a case, the extra-industry R&D expenditure totally replaces the R&D expenditure of each firm. On the contrary, when the absorptive capacity depends on the level of firms' investments this "replacement effect" is extremely attenuated.

Table 5.1. Optimal levels of firm i (i=α, β) accumulated R&D investments. Identical R&D expenditure constraints for firm α and β (m$^\alpha_h \leq 3$, m$^\beta_h \leq 3$).

extra-industry R&D stock	degree of spillovers				
	0.00	0.10	0.50	0.90	1.00
0	5.32456	5.31808	5.29243	5.26715	5.2609
2	5.28185	5.27554	5.25052	5.22586	5.21975
10	5.12114	5.1154	5.09267	5.07024	5.06469
100	3.97214	3.96953	3.95918	3.94891	3.94636

Table 5.2. Comparison between the optimal levels of firm i (i=α β) accumulated R&D investments under different assumptions on the absorptive capacity function. Identical R&D expenditure constraints for firm α and β (m$^\alpha_h \leq 3$, m$^\beta_h \leq 3$).

degree of spillovers	extra-industry R&D stock=0		extra-industry R&D stock=2	
	A.C. < 1	A.C. = 1	A.C. < 1	A.C. = 1
0.00	5.32456	5.32456	5.28185	3.32456
0.10	5.31808	5.02456	5.27554	3.02456
0.50	5.29243	3.82456	5.25052	1.82456
0.90	5.26715	2.62456	5.22586	0.62456
1.00	5.2609	2.32456	5.21975	0.324556

note: A.C.= Absorptive Capacity

Table 5.3. Comparison between the reduction rates of the optimal levels of firm i (i=α, β) accumulated R&D investments under different assumptions on the absorptive capacity function. Identical R&D expenditure constraints for firm α and β (m$^\alpha_h \leq 3$, m$^\beta_h \leq 3$).

degree of spillovers	extra-industry R&D stock=0		extra-industry R&D stock=2	
	A.C. < 1 %	A.C. = 1 %	A.C. < 1 %	A.C. = 1 %
0.00	0.00	0.00	0.00	0.00
0.10	0.12	5.63	0.12	9.02
0.50	0.60	28.17	0.59	45.12
0.90	1.08	50.71	1.06	81.21
1.00	1.20	56.34	1.18	90.24

note: A. C. = Absorptive Capacity

Table 5.4. Optimal levels of firm α accumulated R&D investments. Different R&D expenditure constraints for firm α and β ($m^\alpha_h \leq 3$, $m^\beta_h \leq 1.5$).

extra-industry	degree of spillovers				
R&D stock	0.00	0.10	0.50	0.90	1.00
0	5.32456	5.32131	5.30841	5.29561	5.29243
2	5.28185	5.27859	5.26611	5.25362	5.25052
10	5.12114	5.11827	5.10684	5.09549	5.09267
100	3.97214	3.97083	3.96564	3.96047	3.95918

Table 5.5. Comparison between different optimal levels of firm α accumulated R&D investments under different assumptions on the absorptive capacity function. Different R&D expenditure constraints for firm α and β ($m^\alpha_h \leq 3$, $m^\beta_h \leq 1.5$).

	extra-industry R&D stock=0		extra-industry R&D stock=2	
degree of spillovers	A.C. < 1	A.C. = 1	A.C. < 1	A.C. = 1
0.00	5.32456	5.32456	5.28185	3.32456
0.10	5.32131	5.17456	5.27859	3.17456
0.50	5.30841	4.57456	5.26611	2.57456
0.90	5.29561	3.97456	5.25362	1.97456
1.00	5.29243	3.82456	5.25052	1.82456

note: A. C. = Absorptive Capacity

Table 5.6. Comparison between reduction rates of the optimal level of firm α accumulated R&D investments under different assumptions on the absorptive capacity function. Different R&D expenditure constraints for firm α and β ($m^\alpha_h \leq 3$, $m^\beta_h \leq 1.5$).

	extra-industry R&D stock=0		extra-industry R&D stock=2	
degree of spillovers	A.C. < 1 %	A.C. = 1 %	A.C. < 1 %	A.C. = 1 %
0.00	0.00	0.00	0.00	0.00
0.10	0.06	2.82	0.06	4.51
0.50	0.30	14.09	0.30	22.56
0.90	0.53	25.35	0.53	40.61
1.00	0.59	28.17	0.59	45.12

note: A. C. = Absorptive Capacity

Tables 5.4, 5.5 and 5.6 refer to optimal levels of firm α accumulated R&D investments up to period 1, when firm α and β have different R&D expenditure constraints, i.e., $m^\alpha_h \leq 3$, $m^\beta_h \leq 1.5$ (h=0,1). As in the symmetric case, the optimal levels of firm α accumulated R&D investments decrease both as the degree of intra-industry spillovers increases and as the level of extra-industry R&D stock rises. Furthermore, both these effects are extremely mitigated when the absorptive capacity depends on the level of firms' R&D investments. It is worthwhile noticing that if the degree of intra-industry spillovers is positive, then the optimal levels of firm α accumulated R&D investments are greater than symmetric case ones. This is due to the fact that in the symmetric case both firms have the same R&D expenditure constraint, i.e., $m^i_h \leq 3$ ($i = \alpha$, β; $h = 0,1$), whereas now $m^\alpha_h \leq 3$ and $m^\beta_h \leq 1.5$ ($h = 0,1$). Hence, firm β is forced to reduce its optimal R&D expenditure, so reducing the external R&D that firm α is able to assimilate. Therefore, firm α compensates this by increasing the level of its own R&D investments.

Owing to the particularly restrictive R&D expenditure constraint, the optimal levels of firm β accumulated R&D investments up to period 1, not displayed in the tables, are equal to 3 when the degree of intra-industry spillovers and the level of extra-industry R&D stock are low, whereas they decrease as spillovers and extra-industry R&D stock become so high to make the R&D expenditure constraint inactive. Also for firm β these reductions are extremely mitigated when the absorptive capacity depends on the level of firms' R&D investments.

Table 5.7 shows the effects in the degree of intra-industry spillovers and the level of extra-industrial R&D stock on firm α market share at period 2, under the assumption that the absorptive capacity depends on the level of firms' accumulated R&D investments. Observe that for each level of the degree of inter-industry spillovers the market shares decrease as the extra-industry R&D stock rises. In fact, an increase in the extra-industry R&D stock has a proportionally greater impact on firm β, which is compelled to a smaller amount of internal R&D investments by its R&D expenditure constraint. Moreover, from Table 5.7 it can be seen that for each level of extra-industry R&D stock, the market share of firm α decreases as the degree of intra-industry spillovers rises. In fact, a rise in the degree of intra-industry spillovers produces an increase in the R&D assimilable from the rival firm and this increase has a proportionally greater impact on firm β.

Table 5.7. Market share of firm α at period 2. Different R&D expenditure constraints for firm α and β ($m^\alpha_h \leq 3$, $m^\beta_h \leq 1.5$).

extra-industry	degree of spillover				
R&D stock	0.00	0.10	0.50	0.90	1.00
0	0.516886	0.516705	0.515992	0.515295	0.515123
2	0.516352	0.516176	0.515484	0.514808	0.514642
10	0.514387	0.514231	0.513617	0.513016	0.512868
100	0.503375	0.503321	0.503105	0.502892	0.502839

5.6 Conclusions

The basic findings of the above discussion can be summarized as follows. There is a positive correlation between the degree of spillovers and the impact produced by a firm R&D investment at a given time on its own market share in the short run, whereas this correlation is negative in a longer perspective. Therefore, the total effect is ambiguous. Moreover, if there are intra-industry spillovers, a firm cost function and, consequently, its own market share over time depends on its own R&D investments as well as on those of the rival firms. In this framework, if all the firms have a finite R&D expenditure limit in each period, there exist an optimal R&D strategy over time for all the firms in the industry.

The numerical simulation highlights that the optimal level of firms' R&D investments decreases both as the extra-industrial R&D expenditure increases ('replacement effect'), and as the interindustrial spillover degree rises ('spillover effect'). However, our results show that when the absorptive capacity depends on the accumulated level of R&D investments, according with the model presented in this paper, the replacement and spillover effects are weaker than those forecast by the traditional approach with fixed absorptive capacity.

References

Allen T J (1977) Managing the Flow of Technology. MIT Press Cambridge

Arrow K J (1962) Economic Welfare and the Allocation of Resources for Invention. In: Nelson R R (Ed) The Rate and Direction of Inventive Activity. Princeton University Press Princeton

Basar T and G J Olsder (1982) Dynamic Noncooperative Game Theory. Academic Press New York

Campisi D and A Nastasi (1993) Competitive Pressure and R&D Strategies in an Oligopolistic Industry. Atti del IV Convegno Nazionale AiIG. Roma 29 ottobre

Cohen M W and D A Levinthal (1989) Innovation and Learning: the Two Faces of R&D. The Economic Journal 99:569-596

Friedman J (1990) Game Theory with Applications to Economics. Oxford University Press New York

Grossman G M and C Shapiro (1986) Optimal Dynamic R&D Programs. Rand Journal of Economics. 17:581-593

Kamien M and N Schwartz (1972) Timing of Innovations under Rivalry. Econometrica 40:43-60

Kamien M and N Schwartz (1982) Market Structure and Innovations. Cambridge University Press

Mowery D C (1983) The Relationship between Intrafirm and Contractual Forms of Industrial Research in American Manufacturing, 1900-1940. Explorations in Economic History 20:351-374

Nelson R R (1959) The Simple Economics of Basic Research. Journal of Political Economics 67:297-306

Quirmbach H C (1993) R&D: Competition, Risk, and Performance. Rand Journal of Economics 24:157-197

Reinganum J (1982) A dynamic game of R&D: patent protection and competitive behavior. Econometrica 50:671-688

Spence A M (1984) Cost Reduction, Competition, and Industry Performance. Econometrica 52:101-121

Tilton J H (1971) International Diffusion of Technology: the Case of Semiconductors. Brookings Institutions Washington DC

Tirole J (1988) The Theory of Industrial Organization. MIT Press. Cambridge Massachusetts

6 Innovative Behaviour and Complexity in Organized Systems

Mario Lucertini and Daniela Telmon

L 23

031
032

6.1 Innovation, Continuous Improvement and Performance Measurement in Complex Organizations

Technological innovation, we believe, is increasingly accompanied by a concern with continuous improvement in company performance. This requires constant assessment and, thus, the identification of performance indicators. This assessment is necessary in order to identify the assumptions that underlie a given managerial strategy and possible obstacles in its path.

A first effect of innovation is the need to break with existing company culture and traditions. This can be achieved by means of a constraints analysis (by which we mean the discovery of the causes that prevent the optimal performance of a given production activity).

A typical example of constraint removal resulting from technological innovation is given by telework. This breaks a series of constraints traditionally considered irremovable with the need to gather workers together in the physical space in which work is performed.

Constraints can be of different types. There may be *environmental* constraints, such as public policies regulating practices, products and services performed by the firm; market forces, product characteristics and available technologies. This type of constraint cannot be directly removed by management. There are *structural* constraints, like strategic commitments and systemic processes that are inherent to structural aspects such as the location of the company and its organizational design. Finally, we can speak of *operational* constraints, resulting from the production activity. These are easier to remove. Another classification of constraints is proposed in the last section of this paper.

It is not, of course, possible to measure the constraint itself, but its effect can be measured using criteria (these vary according to the type of constraint).

Innovative companies are characterized by the search for new solutions, either technical or managerial, and for new ways of operating. These new practices are very often suggested by the comparison with other companies. Constraint analysis is fundamental to benchmarking.

Benchmarking has been defined as "a surveyor's mark indicating a point in a line of levels; a standard or point of reference". For Rank Xerox, the first company to adopt this methodology, benchmarking was an effort to take conventional competitive analysis one step further. It was introduced in 1979 in manufacturing operations and encompassed an in-depth study of best competitors, detailed reverse engineering of their products and achievements. Operating capabilities and features of competing products were subjected to a tear down analysis.

As experience was gained and the company learned more about itself and the best practices in industry, the need for change in the whole company became increasingly clear. The focus was extended beyond manufacturing and the definition of goals and benchmarks to an emphasis on the *process* used by the best in class. Benchmarking is now used as a strategic quality tool and is being extended to all aspects of business and integrated into the management process.

Benchmarking today is the "continuous process of measuring products, services and processes against the strongest competitors, or those renowned as world leaders in their field". Other companies' experience helps to set realistic targets in business plans (since the standard is achievable) and to identify the specific actions and resources required to improve performance. The aim is not only to match existing performance levels, but to exceed them and become leaders in areas that provide competitive advantage.

The process is continuous as the high rate of change in the business environment means that benchmarks need to be continually redefined.

In conclusion, we can state that technical and organizational innovation and continuous improvement require:

• a clear model of the system, representing it globally and in all its complexity;
• the definition of the right performance indicators;
• the measurement of performance.

The measures through which performance is assessed may be both internal and external to the company. In the latter case, they involve the comparison with the best companies (benchmarking).

In this paper, we firstly underline the importance of a typology of organizational system modelling that takes into account the flows of the activities inside the system and the organizational structure supporting them. A series of indicators are then proposed. These indicators are determined on the basis of the interaction between the process and the structural dimensions.

6.2 The Network Organization: A Network of Processes

It is becoming increasingly common for environments to consist of networks of interdependent entities which often have different locations. The interactions between entities of different companies are sometimes stronger than interactions between entities of the same company. The term production is used here in a very broad sense. The system may produce objects (manufacturing), continuous flows of goods (process industries, such as nuclear or chemical plants, and some food industries), industrial services (such as logistics and distribution), commercial services (such as shops, maintenance of buildings) and public services (such as transportation and communication).

Network organizations may have different shapes (Bekiroglu 1984; Bonini 1963; Butera 1991; Coase 1937; Dioguardi 1983; Hastings 1993; Lucertini and Telmon 1993; Lucertini and Telmon (forthcoming); Malone 1987; Malone and Smith 1988). A structure which is widely adopted by large firms consists of a corporation of decentralized companies under single ownership, but subdivided into a number of individual structures, or units, that are very similar to autonomous enterprises (divisions, business units, goal-directed work teams, etc). These organization units go back and forth through the boundaries between market and hierarchy with great flexibility and, like consortia, rely heavily on suppliers or outside collaboration.

In fact, a network organization can also be a consortium of juridically independent companies tied together by strong associative links, common facilities and market policies that allow it to operate as a single system.

Corporations adopt political structures and operation systems that enable them to directly influence the whole organization, e.g. the quality standards and the know how of suppliers. The latter can be autonomous for different reasons: because they are already in the market, because they work for different clients and sell their products directly to the final market, etc. The central company integrates its internal structures and the suppliers not so much with hierarchical instruments (organization charts, internal communications, rules and procedures) as with operational instruments (planning systems, management information systems), reporting systems) and integration structures (teams, task forces, committees, etc.), together with corporate culture, management philosophy, etc.

Some of the characteristics of network organizations are that:
- they are often hybrid companies, partly constituted by formal structures, partly by markets;
- the juridical and organizational boundaries of the corporation and the boundaries of the managerial and technical action of the staff may be different;
- the real regulator of the economic and organizational processes is the relationship between constituent companies and not the structure of the single companies.

All these different types of company can be defined and analyzed as a network of processes (Zülch and Grobel 1993). This representation emphasizes the interconnections between the elementary transformation processes that constitute the system (has). Each process is characterized by well defined components: upstream operations, input, a transformation process, output and downstream operations (see Figure 6.1).

Inputs (e.g. materials, information) come from upstream systems or suppliers, and outputs (products or services) are delivered to downstream systems or clients.

The basic processes are aggregated and coordinated at various levels depending on the structure of the organization (Malone and Smith 1987; Saaty 1988). The behaviour of these subsystems varies under the impact of the environment and the control instructions issued. A multilevel arrangement of the system leads to the need for sequential stage-by-stage coordination of activities within the system. The efficiency of such coordination depends on the choice of local controls at each stage. A process of sequential coordination, conceptualized in the form of an iterative procedure, has been demonstrated to be capable in many cases of improving the global performance of a large system (Meraroviç et al 1970).

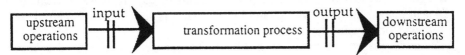

Fig. 6.1. The transformation process

Whereas the basic transformation processes are generally well known and standardized, the behaviour of the whole system can only be analyzed using sophisticated modelling tools. Moreover, although the information provided by modern information systems is, in principle, fairly complete, in practice most of the basic data and the knowledge needed to interpret the data is compartmentalized. This is largely due to the nature of the technology, which is being divided into narrower and narrower specialities, each advancing along a cramped frontier, within strict boundaries and developing its own jargon. But it is also because of the explosion of information within each field, which makes it difficult for any specialist to explore beyond his chosen arena and to communicate across these boundaries.

Many authors have conducted extensive investigations on managers who make decisions with incomplete information. Most non-programmed decisions involve too many variables for a thorough examination of each and managers rarely consider all possible alternatives for the solution to a problem. There exists, in every problem situation, a series of boundaries or limits that necessarily restrict the manager's picture of the world and, therefore, the decision range. Such boundaries include individual limits to any manager's knowledge of all the alternatives as well as such elements as policies, costs and technology that cannot be changed by the decision maker. As a result, the manager seldom seeks the optimum solution but realistically attempts to reach a satisfactory solution to the problem at hand. In other words, instead of attempting to maximize, the modern manager satisfies. He examines the five or six most likely alternatives and makes a choice from among them, rather than investing the time necessary to examine thoroughly all possible alternatives. In decision making studies, many researchers have concentrated on the analysis of alternatives with given constraints. Limited attention has up to now received constraints (Balm 1992; Lucertini and Nicolò 1994) and the most important constraints that can be used as drivers of an improvement process depend on the embedding of a set of interconnected activities or operations in a set of interdependent organizational units.

The conceptualization proposed here aims at identifying i) the quantities to be measured (so that suitable performance indicators, based on such measures and on a model of the whole production process, could be evaluated), and ii) the procedures needed to modify the constraints in order to obtain better performance.

It is common experience in performance improvement (Ostrenga et al 1992) that flows of resources which have been eliminated from some stage of the transformation process appear at some other stage, often in a different form. Our claim is that this sometimes happens because the drivers we have used to achieve the improvement were tailored to one specific aspect of the system traditionally the organization, and more recently the transformation process) and do not integrate information on the three main elements of the production system: the transformation process, the organization and the decision process (Lucertini and Telmon (forthcoming)).

By performance we mean the result of the management of activities in an organization over a certain lapse of time. Performance measurement is essential to planning, as it is difficult to plan if you do not know how to perform. By performance measurement we mean feedback and information on the management of business activities in relation to company strategic goals and customer satisfaction.

6.3 Performance Indicators

An organized system consists of a set of operations and activities (whose interrelations form a process), an organization structure, and the link between the two.

There are traditionally two ways of considering and analyzing an organized system: a structural (sometimes referred to as functional) approach and a process approach. The structural process mainly considers the organization chart. All the company's resources are divided and distributed between business units (divisions, offices, etc.). Business units are created on the basis of specialization and technical know how, and are connected through a hierarchical line (Ostrenga et al 1992).

Traditionally, managers have tended to consider their company as a pyramid organization, i.e. a hierarchical group separated functional entities. This is the consequence of an old model of division of work that has had a great influence on the way in which companies are organized. The typical organizational chart, for instance, is divided into different production units, suggesting that each unit puts together a series of activities that can be managed and measured independently from each other. In this way barriers are created between different units and the flow of the work performed in the whole company ignored.

The process point of view (sometimes called arrow organization) focuses on the work performed and not on the organizational structure that manages it. The process point of view identifies the main set of activities personnel needs to perform for the company to produce and sell its output. These sets of activities are known as processes. To consider the organization as a set of processes and not as a hierarchy of units is a new and important theoretical requirement. *Continuous improvement*, a central element of total quality, involves the measurement of quality and performance in all the company processes, so that action can be taken to improve them. These indicators may be customer satisfaction, the number of design errors made in a month, or other indicators used to characterize a process.

In the literature on complexity in organizations, there have traditionally been two ways of seeing the link between activity and organization structure. The first is the classical organization model. This starts by considering the existing organization structure and tries to find the best combination between activity and organization structure through adjustments to the latter. The second approach, adopted by the total quality movement, starts with the primary process, that is to say by the set of activities that have to be performed, and builds around them the best organization structure.

One of the major contributions of the quality movement is the recognition that the best solution can be found only by considering the system in an integrated way and that there has to be a continuous redefinition of the organization. The goal of continuous improvement focuses on the process to improve customer satisfaction and to reduce the costs associated with its achievement. Therefore, the processes, traditionally seen only as a set of activities, are now seen as a set of activities supported by a dynamic organization structure, that follows the primary process and adapts to it.

Based on this analytical approach, tools have been developed to analyze company processors and costs more accurately (activity based costing) and hence achieve continuous improvement.

To manage a company it is necessary to manage processes. To do this, the link between operations (or activities) and costs becomes essential for cost management. The key to understanding cost dynamics is to establish relationships between activities and their causes, between activities and costs. To consider the organization from the process point of view enables one to manage costs, by managing the activities that generate them. In changing the activities, it is possible to influence the behaviour of costs.

Appropriate tools are needed to identify the process output, the internal and external clients, the activities necessary to produce the output and also the input. having identified these activities, the process is then subject to value analysis (i.e. definition of cycle time, definition of the cost of each activity, determination of value added for each activity). The aim of value analysis is the identification of activities that add time and cost to a process without adding value for the client. This is the basis for identifying non value-adding activities that have to be eliminated in an improvement plan.

Many studies have been carried out on process definition, but much has yet to be said and studied on organization design and management as a support tool for the primary process. The problem is to conceive of the organization structure in a new way, seeing it not simply as the organization structure, something that perpetuates, but something that has to be built around the primary process. In the new approach, processes become the basis for organization design and management: structures and functions become changeable and have to justify their existence and survival as useful contributions to the management of the process.

As far as performance measurement is concerned, much has been done on defining indicators for the first side of the equation (the process), but very little on the second part of it (the organization).

The process to be considered is not only a production flow process but also an information flow process. How do you measure the information flow? How do you measure organization constraints? Let us briefly review the process approach and the performance indicators used to assess it.

A set of new factors have recently put in doubt the conventional company performance measurement system (Johnson and Kaplan 1987; Kaplan 1990; Omachonu et al 1990). The quality movement has contributed to a better understanding of the relations between clients, company processes and business success, and has shown that traditional financially based indicators are insufficient and often also misleading.

In order to evaluate and document company performance, traditional measurement systems adopt the structural and not the process view. Conventional performance criteria for measuring the effectiveness and efficiency of organizations are usually based on monthly financial reports for single units, showing a profit and loss statement, which is then compared with the unit's budget. This information is gathered and used by top management for unit evaluation.

Modelling and measurement of processes has been attempted using the process approach but tends to neglect many structural aspects. In fact, it does not take into account that the company operates in a constrained system. Process indicators can be good for evaluating final business results, i.e. the company output, but they are often not sufficient for determining how structural improvement may increase the output, i.e. evaluating how a constraint affects performance and how it can be changed.

Different kinds of performance criteria are taken into consideration when measurement system is examined.

A first set of criteria, (including indicators to measure throughput, the resources, financial flows, effectiveness, efficiency, productivity, defect rate and so on) are based on measures of the production and support flows. Any implications for the organization are indirect, since it is the effect of the organization on the flows which is measured and not vice versa.

A second set of company wide criteria (including indicators such as organization size, number of personnel, their qualifications, number of different operations the organization performs, quantity of resources available, organizational structure complexity, number of managerial levels) is based on measures of the organizational structure. In this case, the implications for the flows is indirect, as only the effects of the flows on the organization are considered.

A third set of criteria (including indicators such as response or lead time, flexibility, concurrency levels, system complexity) is based on mixed measures. Looking at the process of this organization, it is possible to make an evaluation, often based on a simulation of the whole system behaviour.

We argue that process measures are not sufficient to measure performance and that it is equally important to measure organizational variables and the joint effect of process and organization.

In the following, we propose a set of short definitions of some performance criteria, in order to point out the links with the system representation outlined above.

Effectiveness, efficiency and productivity are classical performance criteria which jointly evaluate production and support flows. They take into consideration the input and output of the transformation process in terms of resources used and quantities produced respectively.

Effectiveness is an 'output side' issue, and is represented by the ratio between actual output and expected output, for a given input. Effectiveness is accomplishing the 'right' things, in terms of timeliness, quantity, quality and cost, for a given amount of resources used. A selected effectiveness measure should explicitly indicate whether the organization is achieving the desired results.

Efficiency is an 'output side' issue, and is represented by the ratio between the resources planned or expected to be used and the resources actually used in producing a given output. If effectiveness is doing the right things, efficiency is doing things right.

Expected productivity is the ratio between the expected output and the resources expected to be consumed (expected input).

Actual productivity is the ratio between the actual output and the resources actually consumer (actual input). Productivity measures are designed to analyze output relative to the inputs. These measures may be developed for each input (labour, capital, energy, materials, data information) or in combination of inputs. Productivity for particular types of inputs and outputs, e.g. for financial flows, takes different names.

Profitability is the measure or set of measures for the relationship between revenues (output belonging, in our representation, to the production flow) and costs (input belonging to the support flow). The same criterion for non profit organizations is *budgetability*, i.e. the measure or set of measures of the relationship between what you said you would do and what it would cost and what you actually did and what it actually cost.

Flexibility (De Groote 1994; Jae-Ho Hyun and Byong-Hun Ahn 1992; Suares et al 1991) can be defined as the capability of the system to adapt to different requests. It has four main dimensions (Upton 1994): operation range (how many different operations can be done), throughput range (how many different production rates are technically and economically acceptable), uniformity (how well all the different operations in the range are performed), mobility (how quickly the system switches from one operation to another). To evaluate operation range, uniformity and mobility it is necessary, in general, to consider both the process and the organization. To evaluate the throughput range we also have to take into account the decision process. In fact, the technical and economic feasibility depends heavily on the strategic and operations level decisions.

Speed can be defined as the time needed by the system to respond to different unforeseen inputs. Time competition is often a strategic orientation for companies. It allows them to be customer oriented, to improve quality and to manage costs strategically, but, above all, time competition focuses on process speed. Examples of time competition are: time necessary to develop the design of a new product, time to develop a software programme, time of arrival of an order from the client to the person responsible for processing it, lead time in production lines. Process analysis is an important technique for determining the cycle time of company processes. Linking these methodologies to activity bases cost analysis, we obtain a combination of cost, quality and time indicators, that allows us to give time an economic value.

It is useful for a company to define the different performance criteria that are taken into consideration for measuring the system, although it should always be borne in mind that a definition suitable for one company may not be suited to another. In defining performance, one can never ignore the specific goals of a particular company.

Table 6.1. Company performance criteria

Company-wide performance indicators

Process indicators

Production flow:	throughput,
	number of operations for each type,
	defect rate.
Support flows:	quantity of resources used for each type.
Production/support flows	effectiveness,
	efficiency,
	productivity,
	number of precedence constraints.

Organization structure indicators

Resources:	organization size,
	quantity of resources available for each type,
	(number of people for each specialization,
	number of jobs,
	number of facilities for each type,
	quantity of financial resources, etc.),
	number of different locations.
Organizational units (OU):	number of OU for each type,
	number of different operations allowed,
	number of task areas,
	differentiation vs. integration level.
Links:	number of structural links among OU,
	link size, time and cost,
	number of managerial levels,
	number of meshing links,
	number procedures and decision rules,
	number of hops for each procedure,
	ratio (number of links/numbers of OUs).

Decision process and mixed indicators

response or lead time,
flexibility,
concurrency level,
system complexity.

It is interesting to notice that these different performance criteria refer to processes and therefore to flows, and that the indicators are generally ratios. Performance criteria provide the basis for measuring tools, or indicators.

In order to understand better how the different indicators relate one to the other, we present a model of a production system. As all its elements separated, it is easier to use it as a basis for new types of performance indicator.

6.4 The Process Flows

The process flows considered here concern the production system and are divided into production flows and support flows.

The *production flows* represent the flow of materials through the transformation processes. By materials we mean not only physical elements (such as components, parts and sub-assemblies), but all elements subject to the primary transformation process characterizing the production system (often in the format of information). For instance, we consider the drawings produced by the engineering department, the invoices issued by the administration department and the inventory update prepared by the materials manager all to be materials. To represent the production flow, two types of information are needed. The first concern the operations (an operation is an elementary work unit of a transformation process). The set of operations required, with the corresponding precedences gives the so-called *operation graph* (Agnetis et al 1990). Operations, in project management or activity-based management terminology, are often referred to as *activities*; in production environment operations they are often clustered into so-called *functions*. The second type of data concerns flows (the number of units of material or information flowing through each section of the operation graph in each time interval). The nodes of the *production network* correspond to the transformation processes (and also include buffers and warehouses), the arcs correspond to the logical (often also physical) links between the nodes. The inputs of the system directly related to the production flow are sources of the production network. The outputs are sinks and the entering flows are required top fit the demand or, more precisely, the assigned factory workload.

The *support flows* include coordination (e.g. the flow of information on the production plan and on the state of the system) and the flow of resources needed for each node of the operation graph (Chang 1984). The resources include both the physical resources, tools and activities needed to carry out the operations (e.g. machines, facilities and effectors, time, money, maintenance, software, etc.) and the information which enables the resources distributed along the production flow to operate (e.g. machines, material handling systems, workers). This includes the information which carries the decisions taken by the different decision makers (on routing, scheduling and choice of operation type, for example) to the personnel in charge of performing the operations. It therefore produces the operations, provided that the necessary resources and the information tools needed are available and correct. In order to build the support flows, we must quantify the work content of each operation in the production flow. The nodes of the *support network* correspond to the transformation processes of the production network, the decision centres, the logistic facilities, the information management centres, etc.; the arcs correspond to the logical (often also physical) links between the nodes. The inputs of the system related to the support flow are sources of the support network; the outputs are typically the nodes of the production network.

It is worthwhile noticing that some flows can be considered part of the production flow or part of the support flow depending on how the system is represented. For example, the flow of money can be part of the support flow if we are considering a production system whose primary goal is to produce goods in given quantities. On the other hand, it can be part of the support flow if we are considering a business unit whose primary goal is to produce added value.

Resources are considered both in the support flows and in the organization structure. The use of the resources for production is included in the support flows, the supply of resources is part of the organization structure.

6.5 The Organization

We now look at the resources, organizational units and the links between them. The aspects of the organization structure considered here concern mainly the elements in charge of carrying out the activity (Huber and Glick 1993; March 1965).

Resources are the set of elements available for production or that can be obtained from outside in the time window considered. They include all factors of production: physical resources, people, information, money, etc. Resources are the main element conditioning decisions. By definition, anything that can be used to help solve a problem is a resource, including time, money, personnel, expertise, energy, equipment, raw materials, information.

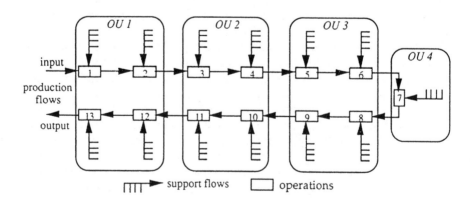

Fig. 6.2. Organizational units in charge of production and support flows

Organizational units (OUs) are the set of elements within the company with organizational responsibility. Each is defined by a mission, specific goals, a series of operations which it performs, the type of decisions it may take, a resource management strategy, etc. OUs should perhaps not be introduced without considering explicitly the allocation of resources to them (in many cases OUs themselves are considered as a set of company resources). But, as our purpose is to analyze how different resource allocation strategies influence system performance, we prefer here to consider resources and OU separately.

Links include all the different types of connection between the OUs. They concern the ways the different units interact one with the other and, in this case, may include some decision aspects, e.g. the so-called automated decisions which are taken on the grounds of given rules and procedures (see Figure 6.2). (Many decisions can be included either in the rule and procedures section, or considered part of the decision process.) The links also include physical connections (such as transportation facilities and service networks), the information network (depending on the way the company information system has been designed and is used), the hierarchical structure (partly included in the rule and procedures section), and all other forms of formal or informal communications. Some links depend only on the OU, others depend also on the resource allocation. Moreover, some links are structural, i.e. do not depend on the decisions considered here, others are the output of the decision process concerning the information exchange.

6.6 The Decision Process

The flows and the organization structure interact to produce different production patterns depending on the decisions taken. The decisions considered are of two types: structural decisions and operations level decisions.

6.6.1 Structural Decisions

Allocation of resources to organizational units, (i.e. who is allowed to operate). An assignment process provides each unit with a set of resources allocated to it for a given time interval. There are several ways to represent the assignment. A simply and effective way is to indicate the set of operations that the resources assigned to each allow it to carry out. Let us consider a bipartite graph $F(O,U,E)$, where O is the set of operations, U the set of OUs and E a set of feasibility edges between O and U (there is an edge from i to j iff unit j can perform operation i). The allocation of resources to units gives the set E.

Resource exchange patterns. For each unit, this consists of the set and the planned time schedule of the resources (especially information) sent to and received from all other units and the environment. This is one of the core aspects of the links between OUs for both physical and human resources, and, above all, information. The way this part of the decision process is dealt with is crucial in the system representation. In fact, although the best decisions can be taken on the basis of fairly complete information (in particular about the future), uncertainties and limited knowledge are usual in the real decision environment. One of the most pernicious aspects of this process is the tendency under stress to discount the future (this is influenced by our emotional state at the time of choice). Side effects are also complex and we often neglect parties who were geographically, socially, institutionally and chronologically remote from the transaction. In a completely rational atmosphere, these conditions should prompt a search for more information, at least to narrow the uncertainties. But there is a price in terms of money and time; information is never free. The real flow of support information, the availability of resources and the flow of decisions must therefore be synchronized and coherent.

6.6.2 Operations Level Decisions

Assignment of operations to organizational units, i.e. who is doing what. For each unit, an assignment process determines the set of operations to be carried out in the time interval considered. Using the bipartite graph introduced above, the problem can be formulated as finding the subset A of E, such that any operation is assigned to exactly one unit. The assignment ascribes a workload to each unit of the production system. It is called *capacitated* if any unit can perform (in the time window considered) all the operations assigned to it without considering the delays introduced by other units or by internal scheduling problems; it is called *balanced* if the workload for each unit is proportional to its capacity, i.e. to the maximum amount of operations that the unit is able to do (with the given resources and in the time window considered). We suppose that, whenever an operation is assigned to a unit, the unit is in charge of the decisions concerning when and how to perform the operation (*how* concerns the case of multiple choice in performing the operation, e.g. when two different machines belonging to the same unit can perform the same operation). The information used to take the decision depends on the information received from the other units.

Scheduling of operations in each organizational unit (i.e. when things are done and with which resources). The aim of the decisions taken by the head of each unit is to find the start and the completion time for each operation, each unit and each resource. It is a resource constrained scheduling problem which derives the behaviour of the unit.

6.7 Constraints as the Key for the Determination of Performance Measurements

Production and support flows represent what has to be done; the organization structure with the assigned resources and operations represent the way the process is carried out, but introduces relevant constraints. Scheduling decisions represent a decentralized decision process in a strongly constrained environment; resource exchange decisions represent a way of making the local decisions possible and effective (Bonini 1963; Huber 1993; Kunar et al 1993).

The organization structure produces different types of constraints on the processes, affecting:

- the routings (a production flow can pass through an organizational unit only if the resources needed for the production exist in the unit),
- the quantities available (an organization unit cannot use, in each time interval and for each type of input, more of a resource than that available. Lack of adequate resources might prove to be a significant constraint),
- the scheduling (an operation cannot be performed by a unit before all the needed inputs are available and later than a deadline after which the input is no longer usable),
- the concurrency (different operations which must be performed, or cannot be performed, at the same time and/or by the same resource),
- the quantities used (the quantity of a given resource used by an operation in an organizational unit, e.g. the operation time length or cost depend on the type of resource used, e.g. the operator's skills, and/or on the global flow allocation, e.g. the information available),
- the range of possible decisions (which depends primarily on the information available).

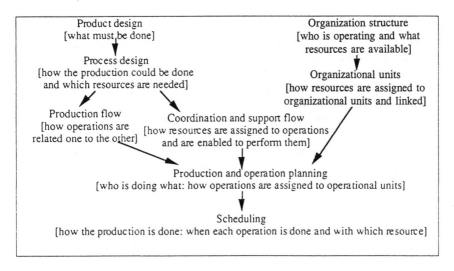

Fig. 6.3. Flows, organization and decisions

From a modelling point of view, the boundaries include both decision variables and constraints. The decisions are taken in a subspace of the real decision space and in a subset of the complete set of feasible decisions in the subspace.

Different authors have developed typologies of organization structures, with boundaries and constraints (Agnetis and Lucertini 1990; Baligh and Burton 1981;

Cyert and March 1963; Daft 1989; Malone and Smith 1988).

Among the constraints there are, first of all, constraints on resources. These may or may not be removable. Non-removable constraints are usually physical, technological or environmental constraints. The constraints on the flows of materials can be in most cases modified only with regard to waste. Logical constraints, such as precedence and concurrency, in many cases can be removed by a different assignment of operations to OUs, although with an additional cost. In fact, an important subset of logical constraints are organizational constraints, that can often be modified by suitably modifying the links between OUs, e.g. company procedures.

The type of constraints influences the type of measures which can be used to evaluate company performance. Operational constraints, for instance, need measurement of efficiency and effectiveness. Process-related constraints need process measurement. Strategic constraints need integrated company-wide measures.

The analysis of constraints is really the core of performance evaluation exercises, and the starting point for further analysis.

6.8 Constraints Analysis and Benchmarking

There are several definitions of benchmarking all based on the idea of evaluating the performance of an organized system by comparing it to exogenous entities. The following is a definition that tries to include the different aspects:

the continuing search, measurement and comparison of products, processes, services, procedures, ways of operating and best practices that other companies have developed with the aim of improving the company performance and obtaining higher output and better global performance.

The concern about performance evaluation has always existed in corporations and has traditionally been carried out on an historical basis (by comparing the performance from one year to the next) and, sometimes, on a competitive basis (by comparing the company to a competitor).

Only in recent times has some attention been devoted to comparisons made on a functional basis, i.e. comparing similar functions in different companies or comparing activities that are similar from a functional point of view, but associated with completely different transformation processes.

Major *one shot* modifications are also considered in the strategy. The problem is that major modifications, even though theoretically sound, could prove impractical from a psychological point of view. To convince management that relevant structural changes can be done, we must give evidence that the new organization could work. A good, wide recognized proof is that other companies have successfully adopted the new organization.

Benchmarking has been applied in many different sectors, but the concepts have been developed independently, with few interactions. Manufacturing processes, data processing systems, accounting systems and company practices are all important areas in which benchmarking has been used (see [Int1] for a more detailed discussion.

In these applications, benchmarking examines practices of leader companies to identify *which* (analogous process parts, i.e. subchains performance indicators), *why* (performance indicators), and *how* (new organizations, interconnections, structure or behaviour) can be transferred.

Benchmarking is more a tool for implementing change, than a tool for merely evaluating company performance. The decision-making process and its links with the set of performance indicators, depicting the company's behaviour, is therefore a cornerstone of benchmarking. This means that it is important to define company goals in order to determine what and where to measure, which are the right indicators and how they relate to measurements.

To put together goals, measures and performance, it is necessary to have a conceptual model of the transformation process that can be used to transform performance evaluation into the performance decisions. In practice, company decisions lie on different levels. Benchmarking focuses only on the type of decisions relating to the intermediate or tactical level. These decisions do not generally concern basic company strategies, such as market selection, process selection, joint ventures or basic make or buy decisions. Nor do they concern operational decisions, such as material routing and operations scheduling.

We will suppose that, given the set of constraints, internal efficiency (i.e. a good solution) is always achieved by the decision-makers at the operational level. All actions at this level can therefore be considered completely determined by the upstream decisions and the environment. For instance, we may suppose that in the constrained optimization of the operational level (where all actions are performed on the grounds of the choices made at the tactical level), the decision process can be represented in an optimization format with a single decision-maker and complete information. The output of the tactical level is assumed to produce the constraints of the optimization problem.

Benchmarking decisions focus on the tactical level, where organization constraints, procedures and practice can be modified. Examples of this can be found in physical material handling, distribution systems, assembly lines, production layout, make or buy, precedence or concurrency constraints (a usual way to represent the relations introduced by the information flow through the organizational units of the system).

Using decision model language, we may characterize the three levels, from operational to strategic, as follows:

Operational level
given: environment, structural constraints and procedures, information flows, operations constraints (different types of technological and organizational constraints), a univocally defined objective function
find: the value of decision variables directly connected to the process
such that: the performance will be optimized (throughput maximization, lead time minimization, etc.)

112

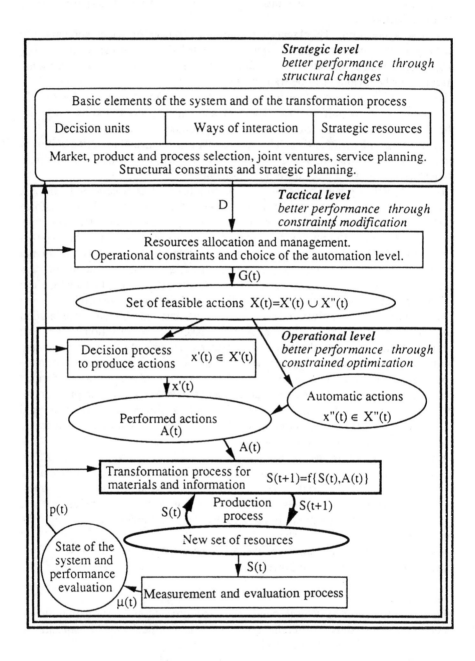

Fig. 6.4. Decision level model

Tactical level
given: environment, structural constraints difficult to modify, company goals, a set of performance indicators
find: operational constraints, information flows, operational procedures and the value of the decision variables
such that: the organization performance will be optimized (flexibility, complexity, etc.)

Strategic level
given: environment, global structural constraints and resources, set of interconnected decision centres
find: company goals and performance indicators
such that: the profitability of investments will be maximized

A first set of decisions concerning organizational constraints are usually the connections between input resources, activities and output resources. Notice that this is the framework for activity based costing analysis. At this point we have to define the relationships between activities, such as precedence constraints, concurrency, etc. Given the resource/activity connections and the relationships between activities, we can define the resource allocation process and plan our activities in time.

The activities can either be the output of a decision making process involving personnel (organized in different decision centres), machinery and information, or the output of an automatic system of decision rules, producing the actions on the basis of a set of measures of the state of the area concerned and, if appropriate, the value of a set of parameters.

The set of decisions produced automatically and the effectiveness of the decision rules are crucial point for the effectiveness of the whole system.

Traditionally, to manage a production system, one begins by defining the activities of the organizational units, then the interconnection network that links them, taking into account the support services only implicitly. From this breakdown into unit activities, the units are aggregated on the basis of the interaction network in order to define linked processes and their coordination needs. But in this way the breakdown becomes an organizational constraint, and the performance optimization process is generally carried out in the framework of a given organization structure. The organization structure, however, is often the main obstacle to improvement, and the optimization of such a structure is a complex problem. Benchmarking is a way of handling this problem.

The activity breakdown is performed on the grounds of a set of different organization structures. The performance can be evaluated for each of the resulting systems. Instead of looking among all possible organization structures, we limit our search to existing ones. To make this effective, it must be possible to compare not only similar companies, but also very different ones. We must, therefore, work on subsystems that have the same input-output functional model. The subsystems to be compared are found bottom up, aggregating the unit activities on the grounds of the relationship among units. The breakdown is performed on the process implemented by the two subsystems.

We therefore compare two companies where the same function is performed in different ways. This means that there are two different breakdowns (i.e. two different organizations and/or two different assignments). These breakdowns are, in our opinion, one of the main aspects of benchmarking.

In order to make an activity breakdown, it is important to describe the functional relationships among input, transformation activities and output, the resources consumed and the resources available for the activities, the aggregation of resources into OUs, the links among OUs and a set of OUs. Given the resource/activity connections and the relationships among activities, we can define the resource allocation process and plan our activities in time.

Lucertini et al (1994) present a modelling approach to benchmarking which attempts to produce a structured methodology based on four types of benchmarking.

Goal benchmarking, which studies the possibility, based on the improvement of performance indicators, of trying to obtain the values of the benchmark.

Organizational benchmarking. which studies the possibility of substituting sets of activities of the whole process with other sets of the breakdown activities of the same process in the best practice company.

Integration benchmarking, which studies the possibility of changing the interconnection pattern for the same activity breakdown (this case is fairly rare on its own because, in general, few interconnections are possible).

Implementation benchmarking, which studies the possibility of redesigning process or support units. In this case, we suppose that the design of some single units can be improved.

A basic assumption underlying the breakdown is *modularity. We can use the breakdown as a decision support tool only* if we can assume that a given subset of activities can be easily replaced by a suitable redesign of the corresponding subset of activities of the benchmarking partner. In practice, this is seldom true today in manufacturing environments. The general trend, mainly because of cost and quality assessment, is however, in the director of increasingly standard subsystems and interfaces.

Generally speaking, although some configurations may prove to be infeasible if the system is fairly modular and the redesign well done, most will produce feasible patterns. For a given configuration, the construction of an efficient way to operate often requires the solution of an optimization problem to define the modes' internal organization and interface procedures correctly. Constraints redesign differentiates optimization procedures from benchmarking procedures. The optimization process evaluates the state of the system and its performance, optimizes the process itself and then formulates improvements. The benchmarking on the other hand, begins with the redesign of the transformation process and continues with the comparison with that of the benchmarking partner. A suitable set of performance indicators are therefore required and constraints redesign is necessary in order to introduce improvement decisions to the process and then to formulate a new optimization problems (see [1nt1; Lucertini et al 1994). It should be noted that to make the comparison effective, we must compare the performance of our process with the performance of the partner's redesigned process, not the original process since this is comparable.

Several tools recently developed for organizational analysis may be of help for benchmarking. We need a structured methodology to find technological coefficients for resource constraints and to express performance indicators, usually defined in terms of products and final outputs, as a function of the activities in which the process breaks down. To do this we must associate products with activities and activities with resource consumption.

First of all, the breakdown required by benchmarking analysis is useful for comparing processes in different companies and may be used to find performance standards for some types of processes. Functional breakdown may also be a good basis for treating information before its introduction into the model.

Activity-based systems aim to correct the traditional costing system's deficiencies in finding technological coefficients for financial resources constraints by introducing three main goals: to assign costs to activities (or actions as we have previously called them, to assign costs to cost objects, and to produce auxiliary information about activities (but not strictly financial).

Conventional cost system analysis presumes that products cause costs. The correct activity-based cost assumption is that the cause of cost is not the product as such, but all the activities necessary to manufacture the product (or produce the service). The product does not directly consume money, but consumes resources. The performance of activities needs a series of resources, and these resources entail costs.

A cost objective is the reason for performing the whole process and its basic activities. It can either be a subproduct or a service being relating to the process being examined. *Activity-based costing* has the goal of measuring the real unit cost, resulting from the sum of all the activities that are necessary to produce a good or service. To have an accurate and fast measure of the use of different activities the definition of *drivers* is also important. The *driver* is a measure through which cost is allocated to the activity; a measurement unit of a *driver* can be, for example, the number of hours worked or the number of parts produced.

Some activity attributes (for instance, the number of hours of inspection, the number of moving parts of a dye, the number of colours in printed parts can be *cost drivers*. A *cost driver* represents the causal factor and hence gives the dominant cause of cost.

The information given by *activity-based costing* can be used to measure performance indicators. A performance indicator may, for example, be the number of pieces refused by the client, or not accepted after a quality inspection. They must be related to measures of how well we perform the activity. Indicators for benchmarking purposes must always be comparable. Thus, an indicator has to be normalized with respect to the complexity of the activity considered, the tools utilized or the place where the activity is performed.

It is important to concentrate on activities that are relevant but also changeable. Their indicators should therefore be either global (and therefore comparable), or ad hoc indicators linked to the alternatives. We can suppose that performance indicators can be expressed as a function of the decision variables that drive the process. In fact, such variables, together with the initial state of the system, determine the state in all the following time frames (because of the assumption of a deterministic system, the state of the system, determines the output and the output determines the performance).

When, in order to achieve better performance, we incorporate some processes of another company into our company, the decision variables can change, but the performance indicators remain the same (even though, hopeful, with a better numerical value).

References

Agnetis A, Lucertini M and F Nicolò (1990) Flexibility-cost Tradeoff in the Design of Production Systems: A Preliminary Approach. in: Carnevale M, Lucertini M and S Nicosia (Eds) Modelling the Innovation. North-Holland

Agnetis A, Lucertini M and F Nicolò (1993) Flow Management in Flexible Manufacturing Cells with Pipeline Operations. Management Science 39 3:294-306

Agnetis A and M Lucertini (1990) Design Criteria for Flexible Production Systems Based on Non-simultaneous Demand Models. Proceedings of the 2nd International Conference on CIM. Troy NY

Balm G J (1992) Benchmarking. Qpma Press Illinois

Baligh H and R Burton (1981) Describing and Designing Organization Structures and Processes. International Journal of Policy Analysis and Information Systems 5 4:251-266

Bekiroglu H (Ed) (1984) Computer Models for Production and Inventory Control. Society for Computer Simulation. La Jolla California

Bonini C (1963) Simulation of Information and Decision Systems in the Firm. Prentice-Hall

Butera F (1991) Il Castello e la Rete. Angeli

Carlsson B (1989) Flexibility and the Theory of the Firm. International Journal of Production Management 3/3

Chang S K (Ed) (1984) Management and Office Information Systems. Plenum Press New York

Cyert R and J March (1963) A Behavioral Theory of the Firm. Prentice-Hall

Coase R H (1937) The Nature of the Firm. Economica 4

Daft R (1989) Organizational Theory and Design. West Publishing St. Paul Minnesota

Dioguardi G (1983) Macrofirm: Construction Firms for the Computer Age. Journal of Construction Engineering and Management. ASCE 1

Groote X de (1994) The Flexibility of Production Processes: A General Framework. Management Science 40 7:993-945

Jae-Ho Hyun and Byong-Hun Ahn (1992) A Unifying Framework for Manufacturing Flexibility. Manufacturing Review 5/4

Hastings C (1993) The New Organization - The Growing Culture of Organizational Networking. McGraw Hill

Huber G P (1990) A Theory of the Effects of Advanced Information Technologies on Organizational Design, Intelligence and Decision Making. Academy of Management Review 15 1:47-71

Huber G P and Glick W H (Eds) (1993) Organizational Change and Redesign. Oxford University Press New York

Johnson H T and R S Kaplan (1987) Relevance Lost: The Rise and Fall of Management Accounting. Harvard Business School Press

Kaplan R S (1990) Measures for Manufacturing Excellence. Harvard Business School Series in Accounting and Control

Kumar A, Ow P S and M J Prietula (1993) Organizational Simulation and Information Systems Design: An Operations Level Example. Management Science 39 2:218-240

Lucertini M, Nicolò F and D Telmon (1994) How to Improve Company Performance from Outside: A Benchmarking Approach. IFIP workshop on Benchmarking: Theory and Practice (invited paper). Trondheim Norway. June 16-18

Lucertini M, Nicolò F and D Telmon (forthcoming) Integration of Benchmarking and Benchmarking of Integration. Journal of Production Economics

Lucertini M And D Telmon (1993) Le Tecnologie di Gestione. I Processi Decisionali nelle organizzazioni Integrate. Franco Angeli

Lucertini M And D Telmon (forthcoming) The Factory in Triplicate

Malone T (1987) Modelling Coordination in Organizations and Markets. Management Science 33 10:1317-1332

Malone T and S Smith (1988) Modelling the Performance of Organizational Structures. Operations Research 36 3:421-436

March J (Ed) (1965) Handbook of Organizations. Rand McNally

McNair C J and K H J Liebfried (1992) Benchmarking: A Tool for Continuous Improvement. Oliver Wight Publications Vermont

Mesaroviç M D, Maco D and Y Takahara (1970) Theory of Hierarchical Multilevel Systems. MacMillan New York

Omachonu V K, Davis E M and Solo P A (1990) Productivity Measurement in Contract Oriented Service Organizations. International Journal of Technology Management 5 6:703-719

Ostrenga M R, Ozan T R, Harwood M D and R D McIlhattan (1992) The Ernst and Young Guide to Total Cost Management. Wiley New York

Saaty T L (1988) Decision Making for Leaders: The Analytical Hierarchy Process for Decisions in a Complex World. RWS Publications

Suares F F, Cusumano M A and Fine C H (1991) Flexibility and Performance: A Literature Critique and Strategic Framework. Working Paper 50-91. International Center for Research on the Management of Technology. MIT Cambridge

Upton D M (1994) The Management of Manufacturing Flexibility. California Management Review 36 2

Zülch G and T Grobel (1993) Simulating Alternative Organizational Structures of Production Systems. Production Planning and Control 4 2:128-138

7 Technological Innovation, Transport and Location: A Comparative Examination of Four Analytical Models

Claudia Azzini, Cristoforo S. Bertuglia and Giovanni A. Rabino

7.1 Introduction

The growing diffusion of the technological/information revolution, which is already having a major impact on the economic development of many industrial sectors and services, is also penetrating the field of transport and influencing the location of urban activities. By 'transport' we refer not only to the physical (material) network for transporting people and goods, but also the network used for the non-material 'transport' of information and for the exchange of knowledge. The enormous increase in the quantity of information and new modes of exchange are generating changes in the economic structure, with consequent modifications in the spatial configuration of urban areas.

This process is producing a transition from monocentric location behaviour, with interactions of the 'one to many' type, i.e. the characteristic radial communications network deriving from the adoption of classical technology (the traditional industrial city), to dispersive location behaviour with 'many to many' interactions, typical of the interconnected communications network deriving from the adoption of new technology (the post-industrial city) (Brotchie et al 1986). A feature of the post-industrial city is the strong interdependence between communications, transport and production activities.

The impact of the technological revolution can be seen in the pervasive introduction of new elements in a whole range of different fields, including telecommunications (e-mail, fax), industry ('just in time' production, new production systems), business (teleconferencing), office work (telework) and services (teleshopping, telebanking). These innovations are generating major changes in: (a) the spatial organization of the activities, (b) the importance of physical mobility, and (c) the nature and perception of distance as a cost. These three phenomena are closely connected and cannot be analyzed separately.

The use of new information technologies is leading to a redefinition of mobility, tending towards the creation of an 'equipotential space' for the location of activities. The distance factor, which previously led to the centralization of economic activities, is becoming less important - to the extent where, in some cases, it is no longer perceived as a cost. This permits a greater distribution of activities, in particular towards peripheral locations, since many interactions previously carried out in the form of physical trips can be replaced by telematic communication.

In reality, these dispersal effects have not until now been as marked as was suggested by early forecasts. There seem to be several reasons. Firstly, the orientation towards the use these new technologies, both in terms of the willingness to create the necessary infrastructure and individual predisposition to the adoption of innovations, is in many cases still relatively weak. In addition, it is perhaps too early for there to have been a profound impact on behaviour. In certain cases, the adoption of innovations has in fact even been found to generate the opposite effects, producing a new forms of centralization and local intensification of mobility.

Examining a number of new technology applications in various different fields, we find examples of these different tendencies. We consider firstly at some of the possible effects on traffic and on residential and office location of the introduction of two innovations in the world of business.

The widespread adoption of teleworking would permit a far greater dispersion of housing and of urban development in general, as a large percentage of people could work from home. Further consequences seem likely to be a reduction in the concentration of offices and management activities in the city centre and a significant decline in the volume of commuting traffic. The overall result would probably be a greater 'spread' of traffic, both geographically and in terms of time (an increase in local traffic for service trips, an intensification of traffic in non-peak periods and possibly also weekend leisure trips). The adoption of teleworking also requires, however, a redefinition of working relationships, with employers (new forms of contract), with colleagues (as less time would be spent in direct contact with other workers) and also at home (because of more time spent in the house). Adoption is at present constrained by the perception of negative implications in these areas.

The introduction of teleconferencing similarly reduces the need for physical trips and, in the sense that it permits the involvement of participants from an unlimited geographical area, creates equipotential in spatial distribution. Physical trips are centralized around those places equipped with teleconference facilities, causing an increase in local traffic. 'Costs' carried by participants include the lack of direct interpersonal contact and the elimination of the journey which, to some, may have provided a form of gratification.

In the area of services, the introduction of teleshopping means that there can be a change in location of sales outlets, which no longer need to be in the city centre. The use of this service is therefore likely to lead to a decrease in private traffic towards shopping centres, but an increase in delivery traffic (though this would probably follow more systematic routes). The consumer pays the price of not being able to inspect the product before purchase. One hypothesis is that while teleshopping may replace regular household shopping, visits to shops will remain as a leisure activity.

The profound conviction found in much of the earlier literature that telecommunications would result in radical and rapid social and economic changes is having to be redimensioned in the light of the many recent empirical studies. In each of the above cases, there is a considerable gap between the expected or potential impact of the introduction of innovations and the actual response. The limited diffusion of certain technologies (e.g. teleshopping) is due, at least in part, to resistance to change in consolidated behaviour. In the case of more general applications of telecommunications, however, this contrast is less marked. The adoption of fax and e-mail, for example, is already widespread and the three effects cited above - dispersed location, substitution of physical journeys and elimination of distance costs - are evident in many places. It seems likely that the impact of telecommunications will spread, not only because of the increasing adoption of these technologies in different sectors, but also because of changes they induce in ways of thinking, and hence in the organization of activities.

Although the limited diffusion effects are due in part to the limits and technical difficulties inherent in the adoption of new technologies (which can with time be overcome), we need to recognize that certain sectors or activities remain strongly constrained by the milieu.[1] An analysis of the potential effects of technological innovation on the centralization/decentralization of firms therefore needs to take into consideration the type of activity involved. Many authors also underline the importance of considering which part of the firm it is possible or appropriate to decentralize. In other words, new technologies do not create a generalized decentralization, as it is not possible to ignore the social, economic and cultural context of which the firm is part. In some studies (e.g. Driver and Gillespie 1993) it is claimed that the phenomenon of decentralization can be attributed more to changes in the transport sector (improvements in the physical transport network) than to new telecommunications technologies. We cannot therefore arrive at valid conclusions concerning the spatial implications induced by innovation if we limit our attention to the technologies themselves (Castells 1989).

The relationships between new technologies, location and transport also need to be examined in the light of those elements which tend to reinforce existing patterns (Bertuglia and Occelli 1995). The new communications technologies can be used by firms to strengthen and extend their spheres of influence, the power of established organizations and hence those 'central places' which already possess a solid economic structure. This can sometimes have the effect of favouring centralization rather than decentralization (Gillespie and Robins 1989).

The aim of the present study is to see how these phenomena are dealt with in four different models selected from current literature. The models are presented in Section 7.2 and then described individually in Section 7.3. In Section 7.4 we make a comparison, attempting to identify the 'points of contact' between them. While the first three models derive from systems analysis, the fourth adopts a neural networks approach. These specific models have been chosen since they are typical of each model type (although slight modifications have been made to adapt them to the particular application, the basic structure is still clearly recognizable).

---

[1]By milieu we refer to those specific qualities and skills of the workforce, entrepreneurship, and the intensity and quality of interpersonal contacts (Dematteis 1994) which distinguish a particular location. The availability of such 'human resources' makes some locations preferable to others for certain types of firm.

7.2 The Models

7.2.1 A Compartmental Model (for Studying the Relations between the Innovation Choice Process and the Location Decision, and the Consequent Effects on Urban Structure)

With the model proposed by Haag and Lombardo (1991) and later developed by Bertuglia et al (1993, 1995) (see also the chapter by Lombardo and Occelli in this book), an attempt has been made to understand and simulate the effects of technological innovation on the spatial organization of urban systems, in particular the impact on metropolitan areas and on the service sector. The service sector is playing an increasingly significant role in economic growth, and the information technologies represent an important tool for improving the services offered. The adoption of innovation, in particular new telecommunications and information technology, influences the organizational structure of human activities, especially in a spatial sense. In the service sector these technologies have an *enabling effect*, extending the degree of freedom of the organizational and spatial structure of firms, thereby giving them greater locational choice. In this sense, at least, it is agreed that information technologies can partially eliminate spatial constraints.

The proposed dynamic model attempts to connect the choice of innovation with the firm's location choice. These two processes are strongly interrelated, since the adoption of new technologies will modify the weight of spatial separation of certain activities and it is also possible that specific locations will be particularly favourable to innovation adoption.

Before describing the model we wish to make some observations on the difficulties of modelling the impact of innovation adoption on location. Firstly, it is extremely difficult to make firm predictions relating to the acceptance and use of new technologies. This is not only due to the fundamental uncertainties regarding the future, but also the large number of exogenous variables which may influence the innovation choice. Secondly, it is necessary to take into consideration that the distribution of innovation will be spatially non-uniform, since some areas will have a higher level of adoption than others. Linked to this is the *self-acceleration effect* (the greater the number of exisitng users, the greater the attraction of an information system).

These observations lead us to a further complication regarding the impact of innovation adoption: the existence, already mentioned above, of contrasting effects. On the one hand, we may expect the reduction in distance costs to cause certain activies to decentralize, leading to an increase in the number of firms locating in peripheral areas where previously there was difficulty in obtaining and managing information in real time. The effect in this case would be a gradual move towards geographical uniformity in the use of technological innovations (virtual equipotential). On the other hand, as pointed out in other studies (e.g. Conway and Kirn 1988, Olson 1989), there may also be a centralization effect due to the role played by the infrastructure. The availability of transport networks, support provided by technical, financial and business services and the existence of telematic networks remain important factors resulting in 'well provided' areas becoming favoured locations. This centralization effect is also encouraged by the reorganization of activities induced by micro-electronic technology. For many firms it is possible to concentrate business services in central areas, while production activities or secondary offices can be located in peripheral areas.

We now describe the compartmental model, highlighting the way in which these various effects are dealt with, looking especially at the innovation and location choice processes.

As far as the innovation decision is concerned, we obtain the dynamic utility function $u_{j\alpha}$ of innovation j in zone α (which expresses the advantage deriving from its use):

$$u_{j\alpha} = s_{j\alpha} + \delta_{j\alpha} \qquad (7.1)$$

where the right hand side expresses the motives pushing the firm to adopt that particular innovation. These are:

- self-acceleration effects ($s_{j\alpha}$) which depend on the degree to which the innovation has already been adopted in a given area. An indicator could be the local share as a proportion of the total market share;
- other effects ($\delta_{j\alpha}$) representing the tendency in area to prefer technology j. This measure includes both subjective and objective evaluations of the quality of innovation j. The indicators can be given by the features of the area (α) and of the innovation (j).

There is a further factor which may influence the choice of innovation. This is the flexibiltiy, v_{ij} of the firm, or its propensity to adopt the innovation. The rate of transition (choice of innovation i rather than j) is given by:

$$w_{ij} = n_{j\alpha} \ v_{ij} \ e^{(u_{i\alpha} - u_{j\alpha})} \qquad (7.2)$$

where $n_{j\alpha}$ is the number of firms in zone α which have adopted innovation j.

The location choice process has been divided into two parts relating to the different possible reasons for modifying the number of offices located in a given zone:

- the decision to expand or reduce the scale of the activity;
- the decision to move to a different area due to the adoption of an innovation.

In the former case, the choice is made for the benefit of the firm. It depends on considerations deriving from: an 'attraction' effect m_{α} for which we use the number of existing offices in the zone; rent prices P_{α} (the higher they are, the less the advantage of moving to the zone) and the supply of infrastructure Z_α (the better the infrastructure, the greater the advantage of the move). The weighted sum of these factors (where α, β and γ are constants which represent the respective weights) gives the expected advantage:

$$V_\alpha = \alpha m_\alpha - \beta P_\alpha m_\alpha + \gamma Z_\alpha \qquad (7.3)$$

from which we can obtain the probability of expanding the number of offices:

$$w_\alpha^+ = \varepsilon_1 \, (k_\alpha - m_\alpha) \, e^{V_\alpha} \qquad (7.4)$$

or of reducing them:

$$w_\alpha^- = \varepsilon_2 \, m_\alpha \, e^{V_\alpha} \qquad (7.5)$$

in which k_α represents the total capacity of zone α, while ε_1 and ε_2 are the different speeds of adjustment of the concentration of offices in each zone.

In this case, the interaction between the innovation and location choices is important. The new technology increases the number of possible locations so the choice is less constrained, although it will still depend on other factors such as rent levels, taxes etc. The transition rate for the firm adopting innovation i to transfer from α to γ is given by:

$$w_{\gamma\alpha} = m_\alpha \, \varepsilon_3 \, e^{V_\gamma - V_\alpha} \, x_{i\alpha} \, Q_{i\gamma} \qquad (7.6)$$

where $x_{i\alpha}$ represents the percentage of firms in zone α which have adopted technology i, V_α and V_γ are respectively the expected advantages from location in α and γ, while ε_3 is the speed of adjustment. In this context, the availability in zone γ of the infrastructure needed for innovation i is important, therefore $Q_{i\gamma} = 1$ if it is available and $Q_{i\gamma} = 0$ if not.

Bringing together all the equations relative to the two processes into a single expression whose form is directly derived from the *master equation*, we have:

$$p(P) = \sum_k p(P+K) \, w(P + K;P) - \sum_k p(P) \, w(P;P+K) \qquad (7.7)$$

where p represents the probability that the system is in the state defined by the vector P, and w is the probability of transition between the states of the system, defined respectively by P and (P + K). We obtain the following first order equations which simulate the most probable behaviour of the firms in relation to the two choice processes:

$$\frac{dn_{j\alpha}}{dt} = \sum_{i=1}^{J} n_{i\alpha} \, v_{ji} \, e^{(u_{j\alpha} - u_{i\alpha})} - \sum_{i=1}^{J} n_{j\alpha} \, v_{ij} \, e^{(u_{i\alpha} - u_{j\alpha})} \qquad (7.8)$$

with:

$$\frac{dm_{\alpha}}{dt} = \varepsilon_1 \, (k_{\alpha} - m_{\alpha}) \, e^{V_{\alpha}} - \varepsilon_2 m_{\alpha} e^{-V_{\alpha}} + \varepsilon_3 \sum_{i=1}^{J} \sum_{\gamma=1}^{A} m_{\gamma} e^{(V_{\alpha} - V_{\gamma})} x_{i\gamma} Q_{i\alpha}$$

$$- \varepsilon_3 \sum_{i=1}^{J} \sum_{\gamma=1}^{A} m_{\alpha} e^{(V_{\gamma} - V_{\alpha})} x_{i\alpha} Q_{i\gamma} \qquad (7.9)$$

$n_{j\alpha}$ = the number of firms in zone α which have adopted innovation j
$u_{j\alpha}$ = the dynamic utility of innovation j in zone α
m_{γ} = the total number of firms in zone φ
k_{α} = total capacity of zone α (maximum number of firms which can locate in the zone)
V_{α} = the expected advantage from location in zone α

$x_{i\gamma} = (\dfrac{n_{i\gamma}}{N})$ percentage of firms in zone γ which have adopted technology i;

N is the total number of firms which use an innovation.

Figure 7.1 shows the logical connections between the various parts of the compartmental model.

7.2.2   An Ecological Model (for the Study of Innovation Diffusion)

Some interesting models have been developed, in particular by Sonis (1984, 1986), to examine the multi-zonal diffusion of innovation from the ecological point of view. The model described here is based on the Volterra-Lotka equation,[2] but produces a *relative distribution* of users, rather than an absolute distribution. The dynamic is represented by a logistic curve, which has a role corresponding to the exponential curve in the absolute version.

---

[2]The Volterra-Lotka equation describes the relative growth of a number of interacting populations and represents the classic case of the prey-predator relationship.

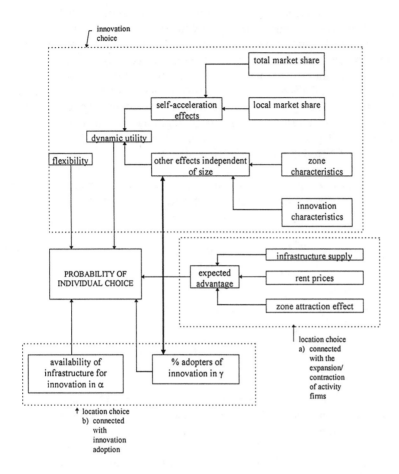

Fig. 7.1. Logical connections of the compartmental model

The models most commonly used in this context, the Logit and Dogit, are static models based on the hypothesis of totally egoistic behaviour (the search for maximum individual utility) and on the assumption that the individual possesses complete information. These highly restrictive hypotheses have been criticized as highly unrealistic. A further weakness is that the individual's decision is considered independently from the decisions of others, ignoring possible interactions and the general social context.

The theory used in the present model is based on the hypothesis of a dynamic choice process with a reciprocal influence not only between individuals but also between the individual and the innovation.[3] It is assumed therefore that there is

---

[3]The characteristics of the time/space diffusion of an innovation reflect the individuals' choice and vice versa, i.e. the greater the frequency of choice, the greater the diffusion of the innovation.

126

competition between innovations,[4] interaction between users, based on the exchange of information relating to utility, and that the environment affects the relations between individuals. The behavioural choices of each individual are represented by the random utility theory.[5] From this theory it follows that, on the supply side, the behaviour of the innovation is measured by the percentage chosen by the users and, on the demand side, it is possible to obtain the solution resolving a log-linear equation. The hypothesis of *competitive innovations* implies that they are considered as alternatives. In the case of a single innovation, the conflict is between adopters and non-adopters. If there are N innovations, the conflict is between adopters of different innovations (including non-adopters, i.e. those who have adopted no innovation).

With the hypothesis of a passive environment (in which, for example, there are no self-organization effects), the change over time in relative distribution of users, y(t), is obtained from the log-linear (competition) equation:

$$\frac{dy_i(t)}{dt} = y_i(t)\sum_{j=1}^{N} a_{ij}y_j(t) \qquad (7.10)$$

under the normalization condition:

$$\sum_{j=1}^{N} y_j(t) = 1. \qquad (7.11)$$

$A = \|a_{ij}\|$ represents the antisymmetrical interaction matrix and a_{ij} describes the influence of innovation i on the adoption of innovation j. Studying A we can deduce both the existence of an equilibrium state E_k and the asymptotic stability of the solution (if $a_{ij} \geq 0$ for each i, j). A simple case of the diffusion of the innovations is represented by the *totally antagonistic behaviour of innovations*. This occurs with the condition:

$$a_{ij} = a_i - a_j \qquad (7.12)$$

where a_i and a_j are the levels associated with the various states ('interaction potentials'). Equation (7.12) is the case in which the interaction depends only on the difference between the characteristic of the final state j (which implies that $a_{ij} + a_{jk} + ... + a_{mi} = 0$, or that if a user changes all the innovations to return to the initial one, there will be no increase in utility). The existence of these interaction

---

[4]The innovations are competitive if: i) they are interchangeable, ii) they are mutually exclusive, iii) the choice is exhaustive.

[5]This theory is based on three hypotheses: i) every user has a choice set I_i, ii) the user aims to maximize utility, represented by u_{ij}, for each choice j from set I_i, iii) as the observer has no knowledge of either u_{ij} or the choice processes, the utility is expressed as an aleatory variable.

potentials allows us to obtain the individual's choice equation:

$$y_i(t) = \frac{y_i(0)e^{a_i(t)}}{\displaystyle\sum_{j=1}^{N} y_j(0)e^{a_j(t)}} \qquad (7.13)$$

which generates the logistic growth and decline curve. The immediate consequence of this interpretation is the *competitive exclusion principle*. If the innovation i is dominant, in the long-term everyone will choose this alternative.

If we wish to take into account an *active environment*, able to influence the behaviour of individuals through the modification of information flows, we can introduce a stochastic matrix[6] $S = \| s_{ij} \|$, so that the new vector solution is $U = S*Y$. Each s_{ij} represents the frequency of adopters who change from i to j under the influence of the environment. The effect on the distribution consists of the redefinition of the extremes of the logistic curve. The upper and lower limits are no longer 1 and 0, but two values which are respectively 0 and 1. The presence of an active environment thus causes a *more balanced distribution of adopters* (the equilibrium is expressed by $S*E_k$, where E_k represents the equilibrium state in the case of a passive environment) and, depending on the specification of S, it will be possible to derive a family of models.

If we take a set P of space/time parameters, in which time is considered as one of the endogenous variables rather than an external reference value (therefore becoming an extra dimension in the R- dimensional space of the system dynamic), it is possible to introduce a potential V_{ij} such that:

$$\frac{\partial V_{ij}}{\partial r} = f_r(y_1 \ldots y_n) \qquad (7.14)$$

for each dimension r of the problem (for $r \in R$), where V_{ij} is the utility of transition from i to j. Given that V_i is the utility of an individual who chooses the alternative i, the solution with a passsive environment is expressed by the *fundamental formula*:

---

[6]This depends on time as well as on Y. This matrix expresses the behavioural barriers of the passage from one alternative to another.

$$y_i(t) = \frac{C_i e^{V_i(t)}}{\sum_{j=1}^{N} C_j e^{V_j(t)}} \qquad (7.15)$$

with $C_i = ln\ V_i = (E_k)$.[7] If, for example, the potential is a function of two parameters (t,s), where t represents time and s is the distance from the diffusion centre, the fundamental formula gives the space/time growth of the innovation. The main feature of this expression is that the fraction of choice $y_i(t)$ of the ith innovation, i.e. the first expression of equation (7.15), coincides with the frequency of individual selection of the same innovation, i.e. the second expression of (7.15), as an alternative. It should be noted that the decision is not static, but evolves with time, as the individual will tend to imitate the behaviour of others.

At time t it is possible to express the choice potential of innovation i in function of the interaction coefficient $a_{ij}(V_i = a_{ij}t)$, producing totally antagonistic behaviour. It can be demonstrated that the adopter seeks marginal utility rather than total utility and, as the individual is assumed to be not completely egoistic, the hypothesis of utility maximization no longer holds. Also in this case, the influence of the environment is introduced with the stochastic matrix S. The dependence of S on the distribution of adopters between the alternatives represents the response of the active environment to the increased adoption of innovations.

As data is, in reality, available for discrete intervals of time, it is necessary to modify the above models accordingly. This produces the paradox, however, that different solutions are obtained with the discrete model and the continuous model, despite the fact that they derive from the same logit model. The behaviour of the model therefore varies according to the interval Δt and the explicit form of dependence of future decisions on past and present choice frequency.[8] The discrete version of the log-linear equation with the introduction of the utility potential C_i for alternative i, or the transition utility from i to j, C_{ij}, makes it possible to resolve the equation identifying two cases: if C_i is constant, we obtain the usual formula for the logistic curve; if C_i represents the increase in utility from time t to time t + 1 we obtain the logit model in incremental form, i.e. with 'discretized' variables. When the same conditions hold for C_{ij} as for a_{ij}, we have totally competitive behaviour and hence the principle of mutually exclusive competition and asymptotic stability.

Figure 7.2 shows the structure of the ecological model.

---

[7] This is similar to the Logit model and therefore V_j can be seen as the systematic component of individual utility.

[8] It is in fact possible to arrive at a discrete form of dy/dt in two different ways: either [y(t+1) - y(t)]/y(t) or [y(t+1) - y(t)]/y(t+1).

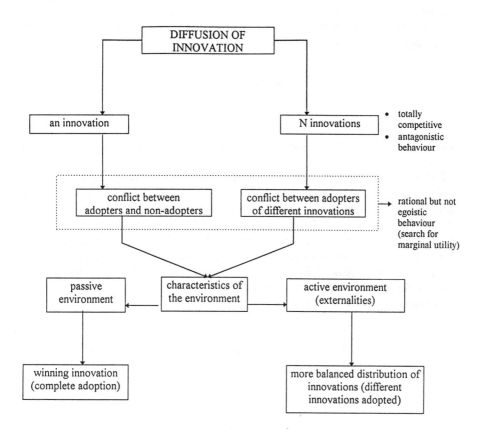

Fig. 7.2. Diagram of the ecological model

7.2.3 Spatial Interaction Model (for the Study of Teleshopping and the Consequent Spatial Redistribution of Retail Activities)

The spatial interaction model considered here (Rabino et al 1992) explores the effects of the introduction of telematic infrastructure on the distribution of retail activities. Teleshopping is considered to represent a new attractor for demand, in competition with existing attractors.

The model simulates the evolution of a Christaller-type system, consisting of a number of nodes, each of which possesses a population $P(i)$ and represents the centre of retail activities. These are divided into different levels, l, on the basis of the type of retail activity carried out there.

The activity (which in this case is teleshopping, but could be any other kind of service) enters the model as another node with zero population. Like the others, it has a distance $D(i,j)$ from nodes j and an attractiveness $M(j,l)$ proportional to the number of sales outlets belonging to level 1 which we call *local units* $U(j,l)$. These can be real outlets or, in the case of teleshopping, fictional equivalents given by:

$$M(j,l) = U(j,l)^{\alpha(l)} \qquad (7.16)$$

where $\alpha(l)$ is a parameter, determined experimentally, which characterizes each level.

The probability $p(i,j,l)$ that a consumer in zone i chooses retail zone (or teleshopping unit) j for services of level l is given by the product of the attractiveness of this zone (or teleshopping unit), $M(j,l)$, and an exponential function of the distance $e^{-\beta(l)D(i,j)}$ normalized with respect to the values in all zones, including that of the service infrastructure. The flows $F(i,j,l)$ are determined for zones i and j as products of the population $P(i)$ and the probability of movement $p(i,j,l)$. The total demand for services in zone j and level l is determined as the sum of flows $F(i,j,l)$ from all zones i.

Assuming the local units per individual for level l:

$$k(l) = \frac{\sum_{j=1}^{n} U(j,l)}{\sum_{i=1}^{n} P(i)} \qquad (7.17)$$

as a characteristic constant, the number of local units for zone j and level l at time $(t+1)$ will be given by $U(j,l)_{t+1} = \text{tot } F(j,l)_t \, k(l)$. We obtain a model which, through successive iterations, simulates the evolution of retail activities in the single centres at different levels of the hierarchy. A number of local units, S, must exist at level $l-1$, if activity of level l is to exist.

Figure 7.3 illustrates the structure of the model described above.

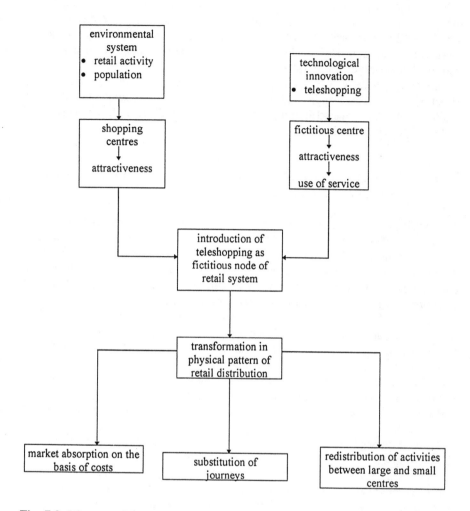

Fig. 7.3. Diagram of the spatial interaction model

The first step in the application of the model to a real situation is the specification of the characteristics of the system: α, β, S ,D. For both α and β, we must choose values which reproduce as closely as possible the current situation in the study area, i.e. the initial state (without the existence of teleshopping). α will have a value <1 as there is a saturation effect in the attraction of shopping centres as the number of local units increases, and β will have a typical value for each level, which decreases as 1 increases (due to the reduced importance of the distance travelled as the level of services offered increases). In specifying the threshold, S, we choose an appropriate value, assuming that the population makes use of local units within the zone of residence.

We also need to decide the way in which teleshopping is introduced. This is done separately for each level, assuming a uniform distribution. Assigning different values to the distances from the fictitious teleshopping centre, we investigate the effects on market shares or absorption, noting in particular the distances for which: i) market absorption is zero, ii) market absorption is total (D = 0) and iii) market absorption is 15%.[9] This last figure is generally considered a realistic market share (Borgers et al 1989).

Once approximate values for D have been obtained, we can experiment with effect of varying these values and with different ways of introducing teleshopping (uniform/non-uniform distribution, joint or separate for different levels, in specific nodes of the same level, etc.).

7.2.4 A Neural Network Model (for the Study of Interzonal Telecommunications Flows)

The application of artificial neural networks as analytical instruments has given rise to considerable interest in the planning field, since the models derived from them constitute an alternative to the more commonly used existing models (in particular, gravity models). This approach offers a new way of interpreting spatial phenomena and appears to be particularly useful for the study of telecommunications flows and their interactions with transport.

A fuller understanding of the spatial structure of interactions in the field of telecommunications is becoming increasingly necessary, especially in Europe, where national policies are tending towards deregulation. If we are to have a clearer idea of the probable complementarity and substitution effects between transport and telecommunications (Salomon 1986), it is particularly important to improve our knowledge of the underlying mechanisms.

The model we examine here was developed by Fischer and Gopal (1994) to represent interregional telecommunications flows in Austria. It is a non-linear model and belongs to the more easily manageable type of neural network models (Hecht-Nielsen 1990). The objective of the authors was to formulate a model of telecommunications flows which makes explicit reference to geographical distance, and to compare the forecasting capacity of the model with that of the classical regression type gravity model. The structure of the neural network model is, nevertheless, based on a conventional gravity type regression model.

The general formulation of the gravity model is:

$$T_{rs} = G(A_r, B_s, F_{rs}) \qquad (r,s, = 1,\ldots,n) \qquad (7.18)$$

where:

T_{rs} indicates the intensity of telecommunications from r to s

---

[9]Other values can of course be adopted and the results compared.

A_r is a factor associated with zone of origin r and represents the intensity of demand for telecommunications of the zone (potential number of calls from zone r);

B_s is a factor associated with zone of destination s and represents the specific attraction factors of the destination (potential number of calls received in zone s);

F_{rs} is a factor associated with the origin/destination pair (r,s) and represents the inhibition effects (impedence) due to the geographical separation between zones r and s.

A and B are factors of mass, F is a separation variable (the higher the value, the less the telecommunications traffic). In general, G is specified in such a way that the model of telecommunications flows is expressed as follows:

$$T_{rs} = KA_r^{\alpha_1} B_s^{\alpha_2} F_{rs} (D_{rs}) \qquad (r,s, = 1,...,n) \qquad (7.19)$$

with:

$$F_{rs}(D_{rs}) = D_{rs}^{\alpha_3} \qquad (7.20)$$

where:

D_{rs} represents the distance from zone r to zone s;

K is a parameter of scale (a constant);

α_1, α_2, α_3 are parameters to be estimated;

n denotes the number of zones.

From this we obtain:

$$\ln T_{rs} = \ln K + \alpha_1 \ln A_r + \alpha_2 \ln B_s + \alpha_3 \ln D_{rs} + \varepsilon_{rs}$$
$$(r, s, = 1,..., n) \qquad (7.21)$$

where ε_{rs} is the error term. (7.21) is the gravity type spatial interaction equation made linear through the passage to logarithms. This model, generally called a 'log-normal' model, is used to evaluate the relative efficiency of the neural network model, even though the term K, being a function of other terms, can only approximately be considered a constant. The general two-level neural network model, characterized by I_1, I_2, and I_3, which indicate respectively input units, hidden units and output units, can be described as follows:

$$X_{2,i_2} = \sum_{i_1=1}^{I_1} W_{1,i_1,i_2} Y_{1,i_1} + W_{1,I_1+1,i_2} \qquad (i_2=1,\ldots,I_2) \quad (7.22)$$

$$Y_{2,i_2} = f\,(X_{2,i_2}) \qquad\qquad\qquad (i_2=1,\ldots,I_2) \quad (7.23)$$

$$X_{3,i_3} = \sum_{i_2=1}^{I_2} W_{2,i_2,i_3}\, Y_{2,i_2} + W_{2,I_2+1,i_3} \qquad (i_3=1,\ldots,I_3) \quad (7.24)$$

$$Y_{3,i_3} = f\,(X_{3,j_3}) \qquad\qquad\qquad (i_3=1,\ldots,I_3) \quad (7.25)$$

where:

I_j is the number of process units in the jth row (j=1,2,3)

i_j are the indices associated with the jth row of the units (j=1,2,3)

X_{j,i_j} is the input of the process unit i belonging to the jth row (j=1,2,3)

Y_{j,i_j} is the output of process unit i belonging to the jth row (j=1,2,3)

$W_{k,i_k,i_{k+1}}$ are the weights (connection weights and parameters) of the kth row between k and (k+1) row, where k=1 indicates the connection weights from the hidden level to the exit level, and k=2 indicates the weights of the connections from the input row to the hidden level

W_{1,I_1+1,i_2} is the 'bias' unit at the hidden level (in the process operated by the network, the bias unit is a constant)

W_{2,I_2+1,i_3} is the 'bias' unit at the output level

the function f is the following logistic function of the hidden units and the output unit, i_2 and i_3:

$$f(X_{k,i_k}) = \frac{1}{1 + \exp(-aX_{k,i_k})} \qquad (k=2,3) \quad (7.26)$$

Combining the previous equations, we obtain the general neural network model in the following compact form for $i_3 = 1,\ldots,I_3$:

$$Y \equiv Y_{3,i_3} = \left\{ 1 + \exp\left[-\left[\left\{ \sum_{i_2=1}^{I_2} W_{2,i_2,i_3} \left\{ 1 + \exp\left[-\left[\sum_{i=1}^{I_1} W_{1,i_1,i_2} Y_{1,i_1} + W_{1,I_1+1,i_2} \right] \right] \right\} \right]^{-1} + W_{2,I_2+1,i_3} \right) \right] \right\}^{-1} \equiv F_w(X) \qquad (7.27)$$

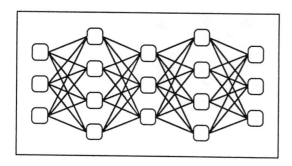

Fig. 7.4. Modular structure of a neural network

(Source: Fabbri and Orsini 1993)

This has a more general functional form compared with the gravity model (the functional input-output relations can be more easily modified) and is expressed from a non-linear regression function of a fairly specific type (White 1989). In addition, it has relatively little sensitivity to the initial conditions.

Models deriving from the artificial neural network approach can play an important role in Regional Science, especially in the analysis and exploration of spatial data. Nevertheless, the definition of the networks is obtained through a standard type network architecture which is characterized by a high modularity and by identical modules (or neural network units) which operate in general in parallel and not sequentially (see Figure 7.4).

The network architecture used in the representation, simulation or forecasting of given spatial phenomena, such as telecommunications flows, has a predetermined structure. It is not deduced or specifically constructed from actual observation - in other words, the architecture of nodes and connections has no specific geographical signficance. The nature of the connections between the various modules (in particular where these are complex neural connections) is established by observing the performance of a network in response to a determined combination of incoming stimuli, and modifying the interconnections until the desired performance is obtained (see Figure 7.5).

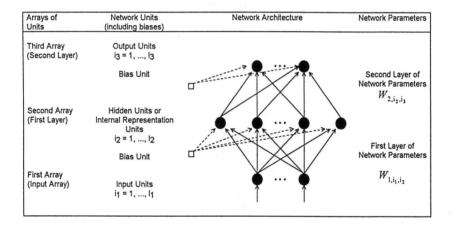

Fig. 7.5. The general two-level neural network model: network architecture (Source: Fischer and Gopal 1994)

The fundamental characteristic of the network architecture proposed by Fischer and Gopal (1994) is that each unit (or node) has connections only with nodes of the level immediately above. The network is therefore a tool for input-output analysis, but is not specifically related to the phenomenon being analyzed. Unlike the previous models, it is not possible to delineate a logical structure, as it cannot really be considered a model in the true sense.

In conclusion, although the study by Fischer and Gopal (1994) does take into consideration some spatial aspects, it is more closely related the methods of econometric analysis which deal with flow/territory relations. It represents the passage from linear to non-linear applications of econometrics rather than to a modelling technique. In the field of Regional Science it is important to distinguish between the theory of territorial networks and network statistics, just as in economics the distinction is made between the network economics and network econometrics.

7.3 Analyses of the Characteristics and Potential of the Different Types of Model

7.3.1 The Compartmental Model

Through the analysis and description of the characteristics and potential of the four different types of model, it is possible to compare the different interpretations of phenomena resulting from the introduction of technological innovations and to identify the particular effects highlighted by each model. These effects, which emerge from simulations conducted with the models (as described in 7.2.3), have physical, social and economic repercussions.

As far as the compartmental model is concerned, the link between technological innovation and location is dealt with in detail. The model takes account of both the 'enabling' effects generated by the adoption of innovations (widening the possibility of location choice and allowing the transfer of activities to the periphery) and the 'centralizing' effects which may occur due to the reliance on specific infrastructure. The model does not take into consideration directly the relationship between innovation and physical mobility.

The relocation of activities towards the periphery is likely to result in a considerable reduction in long distance trips (due in part to their elimination and in part to their substitution) and reduction in city centre congestion. It is possible that the adoption of certain types of new technology can generate a true substitution effect (e.g. teleworking which may eliminate, if not completely at least in part, the need for a journey to work). The effect of new technologies in overcoming distance costs is taken into account only implicitly in the compartmental model, as the location choice is made independently of distance between firms.

7.3.2 The Ecological Model

Initial conditions, such as the characteristics of the local environment or the population, are likely to have a considerable influence on the diffusion and the consequences of the introduction of new technologies. In order to represent these effects in a more realistic way, the ecological model considers an active environment in which the interaction between individuals affects behaviour, since it changes the way in which alternatives are perceived. The innovations are not therefore considered to be mutually excludable and the model is applied to the diffusion of the whole set of innovations.

If we assume that there is total competition between the different innovations and a passive outside environment (i.e. assuming that the user is not affected by any externality and that no new costs are incurred by the introduction of innovations), we would expect complete adoption of the innovations. The most evident result is the elimination of the distance factor: general adoption permits a total dispersion of offices (as a result of telecommunications adoption) and of private housing (as a consequence of teleworking). A further effect is the total substitution of physical mobility. The categories of population affected depend on the type of innovation introduced.

In reality, with an active environment, it is probable that new externalities and new costs for the user are generated. If not all the population are willing to pay these costs, adoption will not be complete and the existing situation will therefore be only partially modified. From the point of view of traffic, for example, it is possible that the effects of adoption could cause a worsening of the current situation. Increased local traffic in residential areas or new journey patterns due to the different distribution of working hours and leisure time could well result in increased congestion.

The contrast in location tendencies between dispersion and centralization effects are similar to the conflict between substitution of physical mobility and its intensification.

7.3.3   The Spatial Interaction Model

Through various simulations the model attempts to analyze how innovation modifies existing equlibria in the territorial system, forcing the formation of new equilibria. As explained above, the model deals with the introduction of teleshopping, but the implications and observations are equally valid for any telematic infrastructure in the service sector.

In the first case, the simulation assumes teleshopping to be introduced uniformly over the study area and separately at each level (each level being characterized by its distance from the teleshopping centre). As distance increases, the teleshopping market share decreases from total to zero. It is possible to identify a distance D for which the substitution represents 15% of the market. The market absorption and therefore the decrease in local units means that some shopping centres do not reach the threshold necessary to have higher level activities (services). This highlights the fact that the adoption of new technology affects not only physical mobility, but also spatial organization, since some smaller shopping centres are downgraded and there is a concentration of higher level retail services in certain specific centres.

This effect is even more marked in the case in which teleshopping is introduced at all levels with the distance value $D_{15\%}$. If the substitution remains on average below 15%, analysing the results carefully, we note that the main higher level shopping centres do not feel the effects of the innovation. Having a high degree of attractiveness, $M(j,l)$, they are able to resist the competition of teleshopping, whereas the smaller centres suffer impact of competition and are downgraded. This produces a greater disparity with the larger centres becoming increasingly dominant. Another significant result is that the average distance travelled for shopping decreases, reducing the total amount of traffic.

It is possible to make further observations deriving from other simulations which adopt more realistic assumptions, such as non-uniform introduction of teleshopping. This assumption reproduces more faithfully the situation in an area where the readiness to adopt new technologies varies according to the local characteristics. By increasing the impact of distance, we go from the situation described above (where the smaller centres find themselves in difficulty) to one where the high costs of adoption in smaller centres means that innovation has less impact and teleshopping obtains a low market share. In the larger centres, the introduction is more effective and reaches high market share.

These examples underline the possibility that in a real situation, where characteristics and potential are not uniformly distributed in space, the impact of technological innovation can result in an increased concentration of activities rather than greater diffusion.

7.3.4 The Artificial Neural Network Model

As the neural network model is qualitatively different from the three previous models, our observations are more general. The model can be more closely compared with the statistical regression methods.

The neural network approach is particularly suited to the identification of empirical patterns and has the advantage that its use is not subject to certain constraints found, for example, in regression analysis. It is also an instrument which makes it possible to minimize estimating errors and to identify patterns even when the data is 'confused'. It allows the representation of processes which do not comply with optimization constraints as well as systems whose internal structure is difficult to analyze. This is because the connections do not have to be defined a priori, but can be modified through a process of self-learning.

The fundamental contribution offered by neural networks is in providing indications of choice when scenarios are continually changing, as in the case of telecommunications, and when the decisions made by the operators (or users) have a limited degree of rationality.

7.4 Points of Contact Between the Models

In this section, in order to be able to make a direct comparison, the formulations of the four models are considered at a more generalized (simplified) level than in the preceding sections. Before analysing the relationships between them, we shall specify for each model the fundamental equations through which the connections and affinities will be established.

In the spatial interaction model, the flows between the zones, $F(i,j,l)$, are calculated through the relation (Rabino et al 1992):

$$p(i,j,l) = \frac{M(j,l)e^{-\beta(l)D(i,j)}}{\sum_j M(j,l)e^{-\beta(l)D(i,j)}} \qquad (7.28)$$

with

$$F(i,j,l) = P(i)p(i,j,l) \qquad (7.29)$$

The compartmental model is based on the following master equation presented in 7.2.1:

$$p(P) = \sum_k p(P+K)\, w(P+K;P) - \sum_k p(P)\, w(P;\ P+K) \qquad (7.7)$$

in which, for ease of calculation, we can substitute the deterministic approximation of average value:

$$\frac{dP(i)}{dt} = \sum_{j=1}^{N} p_j w_{ij} - p_i \sum_{j=1}^{N} w_{ij} \qquad (7.30)$$

which makes it possible to forecast the time/space distribution of the innovation choice and location decision.

With the ecological model, we can define the innovation choice probability as:

$$y_i(t) = \frac{y_i(0)e^{a_i(t)}}{\sum_{j=1}^{N} y_j(0)e^{a_j(t)}} \qquad (7.13)$$

A first point of contact between the first two models is evident if we consider the quantity calculated, i.e. the flows from zone i to zone j, which change in relation to variations in the population of these zones in relation to all the zones in the study area. Similarities can also be found between the terms w_{ij} in (7.2) (the transition coefficients) and $p(i,j,l)$ in (7.28) (the probability that an individual in zone i chooses the level l activity in zone j), between w_{ij} in (7.30) (the mobility parameter), dependent on distance) and the expression $e^{-\beta(l)D(i,j)}$ in (7.28) (distance impedence) and finally between $e^{(u_i-u_j)}$ in (7.8) (utility function) and $U(i,j)^{\alpha l}$ in (7.16) (attractiveness of zone l based on the number of local units U). Although there are differences in the level of analytical expression, the spatial distribution analysis is conducted from the same basic point of view and using similar parameters.

As far as the diffusion of innovations is concerned, there are correlations between the compartmental model and the ecological model. At equilibrium we can obtain either a single solution or multiple equilibria with the equations derived from the master equation. There is therefore a correspondence with the distribution of innovations in relation to the characteristics of the environment (passive environment: single successful solution; active environment: greater redistribution and therefore more than one solution). In the compartmental model these different possibilities also depend on the initial conditions and on the behaviour of the system over time. In the ecological model, we find a dependence on the initial conditions y(0), in the diffusion in continuous time, and on the time trajectories. Different results are therefore obtained according to the time periods chosen.

Being more general, the application of ecological theory includes cases in which a single innovation is introduced (enabling a comparison between those adopting and those not). This means it can be compared with the spatial interaction model which, in our case, analyses the effect of the introduction of teleshopping only. In addition, the basic equation representing the diffusion of innovations, which is obtained by introducing utility V_{ij}, is analytically similar to the equation giving the probability of choice of zone j for a level l activity. In particular, there is a clear analogy between the potential V_i (the utility given by each innovation i) and $M(i,j)$ which represents the attractiveness of each zone j in function of the retail units present.

For reasons explained previously, we have not referred in this section to the neural network model as this model does not lend itself to any direct comparisons with the others.

7.5 Conclusions

There have been many attempts to establish connections between the phenomena induced by the introduction of modern technologies and spatial phenomena. There is a vast literature concerning statistical techniques used for the interpretation of data (often incomplete or confusing) relating to telecommunications and the spatial aspects. These involve simulation techniques and attempts to make forecasts of future communications flows, and examine in particular the effects on the location (or transfer) of firms and the substitution or complementarity between transport and innovations.

The literature concerning models is more limited, although qualitatively of a high standard. One of the main problems is that of establishing appropriate links between the studies of spatial interaction phenomena (e.g. material flows, or mass and distance) and the interpretation of technological innovation phenomena (e.g. non-material flows and size of firms). The studies referred to in this chapter try to identify the nature of such links and to formulate relations of interdependence by concentrating attention on the spatial modifications induced by innovations and the influence of 'space' on the diffusion of these innovations.

As far as the development of new lines of research is concerned, it seems that we can confirm that the models based on neural network theory offer an important contribution, especially in the case of highly complex dynamic systems with numerous feedback mechanisms (such as those which appear at the spatial level following the introduction of technological inovations). This is because the neural networks lend themselves to the representation of interdependences and can be used to model the behaviour of decision-makers and their environment, as well as the interrelations between the two.

Acknowledgement

The authors would like to thank Angela Spence for translating this chapter from the original Italian version.

References

Bertuglia C S, Lombardo S, Occelli S and G A Rabino (1993) Innovazioni tecnologiche e trasformazioni territoriali: il modello Telemaco (Telematica, localizzazione e mobilità: analisi e controllo). Proceedings of the XIV Italian Regional Science Conference, Bologna 2:1053-1085

Bertuglia C S, Lombardo S, Occelli S and G A Rabino (1995) The Interacting Choice Processes of Innovation, Location and Mobility: A Compartmental Approach. In: Bertuglia C S, Fischer M M and G Preto (Eds) Technological Change, Economic Development and Space. Springer-Verlag Berlin:118-141

Bertuglia C S and S Occelli (1995) Transportation, Communications and Patterns of Location. In: Bertuglia C S, Fischer M M and G Preto (Eds) Technological Change, Economic Development and Space. Springer Verlag Berlin:92-117

Borgers A, Gunsing M and H Timmermans (1989) Teleshopping and the Dynamics of Urban Retail System: Some Numerical Simulations. Sixth Colloquium of Theoretical and Quantitative Geography Chantilly (mimeo)

Brotchie J F, Hall P and P V Newton (1986) The Transition to an Information Society. In: Brotchie J F, Hall P and P V Newton (Eds) The Spatial Impact of Technological Change. Croom Helm London:435-451

Castells M. (1989) The Informational City. Basil Blackwell Oxford

Conway R S and T J Kirn (1988) Rural Office Development in Washington State, Its Feasibility and the Role of Telecommunications. Washington State Department of Community Development Olympia

Dematteis G (1994) Global Networks, Local Cities. Flux 15:17-23

Driver S and A E Gillespie (1993) Information and Communication Technologies and the Geography of Magazine Print Publishing. Regional Studies 27:53-64

Fabbri G and R Orsini (1993) Reti neurali per le scienze economiche. Muzzio Padua

Fischer M M and S Gopal (1994) Artificial Neural Networks: A New Approach to Modeling Interregional Telecommunication Flows. Journal of Regional Science 34:503-527

Gillespie A E and K Robins (1989) Geographical Inequalities: The Special Bias of the New Communication Technologies. Journal of Communications 39:7-18

Haag G and S Lombardo (1991) Innovation in Information Technology and Spatial Organization of Urban Systems: A Dynamic Simulation Model. Sistemi Urbani 13:109-124

Hecht-Nielsen R (1990) Neurocomputing. Addison Wesley Reading MA

Olson M H (1989) Telework: Effects of Changing Work Patterns in Space and Time. In: Ernste H and C Jaeger (Eds) Information Society and Spatial Structure. Belhaven Press London:129-137

Rabino G A, Bolognani O and P Alari (1992) Organizzazione territoriale e telecomunicazione, Urban Modelling Thesis. Milan Polytecnic Milan (mimeo)

Salomon I (1986) Telecommunication and Travel Relationship: A Review. Transportation Research 20 A:223-238

Sonis M (1984) Dynamic Choice of Alternatives, Innovation, Diffusion and Ecological Dynamics of the Volterra-Lotka Models. London Papers in Regional Science 14:29-43

Sonis M (1986) A Unified Theory of Innovation Diffusion, Dynamic Choice of Alternatives, Ecological Dynamics and Urban/Regional Growth and Decline. Ricerche Economiche 4:696-723

White H (1989) Some Asymptotic Results for Learning in Single Hidden-Layer Feedforward Network Models. Journal of American Statistical Association 84:1003-1013

8 Telematics, Location and Interaction Flows: The Telemaco Model

Silvana Lombardo and Sylvie Occelli

8.1 Introduction

At the threshold of the 21st century, any policy decision concerning transport, whether strategic, political or economic, must inevitably take into consideration other means of communication, and in particular the New Information Technologies (NIT). These are rapidly changing the ways in which we communicate and interact, influencing the functional and spatial organization of activities and hence patterns of mobility. Such changes are part of a wider framework of social and economic transformation - the globalization of the economy, the growing importance of the service sector and the crisis in the welfare state - which are leading us to revise the way in which urban policies are both conceived and implemented (Amin 1994, Harris 1994, Batten et al. 1995).

In this chapter we discuss the most recent results of an ongoing research project whose purpose is to investigate the existence, nature and extent of relationships between the adoption of new information technologies and processes of change in urban and regional systems. Attention has been paid to the mobility subsystem, in order to be able to evaluate the impact of NIT and assess the possibility of using it as an instrument for the planning and control of residential and office development as well as the transport system.

The study presented in this chapter has two main aims:

a to describe some of the most significant processes responsible for the functional and spatial changes being observed in urban systems today. The analysis focusses on how the adoption of NIT by firms (whether they belong to the service sector or to production sectors affected by tertiarization), influences their location behaviour, the way they interact and their demand for mobility;
b to develop a tool capable of providing analytical/conceptual and operational support in investigating the 'ingredients' which could constitute effective urban policy measures aimed, in particular, at reducing the negative externalities caused by increasing traffic volumes in urban areas.

On the basis of a previous model (Haag and Lombardo 1991, Lombardo 1993), the TELEMACO (Telematics, Location, Mobility, Analysis and Control) model has been formulated (Bertuglia, Lombardo, Occelli and Rabino 1993, 1995). It simulates changes in the demand for land-use and interaction (on both the transport and telematics networks) produced by different urban policies. Policies here are understood to include both planning policies and the strategies of firms relating to functional re-organization and location change.

The model was applied to a fictitious urban system for which a set of scenarios was constructed. Each scenario represents a hypothesized mix of urban policies.

Despite their experimental nature, the simulations made it possible to examine how an urban system could be affected by different urban policies. They also showed that the evolutionary path of the city (both its final spatial/functional configuration and the sequence of the effects over time) was strongly influenced by the speed with which these processes took place. The simulations carried out demonstrated the potential of the model, not only as an analytical tool, but also as an instrument capable of identifying certain elements of urban policy which take into account new possibilities opened up by the diffusion of NIT.

In the next section we recall the structure of the model and illustrate the type of urban policies which can be investigated. We also indicate the kind of results which can be obtained and their usefulness in evaluating policy mix concerning land-use, transport and telematics networks as well as firms' strategies. In Section 8.3 we describe the urban scenarios constructed and provide a brief discussion of the most significant results of their simulation. Section 8.4 outlines the potential of the model and suggests possible directions for future research.

8.2 The Telemaco Model

8.2.1 The Structure of the Model

The formulation of the Telemaco model is based on the convinction that the flexible, adaptive and continuously evolving nature of communication technologies make them 'enabling' rather than determinant factors of urban and regional transformation. In other words, they are capable of increasing the degree of freedom available for planning the organizational and spatial structure of activities and hence play a catalyzing role, in that they allow other variables to interact, often in new ways (Ernste and Jaeger 1989).

The model simulates the self-organization of an urban system as a result of the adoption by firms of new communications technologies. A number of urban scenarios representing different forms of internal and external re-organization of firms as well as impact of policy measures are considered. The reactions to the adoption of NIT are represented as changes in the location structure of the activities (i.e. the tendency to locate in different parts of the city) and in the structure of the interactions i.e. home-to-work and business-to-business contacts/trips (i.e. whether these are effected via the transport network or a telematic network).

We provide here a brief summary of the main features of the model. A more detailed description of the basic hypotheses and equations can be found in Bertuglia, Lombardo, Occelli and Rabino (1993, 1995).

The model belongs to one of the most recent families of dynamic urban models. It follows the compartmental approach, based on the estimation of the mean value of the master equation. The macro behaviour of the model is therefore determined by the interaction of individual behaviours at the micro level (see, for example, Weidlich and Haag 1983, Lombardo and Rabino 1986).

The individual behaviours considered in the model belong to the N firms in the system, divided into H classes. The firms are assumed to carry out activities with medium/high information content (Lombardo, Ambrogio and de Felice 1990) and therefore to be sensitive to the introduction of a set A of telematic services, characterized by K different attributes. The firms are located in an urban system divided into J zones, each characterized by a set of different location attributes and linked by both a transport and telematic network. The model is made up of four dynamic, interacting submodels, each of which simulates a different choice process relating to the firms nhj (see Figure 8.1).

NIT ADOPTION	LOCATION CHOICE
simulation of NIT choices by firms (including the choice not to innovate)	simulation of location/relocation dynamics of firms in the urban system
MODAL CHOICE	FLOW DISTRIBUTION
describing the choice between transport and telematic networks	describing the transport and telematic distribution of flows in the urban system,

Fig. 8.1. Basic structure of the Telemaco model

The main dynamic equations describing the processes dealt with in three of the four interacting submodels are given below.

i The NIT adoption process

$$
n^h_{\alpha j} = \varepsilon_0 \left[\sum_{\gamma=1}^{A} n^h_{\gamma j}\, \upsilon^h_{\alpha\gamma}\, \exp(\, U^h_{\alpha j} - U^h_{\gamma j}\,) - \right.
$$
$$
\left. - \sum_{\gamma=1}^{A} n^h_{\alpha j}\, \upsilon^h_{\alpha\gamma}\, \exp(\, U^h_{\gamma j} - U^h_{\alpha j})\right]
\tag{8.1}
$$

where
α,γ $= 1,..2(A-1)$ is the type of NIT
j $= 1,2,....J$ is the number of zones
h $= 1,2..., H$ is the activity class of firms
$n^h_{\alpha j}$ is the number of firms which have adopted an NIT (or not yet adopted one $\alpha = A$)
$\upsilon^h_{\alpha\gamma}$ is the flexibility of firms for the adoption of a NIT
$U^h_{\alpha j}$ is the utility associated with the adoption of a NIT
ε_0 is the speed of the adoption process

ii The location choice process

$$m^h_j = \varepsilon_1 \, (C_j - \sum_{h=1}^{H} m^h_j) \, \exp(V^h_j) \; -\varepsilon_2 \, m^h_j \, \exp(-V^h_j) \; +$$

$$+ \, \varepsilon_3 \sum_{\gamma=1}^{A} \sum_{i=1}^{J} m^h_j \, \exp(V^h_j - V^h_i) \, n^h_{\gamma i} Q_{\gamma j} \; - \qquad (8.2)$$

$$- \, \varepsilon_3 \sum_{\gamma=1}^{A} \sum_{i=1}^{J} m^h_j \, \exp(V^h_i - V^h_j) \, n^h_{\gamma j} Q_{\gamma i}$$

where

m^h_j is the number of firms
C_j is the zone capacity (i.e. the maximum number of offices)
V^h_j is the expected location advantage
$Q_{\gamma j}$ is the availability of NIT
ε_1, ε_2, ε_3 are respectively the speeds of adjustments for the increase, decrease and
     relocation of offices in the system

iii The modal choice process

$$\psi^h_j = \varepsilon_4 \, [\exp(VF^h_j) \, / \, [\exp(VF^h_j) + \exp(VT^h_j)]]; \phi^h_j = 1 - \psi^h_j \quad (8.3)$$

where
ψ^h_j is the probability of choosing transport interaction
ϕ^h_j is the probability of choosing telematic interaction
VF^h_j is the expected advantage from the use of transport network
VT^h_j is the expected advantage from the use of telematic network
ε_4 is the speed of adjustment

The mathematical formulation of the flow distribution is not given here. It should however be noted that there are two alternative versions. This makes it possible to take into account two different hypotheses relating to changes in the stocks and flows. In one case, the latter change very rapidly and can therefore be assumed to be in equilibrium with respect to the more slowly changing stocks. In the other, it is assumed that the two rates of change are comparable and the flow dynamic is therefore modelled explicitly (a theoretical and experimental study of the latter formulation is made in Lombardo 1986).

The formulation of the submodels is based on the assumption that for each firm there is a given probability, per unit of time, of making choice c within the set C of alternatives available. This probability function derives from certain hypotheses about the factors on which they depend. The factors considered relevant, and therefore included, are summarized in Table 8.1. Some can be considered 'policy elements', in that they make it possible to identify or define urban policies and business strategies (see Section 8.2.2) which take account of the opportunities made available by the diffusion of NIT.

Table 8.1. Effects considered in the Telemaco model and relative indicators

EFFECTS	INDICATORS
Adoption process	(for each telematic innovation)
self-acceleration	- total market share (global acceptance) (EN) - local market share (local synergetic effects)(EN) - market share in specific activity classes (functional synergetic effects)(EN)
preferences	- characteristics of innovation (EX) - level of congestion on transport network (EN)
innovative flexibility	- effects of information diffusion on NIT (EX)
Location choice process	(for each zone)
pull effects	- location choice by other firms (EN)
costs	- rent prices (demand sensitive) (EN) - costs and rates of telematic services (EX)
preferences	- level of congestion on transport network (EN)
advantages	- infrastructure and services available (EX)
labour availability	- number of resident workers (EX)
Modal choice	
interaction requirements	- importance of face-to-face contact (EX)
preferences	- quality of efficiency of telematic connections (EX) - level of congestion of transport network (EN)

Flow distribution

For the three different kinds of flow, the effects considered and the indicators adopted are similar to those used in single constrained spatial interaction entropy models. The probability of choice between transport and telematic netkworks is explicitly dealt with in Telemaco with the necessary modifications to the modelling of flow distribution.

Note: EN=endogenous variables, EX=exogenous variables

Considering a class of activities h, the model therefore assumes that the firms in zone j attribute a certain 'utility', which is a function of these factors, to each choice. This level of utility is the basis for the decision concerning the adoption (or not) of new information technology in the first submodel (the $U^h_{\alpha j}$ term in Equation 8.1), the choice of location/relocation in the second submodel (the V^h_j term in Equation 8.2), the choice of interaction network in the third submodel (the VFh j and VTh terms in Equation 8.3) and the distribution of interactions in the fourth submodel. Obviously, each sub-model the utility levels also depend on the values of variables in the other submodels.

8.2.2 The Model as Simulator of Scenarios

One major area of application of the model is in the investigation of alternative scenarios representing likely or expected future system configurations. In this context, TELEMACO can be used to explore sets of measures or actions to be included in urban policy and business strategies. A number of these are listed below (the dot indicates the variables and parameters actually considered in the definition of the scenarios described in Section 8.3.1).

a *Urban policies (public and joint public/private)*

a1   Policies relating to land-use planning (housing, production, infrastructure, etc.)
- Residential areas: zone capacity in term of resident workers
- Areas for industrial plants
- Service infrastructure: areas destined for various types of services
- Location constraints due to land-use destinations and existence of protected areas

a2   Policies relating to the planning and management of infrastructure and transport services:
- Generalized travel costs
- Transport infrastructure: car parks, intermodal nodes, etc.
- Anti-congestion strategies, e.g. information systems, access limitations, etc.

a3   Technology policies relating to the planning and management of telematic infrastructure and services:
- Infrastructure supply
- Characteristics, performance and quality of the different telematic services
- Cost of telematic services: node and link hardware costs, rates of access and transmisssion

b *Business strategies (public or private firms)*

b1   Office location strategies. These will depend on the sensitivity to the following factors and therefore to the weight they are given in the decision process, as well as on the speed with which the location choices are activated (ε_1, ε_2, ε_3)

Direct costs:
- Rent prices in zones
- Costs of installation and use of NIT in zones

Indirect costs and externalities:

- Infrastructure supply in zones
- Congestion of the transport network
- Location of labour force
- Generalized travel costs

Synergetic effect (imitative or competitive):

- Location choices of firms in the same activity class
- Location choices of firms in different activity class

b2   Organization strategies relating to production and processing and use f information. These will depend on the sensitivity to the following factors and the weight they are given in the decision-making process as well as on the speed of activation of choices (e3, e4).

Operative and structural factors:
- Innovation flexibility depending on the coordination capacity and R&D activities undertaken by the firm, the availability of a skilled and motivated labour force able to take advantage of NIT developments.
- Attributes of the innovation
- Level of diffusion of NIT

Synergetic effects (self-acceleration):
- Level of global adoption by firms in same activity class
- Level of local adoption by firms in same activity class
- Level opf adoption by firms in different activity class.

An evaluation of the effectiveness of urban policies and business strategies could be made by comparing the evolution of certain state variables of the system and the indicators derived from them.

The following list contains some suggested variables and indicators: A) total number of offices in the system, B) spatial distribution of offices, C) density of offices i.e. B/area of zone, D) level of saturation, i.e. B/capacity of local area, E) total number of adopters of NIT, F) spatial distribution of firms which have adopted NIT, G) rent prices in each zone, H) ratio G/level of local infrastructure, I) imitative or competitive effects between firms in different classes, J) advantages expected from different adoption and location choices, K) advantages expected by firms in relation to 'modal' choice between travel and telematic contact, L) distribution of journey-to-work flows on the transport network, M) distribution of flows (business trips) between offices on the transport network, N) distribution of flows between offices on the telematic networks, O) level of congestion of the transport network (general and local), P) relationship between traffic congestion and variation in costs of NIT, infrastructures, etc.

8.3 Simulation of Scenarios

8.3.1 Definition of Scenarios

In order to simulate and compare scenarios representing different paths of urban evolution, it was necessary not only to set up the scenarios, but also to identify a *reference scenario* which would serve as a basis for comparison.

The reference scenario PAT *(Present Assumed Trend)* considered in these experiments, captures some major tendencies which appear to be typical of mature urban systems and economies in most parts of Europe. PAT is distinguished by:
- a relatively weak propensity to use new information technologies;
- a slow pace of economic growth;
- an urban core which still has a dense concentration of functions acting as producers/attractors of a large number of trips, despite the existence of a tendency to spatial diffusion.

Urban changes depicted in the reference scenario variables therefore reflect a slow adoption of NIT and a weak rate of growth of firms. In addition, although decentralization is considered to be a major tendency, firms still have a higher preference for central locations. The urban system is assumed to be evolving slowly and hence relatively low values are assigned to the parameters accounting for the speed of these changes.

Building upon the reference scenario (PAT), an effort was made to set up the scenarios according to likely urban policies and business strategies such as those put forward in 8.2.2. In this application, the scenarios are not mutually exclusive but would correspond to different stages of the evolution of an urban system. Each successive scenario is then developed incorporates the previous one.

In the following, we describe the hypotheses introduced in each scenario and their implications in terms of change of weights (i.e. parameters value) for the different factors. The values of the relevant parameters considered in each scenario are shown in Table 8.2 which also indicates how they are sequentially activated.

1) In the first scenario, AP *(Adoption Pull)*, it is assumed that there exist urban policies aiming at decongesting the city centre. Restricting the location capacity of central areas is the major policy measure. As their re-organization strategies are influenced by such measures, firms are more keen to look for out-of-centre locations. There is also a widespread diffusion of NIT and their global acceptance in the system increases. In addition, imitation effects, competition and functional interdependences will make location choices of firms more sensitive to those of other firms.

2) LP *(Local Pull)*. As a result of the changes in the land-use demand and the emerging of new clusters of offices, location factors - and mainly rent prices - in the different parts of the urban system will be affected. 'Local' factors, including the choices of firms in relation to NIT adoption, become more important in influencing their strategies.

3) DIR *(Decreasing Infrastructure Relevance)*;

4) DCR (*Decreasing Cost Relevance*).The increasing demand for peripheral locations which are often less urbanized, will foster both public and private policies aimed at increasing infrastructure supply (i.e. providing better public transport, car parks, retail stores, etc.) and reducing costs of installation and access to telematic services, in order to encourage the decentralization of offices. These policies aim to create a more uniform distribution of location advantages in the system. As a consequence, both cost and infrastructure factors will have a less important influence on the firms' location decisions.

5) ICR (*Increasing Congestion Relevance*);

6) ILC (*Increasing Land Competition*); IPS (Increasing Pollution Sensitivity). The widening of the set of location choices produced in the preceding phases brings about a change in the ordering of preferences on the part of firms. Sensitivity to costs and negative externalities such as traffic congestion and rent prices is also likely to increase as a result of the trade-off between the latter and the costs of NIT adoption, as well as costs incurred by the firm in re-organizing its activity as a consequence of NIT introduction.

7) ITC (*Increasing Telematic Culture*);

8) ITW (*Increasing Diffusion of Telework*). The telematic culture has by now permeated the whole system and face-to-face contact is reduced to special cases. The widespread diffusion of NIT encourages the choice of telematic interaction and the adoption of teleworking (e.g. neighbourhood offices). The traditional relationship between home and workplace is reversed. This also brings about a demand for new styles of design of whole parts of the urban area.

As different timing in the urban policies and business strategies involved in the above scenarios can affect the evolution of the urban system, different speeds of changes in the propensity to innovate, rate of growth of the firms and propensity to (re)locate were assumed. On the basis of the sensitivity analysis carried out in previous model applications, three hypotheses about the speed of the above system changes were investigated :

i) *slow dynamics*: low parameter values for the speed of NIT adoption, growth rate, and (re)location processes of firm ($\varepsilon_0, \varepsilon_1, \varepsilon_2 = 0.001$, $\varepsilon_3 = 0.0005$);

ii) *fast dynamics*: parameter values ten times greater than the above ($\varepsilon_0, \varepsilon_1, \varepsilon_2 = 0.01$, $\varepsilon_3 = 0.005$);

iii) *mixed dynamics*: low parameter value for the speed of NIT adoption and higher for the speed of growth and (re)location ($\varepsilon_0 = 0.001, \varepsilon_1 \ \varepsilon \ \varepsilon_2 = 0.01$, $\varepsilon_3 = 0.005$).

Table 8.2. Configuration of relevant parameters in the simulated scenarios

	PAT	AP	LP	DIR	DCR	ICR	ILC	IPS	ITC	ITW
Innovation choice process										
G1 Global acceptance	0,1	1	1	1	1	1	1	1	1	1
G2 Local synergetic effects	0,1	0,1	1	1	1	1	1	1	1	1
G7 Sensitivity to traffic congestion	0,01	0,01	0,01	0,01	0,01	0,1	0,1	1	1	1
Location choice process										
G4 Pull effects by firms of the same type	0,1	1	1	1	1	1	1	1	1	1
G5 Infrastructure provision	0,1	0,1	0,1	0,01	0,01	0,01	0,01	0,01	0,01	0,01
G6 Costs of technologies	0,1	0,1	0,1	0,1	0,01	0,01	0,01	0,01	0,01	0,01
G7 Sensitivity to traffic congestion	0,01	0,01	0,01	0,01	0,01	0,1	0,1	0	0	1
G8 Neighbourhood offices	0	0	0	0	0	0	0	0	0	0,00001
β Sensitivity to rent prices	0,001	0,001	0,01	0,01	0,01	0,01	0,1	0,1	0,1	0,1
Modal choice process										
a Face-to-face contact	2,0	2,0	2,0	2,0	2,0	2,0	2,0	2,0	1,5	1,5
b NIT adopters	1,0	1,0	1,0	1,0	1,0	1,0	1,0	1,0	1,5	1,5
G7 Sensitivity to traffic congestion	0,01	0,01	0,01	0,01	0,01	0,1	0,1	1	1	1

▓ parameter activated in each scenario

8.3.2   The Simulations

The simulations relate to a fictitious urban system made up of five zones. Zone 1 corresponds to the central business district (CBD), where the largest number of firms is located and rent prices are highest. Zone 2 is a semi-central district and the remaining three zones are in the outer part of the urban area. In these latter zones rent prices and the density of jobs and housing are lower. In accordance with this pattern, limits were set for the residental/business capacity of each zone. In particular, those in the periphery were assumed to have a higher capacity than in the central zones.

There are two main communications networks: i) a physical network on which the journeys-to-work and firm-to-firm trips are distributed and ii) a non-physical network offering telematic services with characteristics which reflect those of the ITAPAC, RFD and CDN networks (i.e. the main networks in urban areas).

As already explained, the model disaggregates the firms into classes within which firms' behaviour in relation to innovation and location choices is considered homogeneous. In the simulations described here, reference is made to a single class. Effects due to interaction between classes are therefore not considered. Firms can choose between the three types of telematic service mentioned above or opt for the 'non innovation' choice.

All of the simulations use the version of the model in which flows are assumed to be in equilibrium (see 8.2.1).

Two series of simulations were carried out. In the first, the scenarios were run sequentially, assuming that each scenario emerged endogenously in the system at certain points during its evolution. Each scenario was then introduced at a given time interval - every 200 time units - and the output recorded every 100. The whole simulation period consisted of 1800 time units. In these experiments it was assumed that the system changes took place at a slow speed. In order to obtain a deeper understanding of their impacts, in the second series of simulations each scenario was run independently over a relatively long time period and the stationary states achieved by the different scenarios were compared. A simulation period of 500 time units - divided into ten intervals of fifty - was considered in each run. This period proved to be long enough for the system to achieve a stationary equilibrium. A total of thirty runs of the model were made: ten scenarios (including PAT) for each of the three system dynamics (see 8.3.1).

The complex structure of the model yields a large number of results. For reasons of space we present here only a limited number of outputs from the simulations. These are however sufficiently representative of the evolution of the system as well as of the spatial patterns produced in the various scenarios to provide a comprehensive picture.

The results refer to the following system variables (see 8.3.1):
- the level and spatial distribution of offices - A) and B)
- the level and distribution of offices which have adopted NIT - E) and F)
- rent prices in the different zones - G)
- the level and distribution of firm-to-firm flows on the transport network - M)
- the level and distribution of firm-to-firm flows on the telematic network - K).

In describing the results, we highlight the major similarities or differences determined by the scenarios in order to underline how apparently different scenarios can yield similar system configurations and vice versa.

Firstly, we will focus on the path of evolution of the system produced by the sequential impact of the scenarios, examining the system trajectories for offices, rent prices and firm-to-firm flows. We will then discuss the results obtained with the reference scenario PAT, emphasizing in particular the effects produced by the three different system dynamics. This will also allow us to indicate the trajectories followed by certain state variables in reaching their final state at the end of the simulation period.

Such a discussion provides a basis for a more straightforward comparison of the outcomes of the various scenarios. Attention is drawn in particular to the results obtained at the end of the simulation period in all scenarios. Finally, we point out specific features which distinguish the system trajectories in certain scenarios.

8.3.3   Simulation Results

8.3.3.1   The Path of Evolution of the System

In the sequential run of the scenarios two features characterize the overall effects of the simulated scenarios:

a   first, as might be expected, all the scenarios contribute to the decentralization process in the urban system. As a result, a steady decline of the core - a decrease of offices and rent prices in zones 1 and 2 - is produced, partially counterbalanced by a relative growth of the outer zones, Figures 8.2a and 8.2b;

b   second, the increase in contacts implied by the diffusion of the telematic culture in the later scenarios result in higher levels of firm-to-firm interactions, most of which are telematic flows, Figure 8.2c.

Some critical points in the trajectories of the above variables are revealed in correspondence with the introduction of certain scenarios. The most significant is shown in the middle of the simulation period (at time period 11) when the Increasing Land Competition (ILC) scenario is introduced. This determines a definitive reduction in the number of offices in the central zones along with a stagnation in the growth of firms in the system. From this period onwards, price rents in all zones of the system flatten to a relatively low level. A new location pattern appears in the system although its spatial effects are not yet completely evident.

Two other minor critical points can be detected in the first and second part of the simulation period, respectively. Until the Decreasing Infrastructure Relevance (DIR) scenario is introduced (at time period 7), zone 1 offers greater location advantages in spite of the steady development taking place in the peripheral zones. From this scenario and until the critical point in the middle of the simulation period is reached, zone 1 loses its advantages although a growth of offices still occur in all zones - including the central ones.

In the second part of the simulation period, when the decentralization trend is settled, relevant changes continue to occur in the interaction pattern (Figure 8.2c). With the Increasing Pollution Sensitivity (IPS) scenario, in particular, a further reduction of travel flows is accompanied by a considerable increase in telematic flows. These continue to rise also in the subsequent scenarios and by the end of the simulation period a new system configuration has emerged.

156

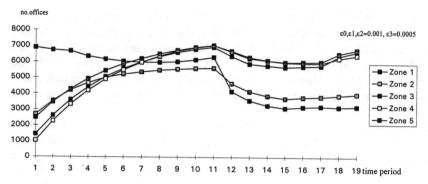

Fig. 8.2a.   Number of offices by zone

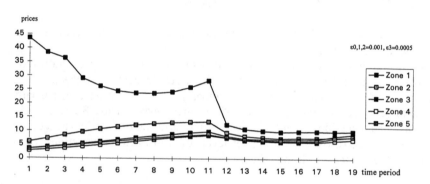

Fig. 8.2b.   Rent prices by zone

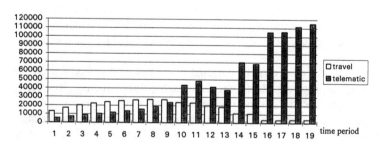

Fig 8.2c.   Firm-to-firm travel and telematic flows

8.3.3.2   The Reference Scenario PAT

In the second series of experiments, each scenario was run independently. The consideration of different system dynamics (i.e. slow, fast and mixed) revealed two significant features which, at least in qualitative terms, characterize the outcome of all the scenarios.

Firstly, and not unexpectedly, an increase in the speed of change affecting the rate growth and (re)location of firms (i.e. higher ε_1, ε_2, ε_3 values for slow and mixed dynamics) led to an early stabilization of the total number of offices in the system, as a consequence of the rapid saturation of the location capacity of zones. As indicated in Figure 8.2, for lower values of ε_1, ε_2, ε_3, the growth capacity in the system at the end of the simulation period, is not completely exhausted, whereas for higher values saturation was already reached in the middle of the simulation period.

Secondly, an acceleration in the NIT adoption process (higher ε_0 values for fast dynamics) favours an increase in the number of offices which innovate, even though the innovation process may not be completed within the simulation period. In the absence of exogenously imposed constraints, this process in fact terminates only when all the firms in the system have adopted innovations. In the reference scenario (see Figure 8.3), a lower value of e0 does not lead to an increase in the proportion of offices which have innovated. At the end of the simulation period, it is slightly lower than at the beginning (the increase in the number of offices adopting an NIT is therefore less than the increase in the total number of offices).

Even though different dynamics may not cause large differences in the final levels reached by the number of offices and rent prices, their zonal trajectories can be very different:

a *slow* dynamics determine a relative smooth change of zones, i.e. the initial spatial pattern is subjected to relatively continuous modifications (see Figures 8.4a and 8.4b). The predominance of the central zone is maintained throughout the whole period, although it is gradually reduced, and despite the steady rise in importance of other zones. This trend characterizes mainly the spatial distribution of offices and is less evident for prices;

b *fast* dynamics involve more rapid spatial changes, especially in the earlier time periods, which slow down later on (see Figures 8.5a and 8.5b). The central zone initially undergoes a significant reduction in the number of offices, then slowly increases again. All the other zones have a marked initial growth which then consolidate in the later time periods. At the end of the simulation period, the spatial distribution of offices is fairly uniform, although it does not appear to be completely definitive. Despite the changes in the relative zone levels, the spatial structure of rent prices on the other hand remains similar to that at the beginning;

c *mixed* dynamics produce system spatial configurations similar to those determined by fast dynamics (see Figures 8.6a and 8.6b). In this case, however, the spatial distribution of both offices and rent prices tend to consolidate more rapidly. At the end of the simulation period, there is a decline in the urban core, i.e. zones 1 and 2 have a lower number of offices than the other zones.

158

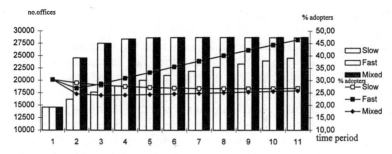

Fig. 8.3. Total number of offices and percentage of NIT adopters in PAT for different dynamics

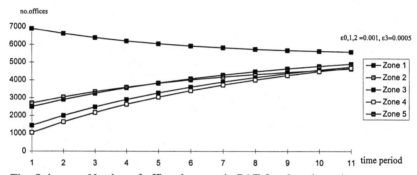

Fig. 8.4a. Number of offices by zone in PAT for slow dynamics

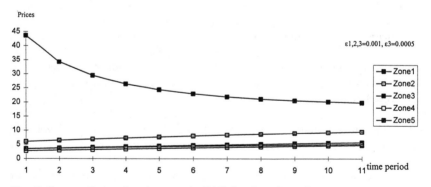

Fig. 8.4b. Rent prices by zone in PAT for slow dynamics

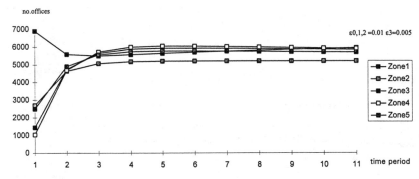

Fig. 8.5a.   Number of offices in PAT for fast dynamics

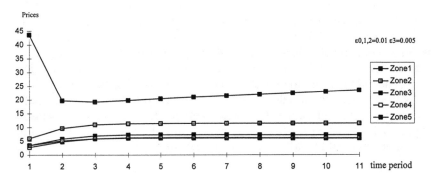

Fig. 8.5b.   Rent prices by zone in PAT for fast dynamics

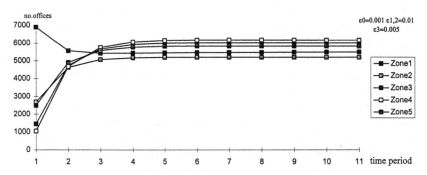

Fig. 8.6a.   Number of offices by zone in PAT for mixed dynamics

160

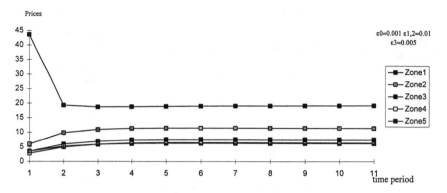

Fig. 8.6b.   Rent prices by zone in PAT for mixed dynamics

As the interaction structure of the system depends on the distribution of firms among the zones, changes in location patterns obviously cause changes in the pattern of flows. Growth in the number of offices in peripheral areas is consequently accompanied by an overall increase in flows of all kinds in these zones. Figures 8.7a and 8.7b show the pattern of firm-to-firm interactions (i.e. physical trips and telematic contacts) at the beginning and end of the simulation period for the different system dynamics. It can be seen that telematic interactions generally have a more uniform distribution than travel flows. This last feature also tends to occur in other scenarios. As would be expected, the increase in telematic flows in terlation to travel flows is more marked when there is an acceleration in the rate of adoption $\varepsilon 0$ for fast dynamics (see Figure 8.7c).

8.3.3.3   A Comparison Between Scenarios

An important feature brought out by the simulations is the clear differentiation between the outcome of the first five scenarios AP-ICR and those obtained by the following group ILC-ITW. Whereas there is a marked increase in the total number of offices by the end of the simulation period in the first five scenarios, in the last four the number decreases (see Figure 8.8). Similarly, in the former the proportion of NIT adopters remains low except - as would be expected - when there is an acceleration in the rate of adoption. In this situation, the higher propensity to innovate and greater sensitivity to functional interdependence lead to full adoption already by scenario AP.

Figure 8.9 shows the number of offices in each scenario, together with an indexx which measures the degree of uniformity of their spatial distribution in the system (calculated as the mean square deviation of the zonal percentage with respect to a uniform distribution). This index indicates that once again, while the distribution of offices in the first five scenarios is relatively uniform and hence not dissimilar from the general tendency to diffusion of PAT, the last four have a very uneven distribution.

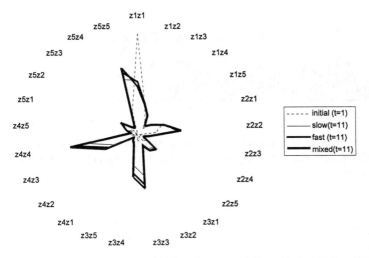

Fig. 8.7a. Pattern of firm-to-firm travel flows in PAT for different dynamics at beginning (t=1) and end (t=11) of simulation period

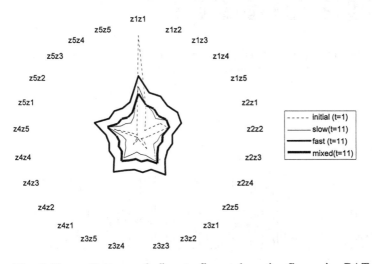

Fig. 8.7b. Pattern of firm-to-firm telematic flows in PAT for different dynamics at beginning (t=1) and end (t=11) of simulation period

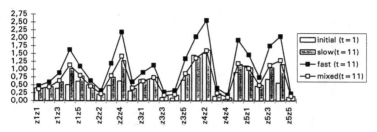

Fig. 8.7c. Ratio between firm-to-firm telematic and travel flows in PT for different dynamics at beginning (t=1) and end (t=11) of simulation period

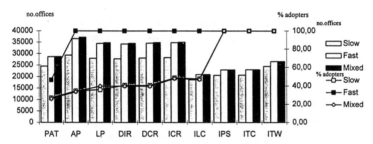

Fig. 8.8. Total number of offices and percentages of NIT adopters by scenarios and dynamics at end of simulation period

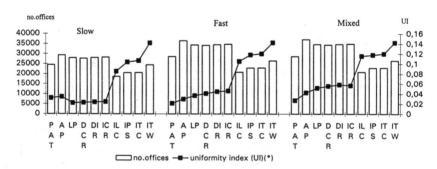

(*) UI: mean square deciation of zone percentage distribution with respect to uniform distribution

Fig. 8.9. Comparison of the total number of offices by scenario and dynamics at end of simulation period

These distinctive features also apply to the spatial pattern of prices, especially in relation to the contrast between central and peripheral zones. As well as illustrating the value of the mean square deviation with respect to PAT, Figure 8.10 also shows the contrast in values between zone 1 (centre) and zone 5 (a peripheral zone). The difference in rent price levels in the last four scenarios is far less than in the first five. As in the reference scenario PAT, the spatial distribution of prices is less affected by system dynamics. The only exception is the scenario AP (Adoption Pull) in which fast dynamics cause a marked relative increase in prices in the central zone.

The distribution of flows within the system is also relatively more uniform than in the first five scenarios, AOP-ICR. Figure 8.11 shows both the mean square deviation with respect to PAT and the values of the entropy index associated with the matrices of flows for each scenario. The relative uniformity for all types of flows and dynamics in the first five scenarios is illstrated by the higher entropy index values. In these scenarios, despite the increase in the total number of offices, the pattern of flows does not idffer dignificantly from that in PAT. Only when there is an acceleration in the NIT adoption process (i.e. for *fast* dynamics) we do not find, as already pointed out above, a greater increase in telematic flows than in physical flows.

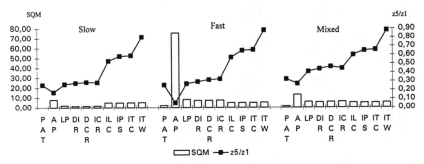

SQM: mean square deviation with respect to PAT(slow dynamics): z54/z1:ratio of prices of zone 5 to 1

Fig. 8.10. Comparison of rent price levels by scenario and dynamics at end of simulation period

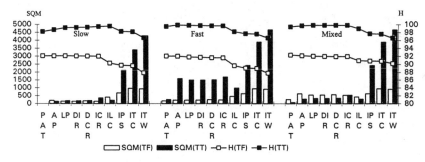

SQM: mean square deviation with respect to PAT (slow dynamics); H: normalized entropy value

Fig. 8.11. Comparison of firm-to-firm travel (TF) and telematic (TT) flow distributions by scenario and dynamics at end of simulation period

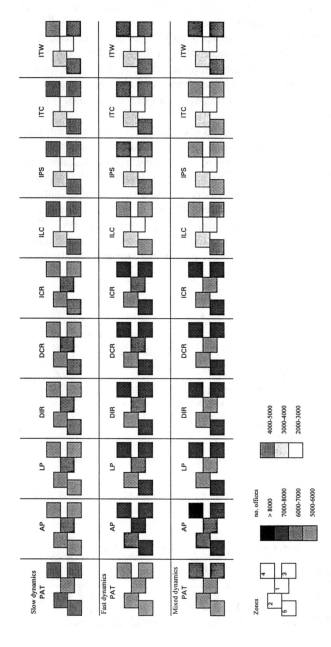

Fig. 8.12a. Zonal distribution of offices by scenario and dynamics at end of simulation period

165

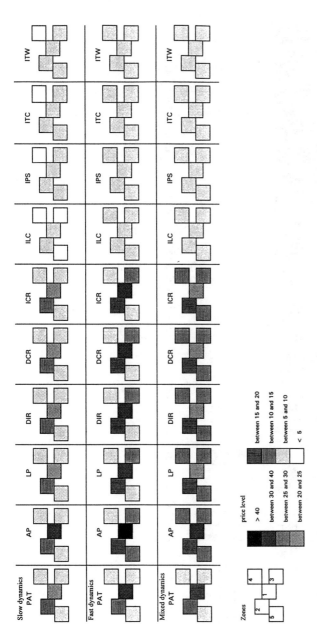

Fig. 8.12b. Zonal distribution of NIT adopters by scenario and dynamics at end of simulation period

166

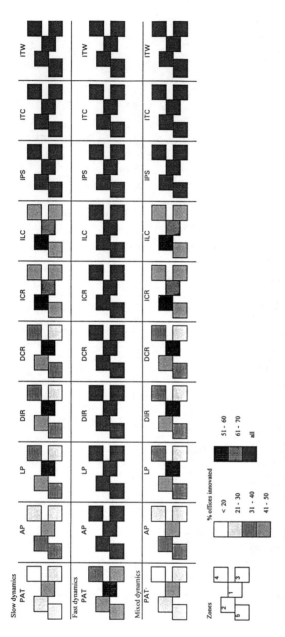

Fig. 8.12c. Zonal distribution of rent prices by scenario and dynamics at end of simulation period

In the scenarios (IPS-ITW) on the other hand, where the total number of firms decrease, but all firms adopt an NIT and tend to prefer a peripheric location, a more marked substitution of physical flows with telematic interaction appears, independently of the speed of system dynamics. This latter process is accompanied by a spatial pattern characterized by a higher density of flows (lower entropy index) especially in the outer zones.

Over the whole simulation time period, the ten scenarios depict an urban system undergoing a significant transformation in its overall configuration. This is clearly shown in Figures 12a-12c which summarizes the zonal distribution of offices, the proportion of NIT adopters and the rent prices. From an initial situation in which relatively few firms have adopted NIT and favour city centre locations where socio-economic development is most concentrated, we gradually arrive at a situation where, as NIT becomes more widely diffused, growth occurs increasingly in the periphery and competition between physical travel and telematic interaction tends to increase.

In this respect, the outcome is the reverse of the process hypothesized in many other studies, such as that of Castells 1989, which suggested that the impact of NIT is likely to be similar to that of telephone diffusion: as the technology becomes more widespread, there will be an initial phase of substitution followed by an intensification of interaction of all kinds.

8.3.3.4   Some Particular Scenarios

Lastly, we wish to examine more closely certain scenarios which proved to be particularly significant and to see of it is possible to draw any implications regarding the mix of urban policies and firm strategies simulated (see 8.3.1 and Table 8.2).

The four scenarios we look at in greater detail are:

- AP (Adoption Pull) which registered the greatest overall differences with respect to the reference scenario, at least in quantitative terms ;
- ILC (Increasing Land Competition) which represents a 'transition' stage between the urban patterns obtained by the scenarios AP-ICR and by IPS-ITW;
- ITW (Increasing Teleworking Diffusion) whose resulting system configuration is very different from that obtained in the reference scenario.

For all the scenarios mentioned, we focus on the results produced by *fast* evolution dynamics and hence a situation in which, by the end of the simulation period, all firms have adopted an NIT.

In AP, the greater propensity to adopt NIT (a higher NIT global acceptance parameter) together with a greater sensitivity to the location choices of other firms (a higher value of pull effect parameter), lead to a uniform spatial distribution of firms. Despite the existence of location capacity constraints, there is an increase in the number of firms in the central zone (an effect produced only in this scenario, see Figure 8.13). Consequently, the level of prices in the cental zone shifts upwards (Figure 8.14).

168

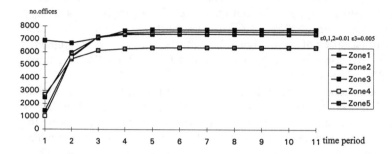

Fig. 8.13.   Number of offices in AP for fast dynamics

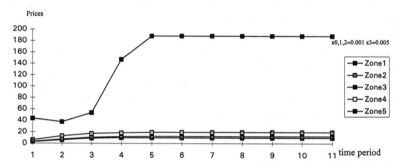

Fig. 8.14.   Rent prices by zone in AP for fast dynamics

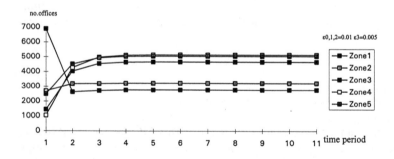

Fig. 8.15.   Number of offices by zone in ILC for fast dynamics

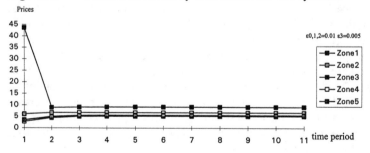

Fig. 8.16.   Rent prices by zone in ILC for fast dynamics

The increase in sensitivity to local conditions - as expressed by higher values of the parameters relative to rent prices and innovation choices - considered in LP (Local Pull), causes an acceleration in the adoption of NIT and only slight growth in the total number of offices. It also triggers a process of office decentralization and a relative fall in the rent price level in the CBD, while the outer zones ungergo a modest rise in prices.

The policies and strategies adopted in the subsequent scenarios DIR, DCR and DCR do not significantly affect these trends.

With the increase in land-use competition introduced in ILC, as expressed through a further increase of the importance of rent, the result is a contraction in the overall number of offices in the system and a decline in the importance of the city centre as a business location, both in relative and absolute terms. At the end of the simulation period, a 'crater effect' is produced: we witness an 'emptying' of the urban core (Figure 8.15) and a consequent levelling out of rent between all zones (Figure 8.16).

The greater sensitivity to environmental conditions (traffic congestion), the diffusion of the 'information culture' and hence the increase in the use of virtual systems of interaction characterizing the next three scenarios (IPS, ITC and ITW) do not significantly modify the spatial distribution of offices and rent prices, even though they affect their levels with respect to scenario ILC. These factors however have a stronger influence on the interaction structure of the system; as the number of travel flows steadily decline telematic flows increase above all in the outer zones (see Figure 8.17).

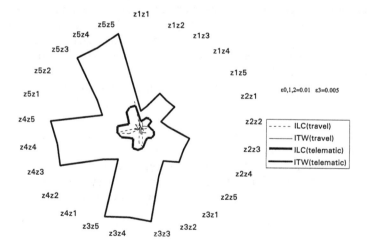

Fig. 8.17. Pattern of firm-to-firm travel and telematic flows in ILC and ITW at end of simulation period for fast dynamics

With the last scenario ITW, the spatial structure at the end of the simulation period is significantly different from that produced in the reference scenario. There is a reduction in the density of business activities in the urban core and hence in its tendency to attract/generate interaction flows. The decentralization of firms results in an increase in telematic interaction (due to the increase in telework system) rather than in the number of trips.

8.4 Conclusions

The simulation results described in this chapter represent the latest stage in a research project whose purpose is to provide a decision-support tool for defining urban policies concerning land-use planning, transport and telematic services which take into account choice process of firms.

In particular, as far as the mobility system is concerned, the TELEMACO model and its methodology contribute to the understanding of the complex interactions which affect urban traffic by explicitly taking into account the diffusion of new information technologies. This factor has until now been given far less attention than more traditional socio-economic factors.

The explicit treatment of NIT diffusion and of the conditions which allow such diffusion provides new element for defining policies able to cope with a conundrum of most urban societies: how to respond to a rising demand for accessibility, while reducing the volume of traffic. Being based on a concept of accessibility which includes both physical and non-physical interaction, this approach makes it possible to identify planning measures which could foster NIT adoption and at the same time alleviate problems of urban areas such as excessive land consumption, existence of specialized districts ('dead' for large parts of the day), overcrowding of functions in urban centres, traffic congestion and pollution.

A useful extension of the application of the TELEMACO model, permitting the full exploitation of the information produced, would be to include it in a wider system of urban models and planning procedures. There are various possibilities, including for example the application of modal choice models to the output of the distribution flow submodels for the transport and telematic network. Another development, already shown to be valuable, is the use of the model output for defining a set of indicators (or 'criteria measures', Lombardo 1992, 1995) to be included in a multi-criteria evaluation procedure.

In the present phase of research, however, attention is being focussed on the following aspects:

- the refinement of the model in the light of the simulation exercises already carried out and further testing;
- experimentation with real data. This will not be easy, because of the difficulties in obtaining appropriate data. For example, as far as the information needed for the NIT adoption submodel is concerned, the only way would appear to undertake a sample survey of relevant firms.

References

Ashin A (Ed) (1994) Post-Fordism. Blackwell Oxford

Batten D. Casti J and Thord R (Eds) (1995) Networks in Action. Springer Verlag Berlin

Bertuglia CS, Lombardo S, Occelli S and Rabino G (1993) Innovazioni tecnologiche e traformazioni territoriali: il modello TELEMACO, Atti del 1

Convegno Nazionale del Progetto Finalizzato Trasporti 2, Rome, 879-910 and Atti della Conferenza Italiana di Scienze Regionali. Bologna 1053-85

Bertuglia CS, Lombardo S, Occelli S and Rabino G (1995) The Interacting Choice Processes of Innovation, Location and Mobility: A Compartmental Approach. In Bertuglia CS, Fischer MM and Preto G (Eds) Technological Change, Economic Development and Space. Springer Verlag Berlin 115-141

Castells M (1989) The Informational City. Blackwell Cambridge

Ernste H., Jaeger C. (1989) Information Society and Spatial Structure. Belhaven Press London

Haag G and Lombardo S. (1991) Information Technologies and Spatial Organization of Urban Systems: A Dynamic Model. In Ebeling W, Peschel M and Weidlich W. (Eds) Models of Self-organization in Complex Systems. Springer Verlag Berlin 29-36

Harris B (1994) Some Thoughts on New Styles of Planning. Environment and Planning B 21 393-398

Lombardo S (1986) New Developments of a Dynamic Urban Retail Model with reference to Consumers' Mobility and Costs for Developers. In Haining R and Griffith DA (Eds) Transformations through Space and Time. NATO ASI Serie. Martinus Nijhoff Boston 192-208

Lombardo S (1992) Processo di Piano, valutazione e modelli di autoorganizzazione: un esempio. Atti della Conferenza di Scienze Regionali 177-200

Lombardo S (1993) Innovazione tecnologica e localizzazione. In Lombardo S. and Preto G (Eds) Innovazione e trasformazione della città. Teorie, metodi e programmi per il mutamento. Angeli Milan

Lombardo S (1995) Nuova modellistica dei sistemi urbani e valutazione. Esemplificazione di una procedura logico-operativai. In Lombardo S (Ed) La valutazione nel processo di piano. Angeli Milan 91-113

Lombardo S, Ambrogio M and de Felice G (1990) Innovazione tecnologica e trasformazioni territoriali. La carta telematica di Roma. E.S.A. Editrice Rome

Lombardo S and Occelli S (1995) Un modello di simulazione per configurare scenari di evoluzione urbana caratterizzati dall'introduzione di nuove tecnologie dell'informazione. Atti del II Convegno del Progetto Finalizzato Trasporti 2 Genoa.

Lombardo S and Rabino G (1986) A Compartmental Analysis of Residential Mobility in Turin. Sistemi Urbani Vol 8 2 263-84

Weidlich W and Haag G (1983) Concepts and Methods of a Quantitative Sociology. The Dynamics of Interacting Populations. Springer Verlag Berlin

9 Innovation and Spatial Agglomeration

Dino Martellato

031

032

R12

R32

9.1 Introduction

The fundamental source of benefits accruing to households and firms in a city are the economies of agglomeration. It seems quite obvious to think that the spatial effects of innovation may be partially understood by considering how innovation can impact on agglomeration economies. This is the precisely the approach followed in this chapter. To pursue our task we need a model in which production takes place under agglomeration economies. The model will be provided in Section 9.3. In the same section we will then present some preliminary thoughts about the impact of innovation on agglomeration economies and the competitive position of cities.

In the following section we will put forward some ideas about the sources and the nature of agglomeration economies for producers in a city and also make references to the literature about the effects of innovation on the spatial structure of the economies.

9.2 Innovation and Spatial Agglomeration

Agglomeration economies emerge from both the consumption side and the production side and have many different components. We restrict our attention here to the agglomeration economies that are conferred to the single firms by their clustering.

The classical argument in this respect has been put forward by Weber (1909) and Marshall (1920) who exploit the concept of scale economies. Hoover (1948), Chipman (1970), Aoki (1971), Segal (1976) and Moomaw (1981), David and Rosenbloom (1990), among others, have made the argument more precise and have applied the original idea.

More recently, taste heterogeneity and product variety has been forcefully advanced (Fujita and Rivera-Batiz 1988, Abdel-Rahman and Fujita 1990, among others) as an alternative to the concept of Marshallian externalities. Since externalities are often deemed to be too vague a concept to be really useful, taste heterogeneity and product variety might be a way of escaping the traditional concept of the compact monocentric city (Anas 1990). Indeed, taste heterogeneity has recently proved able to give a rigorous explanation of agglomeration in a way which overcomes the difficulties of Hotelling's theory of spatial duopoly (Anderson, de Palma and Thisse 1992).

In this chapter, we combine the two principles of Marshallian externality and product heterogeneity by using a variant of a cost function which was originally proposed for describing the technology of a multiproduct firm (Baumol et al. 1982). We assume, in other words, that economies of both scale and scope are present in a city. The reason is that the two principles together - i.e. scale and scope economies - may arguably provide a better insight into the working of a complex system such as a city than is possible with one principle only. In this instance we depart from a recent similar contribution by Abdel-Rahman (1994).

As is well known, innovation generally implies a reduction in the production costs of existing products and the appearance of entirely new goods and services. The quality of existing products can be improved as well. It is likely then that innovation is able to exert a complex impact on the economy and thus on its spatial organization of production through the changes in technology, location and trade.

The reduction of production costs, in particular those concerning transport, has always had discernible effects on the number and dimension of cities. However, the appearance of new goods and services and the improved quality of old products can have their own effects on the composition of firms' output and on many other aspects of the production sector. It seems unlikely that these phenomena will not have their own effects on the location choices of the producers. The observed changes in the location choices indeed seem to reflect more than simple changes in transport costs.

If we adopt the traditional theory (according to which urban agglomerations are the result of reductions in the transfer cost) and focus only on the decrease in transport costs produced by innovation, we completely ignore the impact of any quality improvement and any variety increase and considerably reduce the scope of the analysis.

Innovations in transport and communication technology are deemed to be highly influential on location and mobility patterns, but the effects can be difficult to understand. Any innovation leading to a reduction in transport costs could lead to the relocation of some households and firms. This view has been taken into consideration, for instance, by Orishimo and Beckmann (1988). The latter (Beckmann 1988, 153-160) in a study concerning the impact of transportation cost changes found that the adoption of information technology leads to a reduction in the cost of overcoming distance. The main result of Beckmann's speculation seems to be that the effects will go both ways, in that there will be both centralizing and decentralizing forces at work. Although it may be difficult to understand which force will dominate the other, the investigation is worth pursuing.

More recently, Bertuglia and others (1995) have correctly pointed out that the impact of innovation on the spatial organization of the economy is extremely far reaching as it feeds back not only on location but also on mobility patterns. This poses a serious and complex analytical problem which they solve by adopting a new and promising tool of analysis. The so-called compartmental approach seems to be more adequate than earlier approaches to the study of this complex subject.

However, technological development can induce changes in location patterns in other ways than those deriving from transport costs. Long ago, Hoover (1948, chapter 10) recognized that besides an improved transfer service and a change in the structure of transfer costs, the technical maturing of industries, changes in material requirements and new techniques in the utilization and transmission of energy can also affect the location of production.

The preferred approach in the literature seems to be a microeconomic one as it focuses mainly on changes in transport and communication costs. Kawashima (1993), for instance, reverts to the isodapane analysis of Weber (1909). He finds that a technological innovation favourable for large-scale operation and technological innovation in the transportation sector both tend to enlarge the size of Weber's critical isodapanes (the isodapanes within which the localization economies gained by a change in the location outweigh the increment in transportation costs) thus contributing to the concentration of production activities.

A major recent contribution to the theory of product differentiation and taste heterogeneity (Anderson,de Palma and Thisse 1992) focuses on an entirely different concept. The three authors simply show that firms tend to locate together if the products they sell are differentiated enough.

In this chapter we will follow a similar approach. One which considers externalities and product variety, rather than transport costs, as the basis for urban agglomeration. Instead of looking at the place where the representative firm chooses to produce - that is how far from the city centre and from its competitors the firm will locate - we will look at the amount and the composition of production of a city as the result of the change in the forces that are behind scale and scope economies at the city level.

By considering what an urban area is producing, after the structure has been allowed to react to a given shock, rather than the location chosen by the representative producer or consumer, we are clearly adopting an aggregate approach.

It might be reasonable to assume that the output of some activities within a factory or a firm could be inputs of other activities in the same factory and, in much the same manner, the output of some firms can be considered a shared input of all the remaining firms in the urban area. This will indeed be our basic assumption throughout the chapter.

The handling of information is an obvious example of such a category of activities. Transport and other service activities can also be thought of as a source of externalities. Other obvious examples are: repair and maintenance, legal, financial, engineering and commercial services. Since the production of some information, transport and other examples in the list above is common both at firm level and urban level, it is possible to say that externalities exist outside the firm, but - at the same time - that the firm may be conceived as a way of internalizing them. The existence of such activities and firms produces externalities in their neighbourhood which, in their turn, induce agglomeration of certain other activities and firms in the same neighbourhood.

If a certain innovation changes the preferred location of some firms and some activities it will drive - sooner or later - some activities out of the urban area and drive some other activities into the same area. It is then obvious that only by adopting a model for the whole urban area and the structure of the production in the different urban sectors is it possible to assess the spatial impact of innovation, since this is a useful way to take the locational interdependencies into account.

All the above-mentioned specialized services can be considered as a group of activities with a rather low degree of substitution, which is able to exert a positive impact on the production costs of other traded goods.

The focus will then be on the agglomeration economies rather than on transport costs as we believe that the adoption of new technologies - particularly in the sectors which produce universal services (such as communications and transport) - is able to modify the level of agglomeration economies.

9.3 Shared Inputs and Agglomeration Economics

In this section we will offer a simple model for investigating the impact of innovation on the degree of urban agglomeration. The availability of many differentiated intermediate services can be found in any large city under monopolistic competition. As we will argue, this leads to the existence of scope economies. Marshallian externalities, which imply the existence of monetary scale economies within a sector but external to the single plants, are also a feature specific to a sufficiently large city. In our model the specialization in intermediate services and Marshallian externalities coexist.

There are then two groups of products in a city. The first is made up of the goods and services produced by a number of different factories, and thus sectors, each potentially characterized by Marshallian externalities.

The second group consists of a number of specialized and intermediate services produced under monopolistic competition in a market regime where firms face downward sloping demand curves and where the cross-elasticities of demand for any pair of firms are low, but where there is free entry. Under such assumptions - which are deemed to be a fairly good approximation of the markets for retailing and service industries - each firm behaves like a monopolist, as it can choose any price-quantity combination of its demand curve without provoking a change in the price asked for by its rivals. If the same firm is making profits, other firms may enter the market bringing profits down to zero.

The availability of such a group of services exerts a positive effect on each sector of the first group. This implies the existence of scope economies (or urbanization economies). By increasing the availability of services of the second group - i.e. by increasing the number of services - the city can exploit scope economies. This is quite evident, for instance, in transport, communications, maintenance services and consultancy. An increase of the production in any sector of the first group, yields scale economies i.e. Marshallian externalities.

As will soon be apparent, the resulting equation will be a cost function by which the total production cost in a city is related to the quantity of goods produced in the first group of sectors. The relation between the total production cost and the price vector of both groups of products is only implicit.

As the suggested function is a relation between the total production cost and the output level in a city, we will be able to offer an analytical tool - more general than many existing models of the urban production sector - which is suitable for investigating some aspects of the overall spatial impact of innovation. This will be done considering all the producers in the city, i.e. by adopting an aggregate approach.

The available quantity of specialized intermediate services (u_j with $j=1, J$) is summarized by a quantity index taking the form of the following CES function:

$$S = (\Sigma v(u_j)^r)^{(1/r)} \tag{9.1}$$

where r is a positive parameter: $0 < r < 1$. Also v is a positive constant which is a combination of the elasticity of substitution (to be defined below) and the fixed production cost in the urban service sector. The index is equal to the sum when $r=1$. In this case the diversified services are almost perfect substitutes and variety is not useful. For r close to zero, variety is highly valued and the services are far from being close substitutes. If σ ($\sigma > 1$) is the elasticity of substitution between the different specialized services, we have the following relation between r and σ:

$$\sigma = 1/(1-r)$$

The quantity index has a dual price index of the type:

$$T = (\Sigma p_j)^s / v)^{(1/s)} \tag{9.2}$$

The expenditure T has an interesting interpretation since it can be conceived as the minimal expenditure required for purchasing the available n composite goods, making up one unit of the composite good and given their prices p_j ($j=1,J$).

Summarizing, the three parameters r, s and σ are linked by the following relationships:

$$(1-r)(1-s) = 1$$

$$\sigma = 1/(1-r)$$

$$\sigma = 1-s \tag{9.3}$$

thus: $(1/\sigma)(\sigma) = 1$.

T is an exact or ideal index because S and T share the same weighting formula and because they are consistent, i.e.:

$$ST = E$$

where E is the minimal expenditure for the whole amount of units, S, of the composite good.

As stated above, we assume that factories locate in a city because of the possibility of reducing their production costs:

1 by exploiting Marshallian externalities and
2 by sharing many differentiated intermediate services.

This idea can be translated quite easily into a function relating the total cost of production to the available quantity of specialized intermediate services. To do this we follow Baumol (1982 p 461), who suggested what is known as the shared input model for a multiproduct firm. By adopting such a device, a city - which is an agglomeration of many multiproduct firms - is simply conceived as a multiproduct firm as in Goldstein-Gronberg (1984).

We indicate with C_i the production cost of good i (i=1,I) and we relate it to the quantity q_i produced and to the unit cost of production f_i which is a function of the price vector p. Without any shared input of production, but with Marshallian externalities only, the cost for sector i would be:

$$C_i = (q_i)^{a_i} f_i$$

(9.4)

$$0 < a_i < > 1$$

The parameter a_i allows the presence of economies of scale. The Marshallian externalities - monetary scale economies which are external to the factory, but internal to the sector (because they arise from the development of the sector and accrue to the firms operating in this particular sector and in the city) - are positive when $a_i < 1$. In this case the cost function is concave, otherwise ($a_i > 1$) they are negative.

The availability of a larger number of differentiated intermediate services is specific to a city and has an effect on production in plants of the first group which are located there. This effect is essentially a scope economy. It may be captured by attaching a positive parameter (e in this instance) to the inverse of the quantity index already defined (equation 9.1). Any increase in the quantity of the bundle of specialized services S available reduces the production cost. The total production cost in the city concerned is then obtained by adding the cost of the shared intermediate inputs to the sum of the production costs for the first group, as follows:

$$C = \Sigma(q_i)^{a_i} f_i S^{-e} + ST$$

(9.5)

$$o < e < > 1$$

where ST is the expenditure for the production of specialized services.

The presence of any shared input thus implies a reduction in the production cost in the first group and an intrinsic production cost in the second one. The two effects partially compensate. But, as long as $e > 0$, the sum of all costs corresponding to any combination of products of the first group and specialized intermediate services in a city, is lower than the sum of costs when the different productions are not agglomerated.

For a given number n of specialized services, the second cost component is constant, by definition, as the amount ST does not vary with r. Indeed, as shown above, the parameters r, s and σ must change in a consistent way in order to keep the total expenditure at its minimum. However, this is not true for the first cost component. It depends on the value taken by the two parameters e and r. The higher the value of e, the higher the reduction of the production costs, for a given r. Note that the effect is the same for every good i of the first group. The higher the value of r, the higher the level of S and thus that of C (for a given e).

From a general point of view, it can be said that if a city is identified by the existence of shared specialized services, the agglomeration is profitable whenever the following condition holds:

$$\Sigma(q_i)^{a_i} f_i (1-S^{-e}) + ST > 0. \qquad (9.6)$$

even if Marshallian externalities were totally absent (i.e. if a_i would always be zero).

We assume, however, that the urban agglomeration is the outcome of two forces. The first is provided by the Marshallian externalities by which larger plants and larger clusters of different plants of the same sector imply scale economies. The second is the possibility of sharing many specialized intermediate inputs.

Innovation can probably exert different, but combined, effects on both parameters e and r and thus yield spatially uneven reductions of the total cost borne by different cities. It is obvious that the precise determination of the changes of the values of e and r is a matter of empirical research which will not pursued here.

With equation (9.5) - which actually extends Baumol's shared input model for a multiproduct firm to a city where the factories reduce their costs by purchasing different varieties of specialized services - at least three different paths can be distinguished.

1 By assuming the amount of shared input services S to be given, one can determine the optimal amount of production (q_i).

2 The opposite path, which would seem more interesting involves assuming the location and production decision of firms as given - i.e. by taking q_i to be given - we could determine the quantity S which minimizes the total production cost.

3 Another approach we could adopt is to fix all production levels and investigate the differential effects on total production costs of given changes in the two parameters e and r and also in production prices.

In order to prepare the ground, we elaborate on equation (9.5). We will assume accordingly that the total production cost is minimized by a suitable amount of shared services, all prices and parameters being constant. The following first order condition holds:

$$-e \, \Sigma(q_i)^{a_i} \, f_i \, S^{-(1+e)} + T = 0. \qquad (9.7)$$

This can be used to obtain the optimal amount of shared inputs which minimizes the cost of a proportional production level:

$$S = (T/(e\Sigma(q_i)^{a_i} \, f_i))^{-1/(1+e)} \qquad (9.8)$$

By back substitution in equation (9.5), we get the final formulation of the cost function which gives the production cost corresponding to any assigned production level under the assumption of an optimal supply of services. It reads as follows:

$$C = (\Sigma b_i \, (q_i)^{a_i})^{1/(1+e)} \qquad (9.9)$$

The following definition has been used:

$$b_i = (e^{1/(1+e)} + e^{-e/(1+e)})^{(1+e)} (T)^e f_i \qquad (9.10)$$

The last equation shows that each constant b_i vehiculates both the price effect and the scope effect. The resulting cost function is in some way similar to a standard CES cost function. In equation (9.9) however, the cost depends on the total production of the first group of sectors. The relation with prices is implicit, as revealed by (9.10) and, even more important, the amount of services of the second group is, by virtue of equation 9.8, the optimal one.

It is interesting to consider a few other features of this particular urban cost function. We have already noted that a parameter $0 < a_i < $ or > 1 allows some degree of Marshallian externalities in the sector i concerned.

Since some inputs are shared equally by all sectors of the first group, the parameter e is positive. This implies that $1/(1+e) < 1$. The most interesting situation is obtained when $0 < a_i \, (1+e) < 1$. In this instance there is concavity in the cost function of product i. This property would be assured also by entering fixed costs of production in the same sector i, but we prefer to highlight the combined effect scale and scope economies in a standard CES formulation.

There are also economies of scope between the two groups of products. This occurs when any given amount of production in the first group is matched by an adequate amount of services produced by the second group, with a global cost lower than that of the sum of the two corresponding individual sub-totals.

The dynamic interaction between the changes of Marshallian economies and scope economies gives a better explanation of the large differences in the competitive position of different cities than does path-dependence, which is based on Marshallian externalities alone (David-Rosenbloom 1990). In other words, the coexistence of production of the first group and shared inputs (typically transport and communication) is always positive for a city, but the intensity of cost reductions resulting from the broadening of the scale and scope in production can exhibit large variations between different cities and at different points in time. The reason is that the product composition of the two groups is neither even over space nor constant over time. This is to say that innovation (widely recognized as the principal factor sustaining growth) brings about changes in the parameters a_i's and e which translate into space inequalities.

As is revealed by definition (9.10), the parameters b_i which have been obtained under the assumption of cost minimization, depend on prices as well as on the other critical parameters a_i, e and r (see equations 9.2 and 9.3). This means that, under the maintained cost minimization hypothesis, any innovation which unfolds with changes in: (1) prices, (2) variety in the two groups, (3) scale parameters a_i's, (4) scope parameter e, (5) substitution parameter r has an effect which is extremely difficult to gauge.

It is unlikely that the competitive position of different cities can be left unchanged, but the equation (9.9) can help us isolate the factors at work.

1 Any increase of e (larger scope economies) is followed by an increase in the two factors entering the constant $b_i(T^e$ and $(e^{1/(1+e)} + e^{-e/(1+e))(1+e)})$, but it leads at the same time to a decrease in the general exponent $1/(1+e)$ appearing in the cost equation (9.9). The net result on the level of the minimum production cost is ambiguous.

2 Any increase of the parameter r (higher elasticity of substitution σ) yields a reduction in the parameter s (see 9.3), thus changing the price index $T = (\Sigma p^s/v)^{1/s}$, and the b_i's. However, the index T is not a monotonic function of s (given v) or vice versa. So, also in this case, the effect on the minimum production cost cannot in general be evaluated.

3 Any change in the scale parameters a_i yields a change in the total production cost. If a_i becomes lower, scale economies are greater and the total cost decreases.

4 Any increase in the two categories of prices clearly has a positive effect on the b_i's and thus on the production cost.

As a general conclusion, we can say that changes in prices and technologies combine in a complex way to produce an impact which will not be the same for all cities, since each city specializes i.e. it has different combinations of x_i's and u_j's.

9.4 Summary

This chapter examines the spatial effects of innovation. The subject has been addressed by assuming that urban agglomeration is more influenced by forms of locational-interdependencies between agents than by pure transport cost or accessibility considerations. We have argued furthermore that agglomeration economies stem from the combined effects of both scope economies and of increasing returns to scale at the level of firm and sector i.e. Marshallian externalities. This form of externality accrues to firms because many diversified and thus specialized intermediate services are readily available in a large city.

We present a simple aggregate model where the two principles are made explicit in a Baumol-type cost function for multiproduct firms.

Indeed a city can be considered a multiproduct firm where two groups of products coexist. There is a service sector made up of a large number of specialized intermediate services produced under monopolistic competition and a group of sectors each exploiting its own degree of Marshallian externalities and sharing the available scope externalities generated by the first group.

Although the model is a cost function where prices are treated in an exogenous way, the effects of parameter and price changes attributable to innovation can be captured.

References

Abdel-Rahman H (1994) Economies of Scope in Intermediate Goods and a System of Cities. Regional Science and Urban Economics 24:497-524

Abdel-Rahman H and M Fujita (1990) Product Variety, Marshallian Externalities and City Sizes. Journal of Regional Science 185-183

Anderson S P, Palma A de and J F Thisse (1992) Discrete Choice Theory of Product Differentiation. The MIT Press. Cambridge Massachusetts

Anas A (1990) Taste Heterogeneity and Urban Spatial Structure: the Logit Model and Monotonic City Reconsidered. Journal of Urban Economics 28:315-335

Aoki M (1971) Marshallian External Economies and Optimal Tax Subsidy Structure. Econometrica 39:35-53

Baumol W J, Panzar J C and R D Willig (1982) Contestable Markets and the Theory of Industry Structure. Harcourt Brace Jovanovich. New York

Beckmann M (1988) Information Technology and Location. In: Orishimo I, Hewings G and P Nijkamp (Eds). Information Technology: Social and Spatial Perspective. Springer-Verlag Berlin 153-160

Bertuglia C S, Lombardo S, Occelli S and G A Rabino (1995) The Interacting Choice Processes of Innovation, Location and Mobility: A Compartmental Approach. In: Bertuglia C S, Fisher M M and G Preto (Eds). Technological Change, Economic Development and Space. Springer-Verlag Berlin

Chipman J S (1970) External Economics of Scale and Competitive Equilibrium. Quarterly Journal of Economics 84:347-385

David P and J L Rosenbloom (1990) Marshallian Factor Externalities and the Dynamics of Industrial Localization. Journal of Urban Economics 28:349-370

Fujita M and F L Rivera-Batiz (1988) Agglomeration and Heterogeneity in Space, Introduction. Regional Science and Urban Economics 18:1-6

Goldstein G S and T J Gronberg (1984) Economies of Scope and Economies of Agglomeration. Journal of Urban Economics 16:91-104

Hoover E M (1948) The Location of Economic Activity. McGraw-Hill New York

Kawashima T (1993) Innovation and Location: Spatial Agglomeration and Deglomeration. In: Greenut L, Ohta H and J F Thisse (Eds). Does Economic Space Matter? Essays in Honour of Melvin. MacMillan London 374-385

Marshall A (1920) Principles of Economics. MacMillan London

Moomaw R L (1981) Productivity and City Size: A Critique of the Evidence. Quarterly Journal of Economics 95:675-688

Orishimo I (1988) Development of Informatics and Possible Changes in Urbanisation Processes. In: Information technology: Social and Spatial Perspective. Orishimo I, Hewings G and P Nijkamp (Eds). Springer-Verlag Berlin 250-264

Segal D (1976) Are There Returns to Scale in City Size? Review of Economics and Statistics 58:339-350

Weber A (1909) Uber den Standort des Industrien (Tübingen: J C B Mohr) translated into English by C J Friedrich (1929) Theory of the Location of Industries. UCP Chicago

10 Structural Change in a Metropolitan Economy: The Chicago Region 1975-2011

Geoffrey J.D. Hewings, Philip R. Israilevich,
Michael Sonis and Graham R. Schindler

(U.S.)

Ō33

Ō31   D57

10.1   Introduction

R12

The vast majority of studies addressing issues of the impacts of innovation diffusion or structural change adopt an approach that focuses on one particular sector. In this chapter, a very different approach is used. Firstly, the focus is on a regional *economy* and therefore a set of interdependent sectors. Secondly, the analysis looks backwards from a forecast horizon in an attempt to explore the types of changes that have been predicted in the structure of this set of sectors. The empirical base will be a set of over 30 annual regional input-output tables for the Chicago region (encompassing both an historical and a forecast period) that have been extracted from a general equilibrium model of the region's economy. The exploration of changes in the economy will involve consideration of three approaches to the detective work necessary to uncover what has happened.

The chapter is organized as follows. In the next section, a brief review will be provided of the modelling system that was used to make forecasts of the structure of the Chicago region economy for the period 1993-2018. Thereafter, the three methods will be introduced in turn. The first will explore ways in which the economy, in particular the input-output system, can be decomposed into a *block structure* and to consider whether these block structures themselves have changed over time. The second and third approaches attempt to understand the changes in *structural complexity* that have been forecast over this time period. After the methods and the empirical analysis have been presented, there will be an evaluation of the findings prior to offering some concluding remarks.

Thus, the chapter will provide only a partial picture of the structural changes that have taken place; it will not attempt to give an evaluation of the kind that has been tradition in the past, such as Wolff's (1985, 1989, 1994) work or that of Carter (1970). In spirit, the chapter may be seen to be a first step in the implementation of the conception outlined in Hewings et al. (1988); here, attention was focused on the interdependencies, not only among sectors but between sectors and final demand, that characterized the adoption of new technologies by firms and the consumption of new products by consumers. In this approach, competition between innovation and non-innovation needs to be considered in the production and consumption markets. Furthermore, the data that provides the basis for this work is unusual in that, for the most part, it has been extracted rather than observed.

10.2 The Chicago Region Econometric Input-Output Model (CREIM) and the Forecast of Input-Output Systems

10.2.1 Description of CREIM

The Chicago Region Econometric Input-Output Model (CREIM) generates forecasts of the Chicago economy on an annual basis, with the forecast horizon extending up to 25 years. The model, as its names implies, includes input-output and econometric components. It is a system of linear equations formulated to predict the behaviour of 151 endogenous variables, and consists of 123 behavioural equations, 28 accounting identities and 68 exogenous variables. CREIM identifies 36 industries and three government sectors. For each industry, there are projections of output, employment and earnings.

The input-output module was constructed from establishment-level data obtained from the US Bureau of the Census. Two models have been developed, one based on 1982 and one of 1987 data. Since survey-based systems are prohibitively expensive, researchers developing regions input-output models have relied on a variety of adjustments of national level data. While updates have been made annually, the reliability of these updates is not known. Secondly, the adjustment process in developing regional from national tables relies on a large number of assumptions; the most critical being the one that assumes that the technology at the regional and national levels is identical. Since there has been little survey work done to test this assumption, it often reverts to an assertion. Preliminary analysis with the Census data suggests that differences between national and regional technologies may be significant (see Israilevich, Mahidhara and Hewings 1994).

The adopted approach to table construction avoids many of these problems, since survey data are used to build the manufacturing portions of the tables. Thirty-six sectors were identified for Chicago - essentially, the two-digit SIC manufacturing sectors and somewhat more aggregated sectors for non-manufacturing. While data are available at the individual establishment level, Federal Disclosure Rules preclude the publication of data that would reveal the transactions of individual firms or would enable reasonable estimation from information presented.

In addition to the transactions between sectors, the table also records the purchases made from labour (wages and salaries), capital (profits and undistributed dividends) and imports from outside the state. Complementing the sales made to other sectors are sales to households (consumers), government, investment and exports outside of Chicago. With this table one has, in essence, an economic photograph of the state of Chicago, captured at one point in time. Adding the econometric component enables the analyst to extend this photograph back in time to test the reliability of the system in tracking the changes that have been observed in the end and to redevelop this photograph each year for the next 20 to 26 years producing the annual forecasts.

10.2.2 Extraction of the Input-Output Tables

The methodology is described in more detail in Schindler, Israilevich and Hewings (1994). Only the general scheme will be presented here. Since the general equilibrium model includes a complex set of linear and nonlinear equations, many of which have autoregressive components and lagged variables, the extraction of the forecast input-output coefficients is not a simple task. To account for the problems that might occur, the input matrices were calculated using a sustained shock over a six-year time span to ensure that the full effect of the shock was captured. However, CREIM offers a different path for calculating economic interdependence over time. Recall that individual time series regressions that include input-output relations, make up only 36 of the 150 equations in the system; many of the remaining equations are highly nonlinear and recursive, involving autoregressive (AR) lags of 1 or 2. As a result, the output of one sector is related to the output of other sectors through a complex, multiple chain process in which the input-output relationship is only one component of the chain. Nevertheless, one can relate the output of one industry to the output of another industry within CREIM, This is accomplished through the use of first derivatives. It would be a difficult procedure to derive such derivatives analytically, as a result of the nonlinearities and AR components that are present in many equations. Therefore, these derivatives are calculated numerically; then, tests are conducted to examine their stability.

Essentially, CREIM has an expanded input-output structure - the model is closed in a far more complex fashion that ones that involve making households endogenous - and in many ways retains the appearance and character of a general equilibrium formulation. However, CREIM is solved through adjustments of quantities rather than prices, but market clearing assumptions do hold. Shocks are introduced through the final demand components; given a change in final demand of Δf_j, the model is solved to calculate $\delta X_i / \delta f_j$ for each gross output, X_i. For each Δf_j, n partial derivatives can be obtained; if all Δf_j j = 1,...n (where n is the number of sectors) are considered, then n^2 partial derivatives will result. However, these derivatives represent the elements of Leontief inverse matrix. One vexing question that arose concerned the degree to which the magnitude of the change, Δf_j, might affect the value of the derivative, $\delta X_i / \delta f_j$. If the value is sensitive to the size of the initial shock, then this procedure would be somewhat limited. Fortunately it was demonstrated that significant variations in Δf_j implied the same derivative.

The inverse of the whole matrix of such derivatives then becomes the matrix of direct input coefficients. It should be noted however, that this matrix is not derived from the traditional process of using the direct coefficient matrix derived from observed technologies, nor derived through the application of Shephard's Lemma (neoclassical approach). The process described here is a reverse process. While these input coefficients can be interpreted as traditional coefficients, the world from which these coefficients are derived is created by CREIM. As a result, these input coefficients will be sensitive to changes in some of the macro economic indicators (such as a changed in national GNP, bond rates, or steel exports). The degree of sensitivity has yet to be explored together with an appreciation of the varying degrees of relationships that selected coefficients might have with different external changes. In this regard, the procedure opens up a whole new vision for the definition of error and sensitivity analysis within the context of extended input-output systems (see Sonis and Hewings, 1991). In many price-adjusted computable general equilibrium models, the direct input coefficients are placed at the bottom of a nested production function that admits (via a CES specification) substitution between aggregate inputs (i.e., the sum of intermediate inputs) and value added but usually employs a Leontief technology for individual inputs. CREIM, in essence, allows for substitution between inputs in response to changes in demand generated by shocks to the system. Potentially, underlying this system, there is a dual in which some implicit price elasticities could be calculated.

10.3   Block Decomposition Analysis

While input-output systems are characterized by significant interdependence, this interdependence is not universally present at the same level across all sectors. The purchasing the sales patterns of many individual sectors may reflect similar tendencies while, for others, their patterns may be entirely different. Furthermore, for still other sectors, the purchasing patterns may be idiosyncratic yet the sales patterns may reflect tendencies found in many sectors and so forth. As a result, there has been a long tradition in input-output analysis to explore alternative methods for decomposing the structure of the economy; these decomposition techniques have ranged from attempts to exploit potential block triangular structures, decomposable sub-structures and a variety of hierarchical or slicing procedures (see Hewings 1984; Israilevich 1975; Jackson et al 1990; Sonis and Hewings 1988, 1990, 1991, 1993,1995; Hewings et al 1995).

The main question addressed focuses on the degree to which the sectors have become more or less involved with each other over time; earlier empirical evidence suggested that over the period 1970-1990, the Chicago economy became less interdependent and, concomitantly there was an increase in imports and exports especially involving trading relationships with the rest of the US. Block decomposition analysis offers a macro perspective on the major changes; this technique will be complemented by two others that will add some additional perspectives.

10.3.1   Methodology Employed

The algorithm used draws on the theory of Markov chains. A set of sectors is said to be closed if no sector outside the closed set can be reached from any sector within it. A closed set is irreducible if none of its proper subsets is closed. A set of sectors from which at least one closed set of sectors can be reached is called an open set. Figure 10.1 illustrates these notions; there are seven sectors that form two closed sets {1,2,3} and {4,5} and one open set {6,7}. Arrows in Figure 10.1 represent shipments from one sector to another or shipments within a sector. For example, in the closed set {1,2,3}, sector 2 ships its products to sector 3, sector 1 ships its product to sector 2 as well as consuming its own product. Within this closed set, any produced good is shipped directly from one sector to another (such as the shipment of sector 1 to sector 2) or is used indirectly in each sector of the closed set (good 1 is not shipped to sector 3 but sector 3 uses good 1 indirectly via the consumption of good 2 that, in turn, was produced by sector 2 from the direct consumption of good 1). Goods from the other closed set; {4,5} are not consumed directly or indirectly. On the other hand, goods from the open set {6,7} might be consumed both directly and indirectly by sectors of the closed set. In Figure 10.2 the direct shipments of goods between sectors is shown in matrix form.

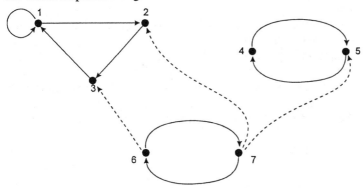

Fig. 10.1. Open and closed sets in graph form

	1	2	3	4	5	6	7
1	1	1					
2			1				
3	1						
4				1			
5			1				
6		1					1
7		1			1	1	

Fig. 10.2. Open and closed sets in matrix form

Four observations can be made using these definitions. First, if all diagonal elements are assigned bo unity, the classification system remains unchanged. Closed sets are formed in such a way that they not only have a zero direct effect between each set but a zero indirect effect as well. For example, the Leontief inverse for Figure 10.2 will have zeros in the following subblocks {1,2,3; 4,5,6,7}, {4,5; 1,2,3} and {4,5; 6,7}. After determining closed sets, one can identify open sets. Finally, the *core* of the subset can be defined: for a given row

I, construct the set:

$$J = (j : a_{ij} > 0 \text{ and } a_{ik} = 0 \text{ for } k \notin J);$$

then construct the set:

$$I = (i : a_{ij} > 0 \text{ and } a_{ij} = 0 \text{ for } k \notin I \text{ and } j \notin J).$$

J is a set of columns that corresponds to a nonnegative element in row I. I is a set of rows with positive coefficients in columns J. The intersection $I \cap J$ represents a set of sectors that ship and receive goods from each other directly. If we group sectors $I \cap J$ in one submatrix, then this submatrix will have no zero cells. Therefore, any irreducible closed set cannot be smaller than $I \cap J$. This intersection, $I \cap J$, is used as a starting set of sectors for the search for the rest of the closed subset.

Based on this observations, the sectors in the input-output table can be ordered in such a way that large coefficients are concentrated on the diagonal subblocks and small coefficients on the off-diagonal subblocks. The algorithm clusters elements into subblocks even if the subsets are not closed. In Figure 10.3, two examples are provided; the matrix on the left has two irreducible closed subsets while the one on the right has one. The asterisked elements on the right *allow* purchases from the closed subsets defined in the matrix on the left.

Assume the following matrix, A:

$$
\begin{array}{c|cccccccc}
 & 1 & 2 & 3 & 4 & 5 & 6 & 7 & 8 \\
\hline
1 & 1 & 1 & 1 & 0 & 1 & 1 & 0 & 1 \\
2 & 0 & 1 & 0 & 1 & 0 & 0 & 1 & 0 \\
3 & 0 & 0 & 1 & 0 & 0 & 1 & 0 & 1 \\
4 & 0 & 1 & 0 & 1 & 0 & 0 & 1 & 0 \\
5 & 1 & 0 & 0 & 1 & 1 & 1 & 1 & 0 \\
6 & 0 & 0 & 1 & 0 & 0 & 1 & 0 & 1 \\
7 & 0 & 1 & 0 & 1 & 0 & 0 & 1 & 0 \\
8 & 0 & 0 & 1 & 0 & 0 & 1 & 0 & 1 \\
\end{array}
$$

$$
\begin{bmatrix}
1 & 1 & 1 & 0 & 0 & 0 & 0 & 0 \\
1 & 1 & 1 & 0 & 0 & 0 & 0 & 0 \\
1 & 1 & 1 & 0 & 0 & 0 & 0 & 0 \\
0 & 0 & 0 & 1 & 1 & 1 & 0 & 0 \\
0 & 0 & 0 & 1 & 1 & 1 & 0 & 0 \\
0 & 0 & 0 & 1 & 1 & 1 & 0 & 0 \\
1 & 0 & 0 & 1 & 1 & 1 & 0 & 1 \\
0 & 1 & 1 & 0 & 1 & 0 & 1 & 1
\end{bmatrix}
\qquad
\begin{bmatrix}
1 & 1 & 1 & 0 & 0 & 0 & 0 & 0 \\
1 & 1 & 1 & 0 & 0 & 0 & 0 & 0 \\
1 & 1 & 1 & 0 & 1^* & 0 & 0 & 0 \\
1^* & 0 & 0 & 1 & 1 & 1 & 0 & 0 \\
0 & 0 & 0 & 1 & 1 & 1 & 0 & 0 \\
0 & 0 & 0 & 1 & 1 & 1 & 0 & 0 \\
1 & 0 & 0 & 1 & 1 & 1 & 0 & 1 \\
0 & 1 & 1 & 0 & 1 & 0 & 1 & 1
\end{bmatrix}
$$

Fig. 10.3. Clusters of coefficients presented by two irreducible subsets

The algorithm first assigns 1 to zero diagonal elements; Table 10.1 illustrates sets I and J and INT (refer to observations noted earlier; INT is the set of intersection of I and J).

Table 10.1. Sets presented by Matrices MI, MJ and MINT

MI	MU	MINT
1 0 0 0 *	1 2 3 5 6 8 *	1 0 0
2 4 7 0 *	2 4 7 0 0 0 *	2 4 7
1 3 6 8 *	3 6 8 0 0 0 *	3 6 8
5 0 0 0 *	1 4 5 6 7 0 *	5 0 0

Row 1 of the Matrix A has positive elements in columns 1, 2, 3, 5, 6 and 8. These columns determine the set J. To determine the set I, the algorithm searches for rows that have positive elements in at least as many places as indicated by set J. Since no other row of the matrix A has this feature, set I will consist of only row 1. This is shown in Table 10.1. Row 2 has positive elements in column 2, 4 and 7, forming another J set. Two more rows have positive elements in the same set of columns (rows 4 and 7). Therefore, the corresponding I set will be rows 2,4 and 7. The algorithm proceeds in a similar fashion; three matrices can be derived, MI contain the sets I, matrix MJ containing the sets J and matrix MINT consisting of the sets INT.

Therefore the algorithm chooses the row in Table 10.1 that has the greatest number of elements in INT. In the example, two choices present themselves, rows 2 or 3. Row 2 is chosen as it appears first; next, it test for the closeness of the selected set INT (INT=MINT [2;]):

INT [2;] ∩ INT [j ≠ 2;] = 0 and MJ [2;] ∩ INT [j ≠ 2;] = 0, therefore the set {2,4,7} is a closed set. Sector 1 ships products to sector 2 and sector 5 ships products to sector 7; thus, sectors {1,5} are suggested as an open set in relation to the set {2,4,7}. This is the first cycle. The second cycle is INT [3,6,8]; the set {3,6,8} is closed and sector {1,5} present an open set. The third cycle is

INT=[1]. Hence MJ [1;]={1,2,3,5,6,8}, INT [5;]={5}. The last closed set is {1,5}, yielding the following set definitions:

Closed Set   {2,4,7}

Suggested Open Set   {1,5}

Closed Set   {3,6,8}

Suggested Open Set   {1,5}

Order of Sectors   {2,4,7,3,6,8,1,5}

The permutation of sectors will yield

$$
\begin{bmatrix}
1 & 1 & 1 & 0 & 0 & 0 & 0 & 0 \\
1 & 1 & 1 & 0 & 0 & 0 & 0 & 0 \\
1 & 1 & 1 & 0 & 0 & 0 & 0 & 0 \\
0 & 0 & 0 & 1 & 1 & 1 & 0 & 0 \\
0 & 0 & 0 & 1 & 1 & 1 & 0 & 0 \\
0 & 0 & 0 & 1 & 1 & 1 & 0 & 0 \\
1 & 0 & 0 & 1 & 1 & 1 & 0 & 1 \\
0 & 1 & 1 & 0 & 1 & 1 & 1 & 1
\end{bmatrix}
$$

A percentage rule was adopted to assist in the identification of clusters and to avoid problems with small coefficients undermining tendencies for cluster to appear. This percentage rule simply designates a critical value, δ, for a particular matrix and sets to zero all elements less than this critical value. For the matrices of direct coefficients, A, the rule is as follows:

$$\delta_t = \frac{\sum_{ij} a_{ij}}{\mu^*} \tag{10.1}$$

where $\mu^* = 0.45$, i.e., the percentage rule eliminates about 45% of the value of the total sum of elements from a particular matrix. For the inverse matrices, B, the rule is as follows:

$$\delta_t = \frac{\sum\limits_{ij} a_{ij}}{\nu^*} \qquad (10.2)$$

where $\nu = 0.135$.

10.3.2 Empirical Findings

The clusters uncovered by the block algorithm are shown in Figures 10.4 and 10.5 for the 36-sector version of the model (see Table 10.4 for sector definitions); Figure 10.4 provides clusters based on the direct coefficients (A); for any particular column, the sectors within the clusters exhibit more interlinkages as one moves down the column. Sectors within any given block are said to reveal the same degree of sectoral linkages. The elements within the first block are referred to as absorption sectors, that is, after elimination of the small elements and after inversion, there is only one non-zero element within the row. The interpretation is that these sectors make important intermediate purchases only from themselves. It turns out that these (self) absorbing sectors remain so for the entire time period, although the entries vary from a low of 17 in 1981 to 20 over the period 1978-1990. The implication here is that the set of non-absorbing sectors (i.e., potentially eligible for entry into clusters) is relatively stable over time. The Chicago economy has not, it would seem, a dynamic that would suggest some significant degree of structural transformation in the manner in which the sales and purchase relationships reposition themselves over time.

For the clustering sectors, the group may be split into between two and four groups. Throughout the entire time period, there is one fairly steady small cluster that comprises from three to five sectors. The reasons for the clustering is not intuitively obvious, since the set of sectors includes paper, rubber, utilities and the restaurant sectors. It should be recalled, however, that the clustering is based on similarities of purchases and does not necessarily imply that the sectors are linked to each other. The most important and prominent cluster is a complex that one might expect to be associated with the Chicago economy - business services, trade, health, fabricated metals, industrial machinery, food processing and construction.

Over the time period, there is little movement between the cluster/non clusters; one notable exception is the primary metal sector (17). Prior to 1986, it was part of the most important cluster but after this time, it joined the set of self-absorbing sectors. While these findings should not be surprising, they do not reflect the amount of change that is observed in individual coefficients (see Israilevich et al 1994). However, the clustering technique focuses less on individual element changes than in the system-wide patterns of change. One might infer that the changes observed in individual coefficients continue to support a macro-level structure of relationships that is changing very slowly if at all. Although some of the pair-wise changes in coefficients have proven to be large, they have not disturbed the overriding structure of the block decomposition of the economy.

Fig. 10.4. Clustering of sectors based on the direct coefficient matrix (columns represent years from 1974-2010)

Fig. 10.5. Clustering of sectors based on the inverse matrix (columns represent years from 1975-2010)

194

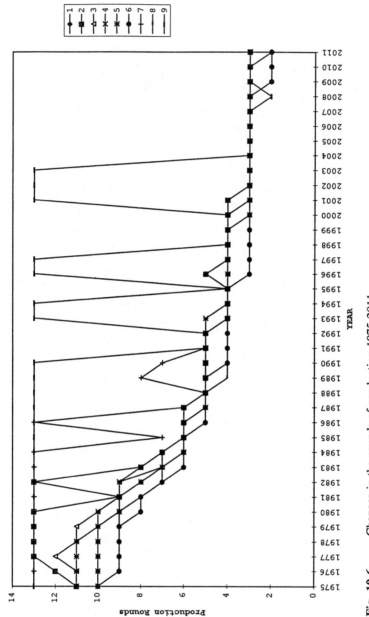

Fig. 10.6.   Changes in the rounds of production 1975-2011

Table 10.2. Production Rounds for Chicago, 1975-2011

Year/Sector	Number of Rounds of Production for Chicago 1975-2011								
	1	2	3	4	5	6	7	8	9
1975	10	11	11	10	11	10	13	10	13
1976	10	12	11	10	11	9	13	9	13
1977	10	13	12	10	11	9	13	9	13
1978	10	13	11	10	11	9	13	9	13
1979	10	13	11	10	10	9	13	9	13
1980	9	13	10	9	10	8	13	9	13
1981	9	9	9	9	9	8	13	8	13
1982	8	13	9	8	9	7	13	7	13
1983	7	8	7	7	8	6	13	7	13
1984	7	7	7	6	7	6	13	6	13
1985	6	6	6	6	6	6	7	6	13
1986	6	6	6	6	6	5	13	5	13
1987	6	6	6	5	6	5	6	5	13
1988	5	5	5	5	5	5	5	5	13
1989	5	5	5	5	5	5	8	4	13
1990	5	5	5	5	5	4	7	4	13
1991	5	5	5	5	5	4	5	4	5
1992	5	5	5	5	5	4	5	4	5
1993	4	4	4	4	5	4	4	4	13
1994	4	4	4	4	4	4	4	4	13
1995	4	4	4	4	4	4	4	4	4
1996	4	5	4	4	4	3	5	3	13
1997	4	4	4	4	4	3	4	3	13
1998	4	4	4	4	4	3	4	3	4
1999	4	4	4	4	4	3	4	3	4
2000	4	4	4	3	4	3	4	3	4
2001	3	4	3	3	4	3	3	3	13
2002	3	3	3	3	3	3	3	3	13
2003	3	3	3	3	3	3	3	3	13
2004	3	3	3	3	3	3	3	3	3
2005	3	3	3	3	3	3	3	3	3
2006	3	3	3	3	3	3	3	3	3
2007	3	3	3	3	3	3	3	3	3
2008	3	3	3	3	3	3	3	3	2
2009	3	3	3	3	3	2	3	2	3
2010	3	3	3	3	3	2	3	2	3
2011	3	3	3	3	3	2	2	2	3

Table 10.3. Frequency of Production Rounds for Chicago, 1975-2011

Year/Sector	Frequency of Production Rounds: Summary for Each Year												
	1	2	3	4	5	6	7	8	9	10	11	12	13
1975										4	3		2
1976									2	2	2	1	2
1977									2	2	1	1	3
1978									2	2	2		3
1979									2	3	1		3
1980								1	3	2			3
1981								2	5				2
1982							2	2	2				3
1983						1	4	2					2
1984						3	4						2
1985						7	1						1
1986					2	5							2
1987					3	5							1
1988					8								1
1989				1	6			1					
1990				2	5		1						1
1991				2	7								
1992				3	6								
1993				7	1								1
1994				8									1
1995				9									
1996			2	3	2								1
1997			2	6									1
1998			2	7									
1999			2	7									
2000			3	6									
2001			6	2									1
2002			8										1
2003			8										1
2004			9										
2005			9										
2006			9										
2007			9										
2008		1	8										
2009		2	7										
2010		2	7										
2011		3	6										

Table 10.4. Sector definitions

Sector Description	36-Sector Number	9-Sector Number
Livestock, Agricultural products	1	1
Forestry, Fisheries, Agric. Services	2	1
Mining	3	1
Construction	4	2
Food and Kindred Products	5	3
Tobacco	6	3
Apparel and Textiles	7	3
Lumber and Wood Products	8	4
Furniture and Fixtures	9	4
Paper and Allied Products	10	3
Printing and Publishing	11	3
Chemicals and Allied Products	12	3
Petroleum	13	3
Rubber and Plastics	14	3
Leather	15	3
Stone, Clay and Glass	16	4
Primary Metals	17	4
Fabricated Metals	18	4
Non Electrical Machinery	19	4
Electrical Machinery and Electronic	20	4
Transportation Equipment	21	4
Instruments	22	4
Miscellaneous Manufacturing	23	4
Transportation	24	5
Communications	25	5
Utilities	26	5
Trade	27	6
Finance, Insurance	28	7
Real Estate	29	7
Hotels, Repair Services	30	8
Eating and Drinking Places	31	8
Auto Repair and Services	32	8
Amusements and Recreation	33	8
Health and Nonprofit	34	8
Federal Government Enterprises	35	9
State and Local Govt. Enterprises	36	9

To complement the attention on the direct coefficients, analysis of a similar kind was also directed to the B matrix. As with the A matrix, there is a fair degree of stability in the separation of absorption and clusterable sectors (see Figure 10.5). Absorption sectors vary from 21 to 25, leaving fewer sectors to form clusters. In the context of the B matrix, the sectors that are absorption states are the least susceptible economically to an external shock. As one would expect, given their definition, these are sectors that react strongly only when the shock is applied to the sector itself. Included in this group are primary and fabricated metals, restaurants and government enterprises.

There are three groups within the clusterable sectors. The bottom group, with the strongest degree of interdependence, consists of the petroleum and real estate sectors - an unexpected cluster! Above them, in the next strongest group, are sectors such as construction, food products, industrial machinery, trade and business services. On interesting transformation occurs in the health sector (34); it begins the period as an absorbing sector but, in the third year, it moves into the first cluster and by the end of the period it is almost at the top of the hierarchy. In comparing the two clustering systems, it should be noted that the majority of absorption states in A are also absorption states in B; hence, while the clusterable sectors are similar in both matrices, the clusters themselves bear little relationship to each other. Part of the reason for this, of course, stems from the complex process of interaction that occurs as matrix A is transformed into B.

10.4   Structural Complexity and Production Rounds

The perspective offered here is one that considers the change in the structural complexity of an economy over time. There is little in the way of received theory on which to base empirical testing, although Hewings et al. (1989) proposed a logistic function in which an economy would develop in terms of some index of complexity (associated with the strength and number of connections between industrial sectors, for example). The approach adopted here draws on some methodology initially developed by Robinson and Markandya (1973) and empirically tested by Hewings et al. (1984). The method exploits the power series version for the estimation of the Leontief inverse:

$$B = I + A + A^2 + A^3 + \dots \qquad (10.3)$$

Each round (power of A) is considered to represent a production round; complexity is measured as a function of the number of rounds it takes for a level of final demand to be satisfied. Since the series is of infinite length, Robinson and Markandya (1973) proposed that production was assumed to have been satisfied in a sector one round after:

$$\sum_z X_{i(z)} \geq 0.8X_i \qquad (10.4)$$

where $\sum_z X_{i(z)}$ is the amount of X_i produced through the zth production round.

If (10.4) is satisfied at round z, then sector i is assumed to require $z+1$ production rounds to meet final demand requirements. The choice of 0.8 is somewhat arbitrary, but was applied consistently to all the input-output tables. In this application, a 9-sector version of the tables used (see Table 10.4 for sector definitions). It should be noted that in the application of the power series expansion (Equation 10.4), the elements of the matrices, A, are the same except for sectors that finished their production in the prior round. In this case, the column is set to zero. At some point, $A^z \rightarrow 0$ since all sectors will have completed their production rounds to meet the vector of final demands.

Tables 10.2 and 10.3 and Figure 10.6 present the results for Chicago. A nine-sector aggregation of the 36 sectors shown in Table 10.2 was used to illustrate the methodology and to attempt to draw some impressions from the overall trends. Clearly, the economy is moving to one that requires far fewer rounds of production to meet final demands. The decrease in production rounds occurs almost immediately and stabilizes at between two and three by 2004. There would appear to be some within-period volatility for sectors 7 and 9 (see definitions in Table 10.4).

The apparent transformation of the economy can be captured graphically with the final technique; the *multiplier product matrix*, which provides a visualization of the economic landscapes derived from the input-output structure. The methodology is described in the next section.

10.5   Comparison of Economic Landscapes

10.5.1   The Multiplier Product Matrix [MPM]

The definition of the multiplier product matrix as follows: let $A = \|a_{ij}\|$ be a matrix of direct inputs in the usual input-output system, $B = (I - A)^{-1} = \|b_{ij}\|$ be the associated Leontief inverse matrix and let $B_{\bullet j}$ and $B_{i\bullet}$ be the column and row multipliers of this Leontief inverse. These are defined as:

$$B_{\bullet j} = \sum_{i=1}^{n} b_{ij}, \quad B_{i\bullet} = \sum_{j=1}^{n} b_{ij} \qquad (10.5)$$

Let V be the global intensity of the Leontief inverse matrix:

$$V = \sum_{i=1}^{n} \sum_{j=1}^{n} b_{ij} \qquad (10.6)$$

Then, the input-output multiplier product matrix (MPM) is defined as:

$$M = \frac{1}{V} \| B_{i\bullet} B_{\bullet j} \| = \frac{1}{V} \begin{bmatrix} B_{1\bullet} \\ B_{2\bullet} \\ \vdots \\ B_{n\bullet} \end{bmatrix} \begin{pmatrix} B_{\bullet 1} & B_{\bullet 2} & \cdots & B_{\bullet n} \end{pmatrix} = \| m_{ij} \| \qquad (10.7)$$

The properties of the MPM will now be considered in the context of the following issues: (1) the hierarchy of backward and forward linkages and their economic landscape associated with the cross-structure of the MPM and (2) the maximum entropy properties of the MPM.

10.5.2 Economic Cross-Structure Landscapes of MPM and the Rank-Size Hierarchies of Backward and Forward Linkages

The concept of key sectors is based on the notion of backward and forward linkages and has been associated with the work of both Rasmussen (1956) and Horseman (1958). The major thrust of the analytical techniques, and subsequent modifications and extensions, has been towards the identification of sectors whose linkages structures are such that they create an above-average impact on the rest of the economy when they expand or in response to changes elsewhere in the system. Rasmussen (1956) proposed two types of indices drawing on entries in the Leontief inverse:

1. Power of dispersion for the backward linkages, BL_j, as follows:

$$BL_j = \frac{1}{n} \sum_{i=1}^{n} b_{ij} \bigg/ \frac{1}{n^2} \sum_{i,j=1}^{n} b_{ij} =$$
$$= \frac{1}{n} B_{\bullet j} \bigg/ \frac{1}{n^2} V = B_{\bullet j} \bigg/ \frac{1}{n} V \qquad (10.8)$$

and

2. The indices of the sensitivity of dispersion for forward linkages, FL_i, as follows:

$$FL_i = \frac{1}{n}\sum_{j=1}^{n} b_{ij} \Big/ \frac{1}{n^2}\sum_{i,j=1}^{n} b_{ij} =$$
$$= \frac{1}{n}B_{i\bullet} \Big/ \frac{1}{n^2}V = B_{i\bullet} \Big/ \frac{1}{n}V$$

(10.9)

The usual interpretation is to propose that $BL_j > 1$ indicates that a unit change in final demand in sector j will create an above average increase in activity in the economy. Similarly, for $FL_i > 1$, it is asserted that a unit change in all sectors' final demand would create an above average increase in sector i. A key sector, K, is usually defined as one in which both indices are greater than 1. It would be noted here that similar ideas were developed by Chenery and Watanabe (1958) for the definition of backward and forward linkages based on the matrix of direct inputs, $A = \|a_{ij}\|$.

The definitions of backward and forward linkages provided by (10.8) and (10.9) imply that the rank-size hierarchies (rank-size ordering) of these indices coincide with the rank-size hierarchies of the column and row multipliers. It is important to underline, in this connection, that the column and row multipliers for MPM are the same as those for the Leontief inverse matrix:

$$\sum_{j=1}^{n} m_{ij} = \frac{1}{V}\sum_{j=1}^{n} B_{i\bullet}B_{\bullet j} = B_{i\bullet}$$
$$\sum_{i=1}^{n} m_{ij} = \frac{1}{V}\sum_{i=1}^{n} B_{i\bullet}B_{\bullet j} = B_{\bullet j}$$

(10.10)

Thus, the structure of the MPM is essentially connected with the properties of sectoral backward and forward linkages.

The structure of the matrix, M, can be ascertained in the following fashion. Consider the largest column multiplier, $B_{\bullet j}$ and the largest row multiplier, $B_{i\bullet}$ of the Leontief inverse. Then, the element $m_{i_0 j_0} = \frac{1}{V}B_{i_0\bullet}B_{\bullet j_0}$, and is located in the place (i_0, j_0) of the matrix, M. Moreover, all rows of the matrix, M are proportional to the i_0^{th} row, and the elements of this row are larger than the corresponding elements of all other rows. The same property applies to the j_0^{th} column of the same matrix. Hence, the element located in (i_0, j_0) defines the center of the largest cross within the matrix, M. If this cross is excluded from M, then the second largest cross can be identified and so on. Thus the matrix, M, contains the rank-size sequence of crosses. One can reorganize the locations of rows and columns of M in such a way that the centers of the corresponding crosses appear on the main diagonal. In this fashion a descending *economic landscape* will be apparent.

This rearrangement also reveals the descending rank-size hierarchies of the

Hirschman-Rasmussen indices for forward and backward linkages. Inspection of that part of the landscape with indices > 1 (the usual criterion for specification of key sectors) will enable the identification of the key sectors. However, it is important to stress that the construction of the economic landscape for different regions or for the same region at different points in time would create the possibility for the establishment of a taxonomy of these economies. Moreover, the superposition of the hierarchy of one region on the landscape of another region provides a clear visual representation of the similarities and differences in the linkage structure of these regions (see Sonis et al. 1994 for an application to Chinese urban areas).

10.5.3   Maximum Entropy Properties of the MPM

Consider all positive matrices, $\Psi = \| \psi_{ij} \|$ with the property that the row and column multipliers are equal to those of the Leontief inverse:

$$\sum_j \psi_{ij} = B_{i\bullet}, \quad \sum_i \psi_{ij} = B_{\bullet j} \tag{10.11}$$

Obviously $\sum_{i,j} \psi_{ij} = V$. Consider, for each matrix, V, the following Shannon entropy:

$$Ent\Psi = -\sum_{i,j} \frac{\psi_{ij}}{V} \ln \frac{\psi_{ij}}{V} \tag{10.12}$$

It is possible to prove that the matrix M will provide the maximal value of the entropy (10.12); thus the matrix M may be considered to represent *the most homogeneous* distribution of the components of the column and row multipliers of the Leontief inverse B. A further perspective may be offered. In the case of equal column and row multipliers, the economic landscape will be a flat, horizontal plane.

The MPM depends solely on the column and row multipliers, and therefore represents only the aggregate charateristics of the interactions of each sector with the rest of the economy. Thus, MPM does not take into account the specifics of the pair-wise sectoral interactions and can be considered as an aggregate representation of some sector *equalization tendency* in the economic interaction between sectors. For the measurement of the realizations of this tendency in actual input-output systems, the vector of gross outputs, $X = \begin{bmatrix} X_1 \\ X_2 \\ \vdots \\ X_n \end{bmatrix}$ derived from

the usual application of final demand, f, with the Leontief inverse matrix, B, i.e.,

Bf, with the vector $X^* = \begin{bmatrix} X_1^* \\ X_2^* \\ \vdots \\ X_n^* \end{bmatrix}$ calculated as $X^* = Mf$. In so doing, the

following evaluation can be revealed:

$$
\begin{aligned}
X_i^* &= \sum_j m_{ij}\, f_j = \frac{1}{\bar{V}}\sum_j B_{i\bullet}B_{\bullet j}\, f_j = B_{i\bullet}\sum_j \frac{B_{\bullet j}}{\bar{V}}\, f_j = \\
&= B_{i\bullet}\sum_j \frac{1}{\bar{V}}\left(\sum_s b_{sj}\right) f_j = \frac{B_{i\bullet}}{\bar{V}}\sum_s \sum_j b_{sj}\, f_j = \\
&= B_{i\bullet}\sum_s \frac{X_s}{\bar{V}} = \overline{X}FL_i
\end{aligned}
\tag{10.13}
$$

where

$$
\overline{X} = \frac{1}{n}\sum_i X_i
\tag{10.14}
$$

is the average sectoral gross output.

Thus, the vector

$$
X^* = \overline{X}\begin{bmatrix} FL_1 \\ FL_2 \\ \vdots \\ FL_n \end{bmatrix}
\tag{10.15}
$$

represents the part of grow output associated with the tendency of sectoral equalization. The weight of this maximum entropy gross output can be defined as the largest number, p, such that $X > pX^*$. This implies that:

$$p = \min X_i / X_i^*$$ (10.16)

Further, the following decomposition holds:

$$X = pX^* + (1 - p)R^*$$ (10.17)

where R^* is the residual gross output, representing the action of sectorally specific tendencies and their impact on gross output (i.e. the deviations from homogeneous tendencies).

10.5.4   Application to Chicago Tables

Figures 10.7 through 10.14 show the economic landscapes derived from the MPM applications to selected Chicago matrices for the period 1975 through 2010. The ordering obtained in 1975 was used as a *numeraire* for the remaining years so that changes would be readily apparent. Over time, the forecasts of structural change reveal two important processes not picked up explicitly by the other methods. First, the economy would appear to be moving towards a state of maximum entropy, i.e. an economic landscape in which there is little differentiation between sectors. Compare, for example, the landscape for 2010 (Figure 10.14) with that for 1975 (Figure 10.7). The relative impacts of all sectors would appear to be converging to a relatively flat homogeneous plane. Secondly, the *height* of this plane reflects a continuation of earlier trends towards a greater dependence on external-to-the-region purchases and sales, since the landscape is much lower at the end of the period than in earlier years. While this process of sector equalization was picked up by the structural complexity measure, neither this measure nor the block decomposition approach was able to shed any light on the nature of changes in regional dependence. An examination on a coefficient-by coefficient basis might have revealed this phenomenon, but the economic landscapes produced by the multiplier product matrix offer a rapid visual interpretation.

Finally, Figure 10.15 provides a summary of the changes in the rankings of the sectors (the rows and columns of the MPM). The forward linkage patterns seem more stable with only small changes in the rankings over the period 1975-2010. During the period 1985-2000, there are some important changes in the rankings of the sectors based on backward linkages, with the construction sector decreasing in importance while non-durable manufacturing increases in importance. (It should be remembered, however, these changes reflect relative rather than absolute changes.)

205

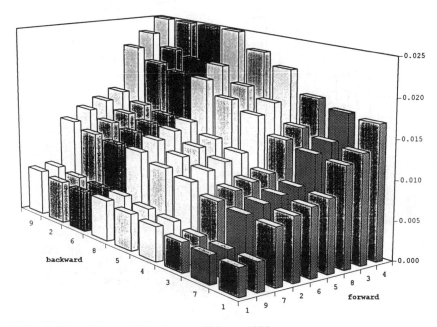

Fig. 10.7.   Economic landscape, Chicago 1975

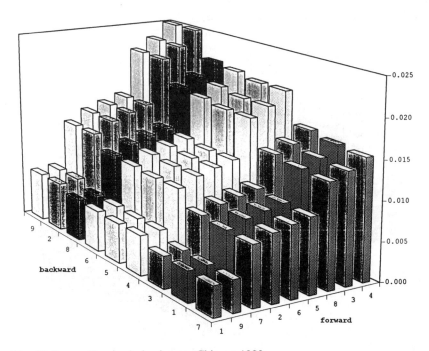

Fig. 10.8.   Economic landscape: Chicago 1980

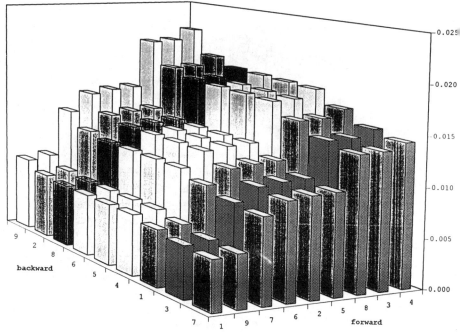

Fig. 10.9.   Economic landscape, Chicago 1985

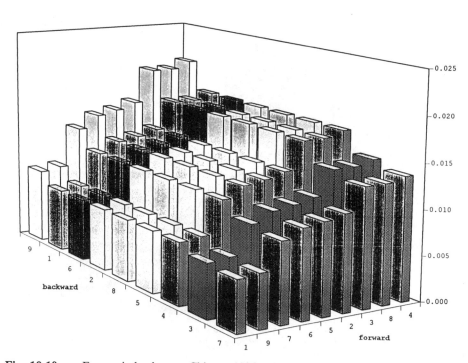

Fig. 10.10.   Economic landscape: Chicago 1990

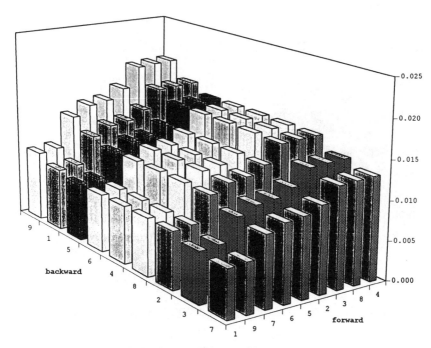

Fig. 10.11. Economic landscape, Chicago 1995

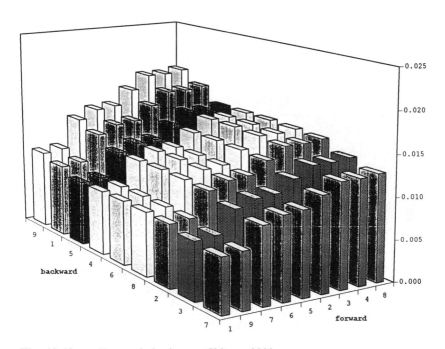

Fig. 10.12. Economic landscape: Chicago 2000

208

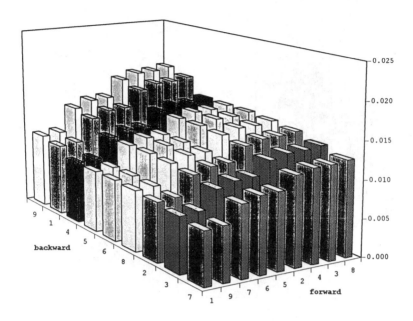

Fig. 10.13.   Economic landscape, Chicago 2050

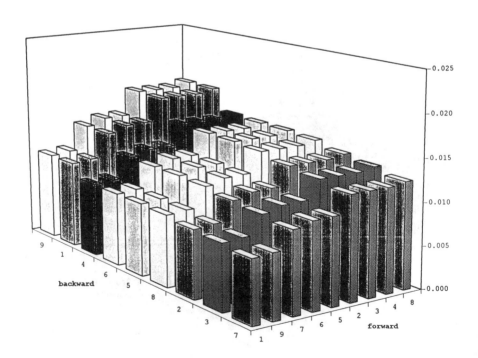

Fig. 10.14.   Economic landscape: Chicago 2010

Forward Linkages

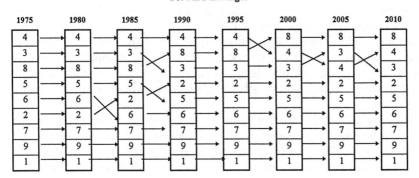

Backward Linkages

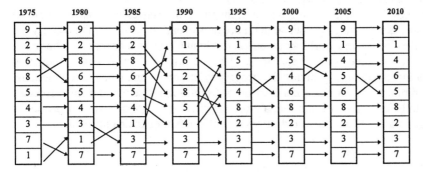

Fig. 10.15.   Backward and forward linkage rankings

10.6   Evaluation

All three methods applied to the set of Chicago input-output tables derived from a general equilibrium model suggest that the changes in the economy over the next two decades are likely to reflect more subtle variations around trends established in earlier periods. While this should not be surprising in one sense, given the nature of forecasting models, the changes are not monotonic. The behaviour of individual sectors and groups of sectors exhibit patterns of change that are not necessarily straightforward. None of the methods explored here attempts to provide an explanation of these changes. Certainly, a more complete explanation would lie in the nature of the changes evident in the rest (non input-output components) of the model - particularly, the changes in regional and extra-regional demand.

The methods do provide the basis for an appreciation of the way in which changes can be portrayed in an economy and thus establish the *macro basis* for more *micro* studies of the process of innovation diffusion of new technologies. The analysis presented here reaffirms the need to explore these types of diffusion processes in an interdependent framework since subtle changes in the sector can create important system-wide effects.

The most important finding would seem to be the suggestion that the regional economy is following out, becoming more dependent on the rest of the country (or even the rest of the world as a source for inputs and as a market for outputs). Structural change would appear to be creating not only new production technologies but also changing the geography of supply and demand interactions. The implications for these changes on transportation systems and infrastructure need to be explored. In addition, a more extensive evaluation of similar processes in other parts of the US would provide confirmation of their universality.

References

Carter A P (1970) Structural Change in the American Economy. Harvard University Press Cambridge

Chenery H B and T Watanabe (1958) International Comparisons of the Structure of Production. Econometrica 26:487-521

Hewings G J D (1982) Holistic Matrix Analysis of Regional Economic Systems. Working Paper. Department of Economics. University of Queensland Australia

Hewings G J D, Merrifield J and J Schneider (1984) A Regional Test of the Linkage Hypothesis. Revue d'Economie Regionale et Urbaine 25:275-290

Hewings G J D, Sonis M and R C Jensen (1988) Fields of Influence of Technological Change in Input-Output Models. Papers of Regional Science Association 64:25-36

Hewings G J D, Jensen R C, West G R, Sonis M and R W Jackson (1989) The Spatial Organization of Production. An Input-Output Perspective. Socio-Economic Planning Sciences 23:67-86

Hewings G J D, Sonis M, Lee J K and S Jahan (1995) Structure of the Bangladesh Interregional Social Accounting System: a Comparison of Alternative Decompositions. In: Hewings G J D and M Madden (Eds) Social and Demographic Accounting. Cambridge University Press: 81-110

Hirschman A (1958) The Strategy of Economic Development. Yale Univerisity Press, New Haven

Israilevich P R (1975) Unpublished Ph.D Dissertation. University of Pennsylvania, Philadelphia

Israilevich P R, Hewings G J D, Sonis M and G R Schindler (1994) Forecasting Structural Change with a Regional Econometric Input-Output Model. Discussion Paper 94-T-1. Regional Economics Applications Laboratory. Urbana Illinois

Israilevich P R, Mahidhara R and G J D Hewings (1994) Forecasting with Regional Input-Output Tables. Accepted subject to revision in Papers in Regional Science

Rasmussen P (1956) Studies in Inter-Sectoral Relations. Einar Harks Copenhagen

Robinson S and A Markandya (1973) Complexity and Adjustment in Input-Output Systems. Oxford Bulletin of Economics and Statistics 35:119-134

Schindler G R, Israilevich P R and G J D Hewings (1994) Three-Dimensional Analysis of Economic Performance. Discussion Paper 94-P-6. Regional Economics Applications Laboratory. Urbana Illinois

Sonis M and G J D Hewings (1988) Superposition and Decomposition Principles in Hierarchical Social Accounting and Input-Output Analysis. In: Harrigan F and P McGregor (Eds) Recent Advances in Regional Economic Modelling. Pion London 46-65

Sonis M and G J D Hewings (1990) The 'Matrioshka Principle' in the Hierarchical Decomposition of Multiregional Social Accounting Systems. In: Anselin L and M Madden (Eds) New Directions in Regional Analysis. Integrated and Multiregional Approaches. Pinter London

Sonis M and G J D Hewings (1991) Fields of Influence and Extended Input-Output Analysis. A theoretical Account. In: Dewhurst J J L, Hewings G J D and R C Jenson (Eds) Regional Input-Output Modelling. New Developments and Interpretations. Avebury, Aldershot

Sonis M and G J D Hewings (1993) Hierarchies of Regional Sub-Structures and their Multipliers within Input-Output System. Miyazawa Revisited. Hitotsubashi Journal of Economics 34:33-44

Sonis M and G J D Hewings (1995) Matrix Sensitivity, Error Analysis and Internal/External Multiregional Multipliers. Hitotsubashi Journal of Economics 36:71-60

Sonis M, Hewings G J D and J Guo (1994) Comparative Analysis of China's Metropolital Economies. An Input-Output Perspective. In: Chatterji M (Ed) Regional Science Applications to China. Macmillan. forthcoming

Wolff E N (1985) Industrial Composition, Interindustry Effects and US Productivity Slowdown. Review of Economics and Statistics

Wolff E N (1989) Productivity and American Leadership. The Long View

Wolff E N (1993) Competitiveness, Convergence and International Specialization
Wolff E N (1994) Productivity Measurement within an Input-Output Framework.
 Regional Science and Urban Economics 24:75-92

11 Endogenous Technological Change, Long Run Growth and Spatial Interdependence: A Survey

Peter Nijkamp and Jacques Poot

033 041
R12

11.1   Introduction

After extensive development of theories of economic growth in the 1950s and 1960s, the subject received relatively little attention in the 1970s and most of the 1980s. Yet a new wave of interest emerged in the late 1980s. The influential articles by Paul Romer (1986) and by Robert Lucas (1988) led to several theoretical and empirical research programmes. Surveys of the 'New Growth Theories' and their relationship with the standard neoclassical model (Solow 1956) have already emerged in special journal issues (e.g. Ehrlich 1990, Stern 1991, Romer 1994), while a range of models can also be found in the books by Grossman and Helpman (1991), Barro and Sala-i-Martin (1995) and Pasinetti and Solow (1994). These theories emphasize the role of externalities in technological change (Romer 1986), specialization and trade (Grossman and Helpman 1990a), monopoly rents from innovation and "creative destruction" (Aghion and Howitt 1992), human capital (Becker et al. 1990) and government policy (Rebelo 1991).

Empirical work has benefited from new comparable data available for a wide range of countries in the world (e.g. Summers and Heston 1991, International Monetary Fund, various years, World Bank, various years), although the statistical inference from such data appears as yet rather fragile (Levine and Renelt 1992). While the neoclassical new growth theories have not gone unchallenged (Scott 1989), a general consensus has emerged that Schumpeterian forces of entrepreneurship, innovation and evolution are important (Grossman and Helpman 1994). This is already reflected in new directions in regional policies (Nijkamp 1991) and policies to encourage trade-driven national growth (Porter 1990).

In this paper we discuss some of the salient features of the current literature on technological change and growth. No attempt is made to provide a full survey of the literature, but the emphasis is on how growth is affected, through technological change, by spatial linkages such as trade, factor mobility or the diffusion of ideas. A simple model is used to highlight the issues. The fact that technological change itself may be stimulated by economic growth creates the possibility of a positive feedback in the economic system, which generates increasing returns. Positive feedback in the economy opens up the possibility of multiple equilibria, diverging growth paths and a sensitivity to initial conditions (e.g. Arthur 1990). In a spatial context, it can also lead to patterns of industrial localization and geographic diversity (Krugman 1991).

The general equilibrium models of the new growth theories allow a broader discussion of diverging growth than the demand-driven Kaldorian models of cumulative causation, which are still commonly used in the regional development debate (e.g. McCombie 1988a, 1988b). We show that the models of endogenous technical change can generate results which are consistent with some of the predictions of the Kaldorian framework. Like the Kaldorian approach, they also point to a need for regional economic policies, although the new theories suggest that such policies would need to be supply-oriented and hence focus, inter alia, on infrastructure, innovation and ecological sustainability rather than the traditional tools of demand stimuli and subsidies.

The next section reviews the neoclassical model. Section 11.3 discusses endogenous technical change in a closed economy. For the purposes of illustration, we formulate a simple model which has some of the features of models proposed by Lucas (1988) and Romer (1986). Section 11.4 discusses the consequences of spatial diffusion of technology and other forms of spatial interaction, such as migration and trade. Section 11.5 briefly outlines the conventional Kaldorian demand-driven approach to regional growth and reconciles its conclusions with the new insights. The last section provides some final comments.

11.2   The Neoclassical Growth Model

The standard neoclassical model of economic growth was formulated in 1956 by Robert Solow and independently by Trevor Swan. Lucas (1988) and Romer (1989) noted that this model was consistent with intertemporal optimization in an Arrow-Debreu competitive equilibrium framework, although Cass (1965) had already shown much earlier that from any starting-point, optimal capital accumulation converges to the balanced Solow-Swan growth path. The Solow-Swan model implies that countries or regions with the same preferences and technology will converge to identical levels of *per capita* income. Trade and/or factor mobility would accelerate convergence. While there is evidence of such convergence among groups of countries or regions (e.g. Baumol 1986, Dowrick and Nguyen 1989), the process tends to be very slow (Barro and Sala-i-Martin 1992) while others argue that a reorganizing of the data may in fact show up patterns of divergence (e.g. De Long 1988).

It is well known in explaining such patterns of growth that the accumulation of human capital and technological change have a more important role to play than growth in capital per worker (e.g. Solow 1988). The neoclassical growth model for closed economies provides a simple explanation of labour productivity growth in terms of capital accumulation (which tends to grow faster than aggregate hours supplied) and other factors, referred to collectively as technological change. If closed economies are similar with respect to preferences and technology, diminishing returns to reproducible capital will generate a long-run convergence in levels of per capita income. Before considering implications of endogenous technical change and spatial interactions, it is useful to review the key features of the standard model (but see also e.g. Lucas 1988).

Consider a closed economy with competitive markets and a constant returns technology. At date t, labour supply is L(t). The exogenously given rate of growth of L(t) is n. Real production Y(t) is assumed to result from combining inputs according to

$$Y(t) = F (K(t), L(t) e^{gt}$$ (11.1)

where K(t) is the stock of capital at time t and e^{gt} represents the effect of exogenous labour-augmenting technical progress with a growth rate g. Neglecting labour-leisure choices and assuming full employment, population and labour force become equivalent concepts and both growth at rate n. Equation (11.1) can be rewritten as

$$\hat{y} = f(\hat{k})$$ (11.2)

where the symbol ^ denotes a quantity per effective unit of labour L(t)egt and where we shall assume that f(.) has the usual 'well-behaved' properties, formalized in the Inada (1963) conditions. If the rate of depreciation of capital is a fraction δ of the stock, net investment is given by

$$\dot{K} = Y - C - \delta K$$ (11.3)

where · denotes a derivative with respect to time and C is the rate of consumption. Hence, k̂ evolves in accordance with

$$\dot{\hat{k}} = f(\hat{k}) - \hat{c} - (n+g+\delta) \hat{k}$$ (11.4)

Households seek to maximize lifetime utility[1] given by

$$W = \int_{t=0}^{\infty} u(c) \ e^{nt} \ e^{-\rho t} dt$$ (11.5)

where c = C/L and ρ is the constant rate of time preference. Note that household utility rather than individual utility is in the welfare criterion since per capita

---

[1]Note that the usual 'infinite horizon' assumption has been introduced for simplicity.

utility is multiplied by household membership which grows at rate n. Assuming that the utility function has the form

$$u(c) = \frac{c^{1-\sigma} - 1}{1-\sigma} \qquad (11.6)$$

marginal utility $u'(c)$ has the constant elasticity $-\sigma$ with respect to c.[2] To find the consumption path $c(t)$ which maximizes (11.5) subject to (11.4) is a standard optimal control problem (see e.g Cass 1965). It can be shown that on the optimal time path for consumption

$$\tilde{c} = [f'(\hat{k}) - \delta - \rho]/\sigma \qquad (11.7)$$

where $\tilde{\ }$ refers to a rate of growth, i.e. $c \,/\, c = \tilde{c}$. Given initial resources $K(0)$ and $L(0)$, the optimal path will converge asymptotically to the balanced, or steady-state, path. In the steady-state, the effective quantities $\hat{y}$, $\hat{k}$ and $\hat{c}$ do not change. Thus, income, capital and consumption per capita each grow at the rate of technological progress, g. The absolute quantities Y, K and C grow at the rate $g+n$. The long-run rate of return to capital is $f'(\hat{k}^*)$ where $\hat{k}^*$ is the steady-state effective capital intensity found by setting the rate of growth in per capita consumption in (11.7) equal to g. Hence,

$$f'(\hat{k}^*) = \delta + \rho + \sigma g \qquad (11.8)$$

If the production function is of the Cobb-Douglas type, with α denoting the share of profits in income, i.e.

$$f(\hat{k}) = \gamma \hat{k}^\alpha \qquad (11.9)$$

then it can be easily seen that

$$\hat{k}^* = \left[\frac{\delta + \rho + \sigma g}{\alpha \gamma} \right]^{\frac{1}{\alpha-1}} \qquad (11.10)$$

and the optimal propensity to save in the steady-state is also constant and equal to

$$s^* = \frac{(n + g + \delta)\, \hat{k}^*}{f(\hat{k}^*)} = \frac{\alpha\,(n + g + \delta)}{\delta + \rho + \sigma g)} \qquad (11.11)$$

---

[2] If $\sigma = 1$, $u(c) = \ln c$, i.e. the intertemporal elasticity of substitution is one.

It can be seen from (11.10) and (11.11) that a low discount rate ρ and a high intertemporal elasticity of substitution (i.e. low σ) increase $\hat{k}^*$ and s^*.[3] This demonstrates the well known prediction of the neoclassical model that a thrifty society will wealthier in the long-run be than an impatient one, but does not grow faster. We shall show in Section 11.3 that in the presence of endogenous technical change, not just per capita income but also the long-run growth rate is affected by savings behaviour.

The model outlined above is easily open to criticism. For example, the model focuses on real capital accumulation in a closed economy and abstracts from monetary considerations. This may be a serious omission in explaining growth differences between countries, although Lucas (1988) pointed out that we do not know as yet how serious this omission is. Monetary considerations are less important at the regional level, since monetary conditions (e.g. interest rates and inflationary expectations) may be assumed largely constant across regions and are thus unlikely to generate differential growth effects. However, the openness of the economy cannot be ignored at the regional level and the terms of trade, factor mobility and the generation and diffusion of technological change then become important issues.

On balance, the empirical evidence rejects the standard model. Barro and Sala-i-Martin (1992) found evidence of convergence across the states of the USA, but the speed of convergence is so slow that it can only be compatible with unrealistically high values for the share of profits in income.[4] Alternatively, capital must be interpreted as being broadly defined and must include human capital. The persisting differences in steady-state growth rates tend to be explained in a rather ad hoc fashion by school enrolment rates and government consumption expenditure (excluding education and defence). A large cross-section of countries shows an even lower tendency to convergence.

Similarly, Mankiw et al. (1992) show that the textbook Solow model needs to be augmented to become useful in explaining differences in income per capita across countries. They find that introducing human capital accumulation explicitly (measured by secondary school enrolment rates) has the same type of positive effect on income per head as the savings ratio s^* in the standard Solow model. The augmented model provides some evidence of inter-country convergence, although again at a slow rate.

Regional openness and interconnectedness may of course be responsible for the

---

[3]The substitution of "stylised", but plausible, values of the parameters for developed economies in (11.11) suggests an order of magnitude for s^*. Let e.g. $\alpha = 0.25$; $n = 0.01$; $g = 0.02$; $d = 0.02$; $\rho = 0.05$ and $\sigma = 1$, then (11.11) suggests that the optimal propensity to save is about 14 percent.

[4]There are two notions of convergence. *Weak* convergence takes place when low income regions grow faster than high-income ones, all else being equal. *Strong* convergence takes place when the standard deviation of the distribution of income across regions declines. Barro and Sala-i-Martin's evidence points to weak convergence rather than strong convergence as there are periods during which the interregional and international dispersion of incomes increased (notably the 1970s).

somewhat more convincing patterns of convergence observed at the regional level in the USA than in cross-country comparisons. For example, Barro and Sala i Martin (1992) point to diffusion of technological change having the potential of generating convergence even if the marginal product of capital is not declining. Nonetheless, despite well known historical evidence of convergence of incomes across states of the USA (e.g. Easterlin 1960), there are fairly lengthy periods during which one can observe divergence[5] and the evidence that factor mobility operates as an equilibrating process is also rather inconclusive.[6] Richer models are needed to explain such observations. The next section addresses the determinants and consequences of the process of technological change in the neoclassical growth model.

11.3   Endogenous Technical Change

Economic restructuring, technological change and the shifts in regional growth patterns have exerted a far reaching impact on resource allocation and welfare. National and regional economic systems have become more interdependent. In addition, public policymakers have become aware of the need to stimulate competitive behaviour and their policies are increasingly oriented towards deregulation, devolution and a reliance on market signals. Among these, regional policies reflect responses to a permanent conflict between the relatively efficient use of scarce resources in core regions and the resulting equity discrepancies with respect to peripheral regions. Production-oriented policies can have strong spatial and sectoral impacts. For instance, a regional innovation policy promoting the microelectronics industry or the telecommunications sector targets areas with a favourable 'seedbed' potential for these sectors. Technological innovation is therefore not 'manna from heaven' as in the standard neoclassical model, but can be generated by well-focused public policies. Regions are competitive geographical units which will try to obtain an economic advantage by encouraging technologically advanced products (or processes). Thus, we need to ask what type of spatial selection environment (Kamann 1988) induces such 'technogenesis' (see also Davelaar and Nijkamp 1989). A clear analytical framework which integrates economic growth, spatial interdependencies and the creation of new technology as an explicit production process is required to formulate production-oriented regional policies.

Malecki and Nijkamp (1988) argued that the blend of entrepreneurial spirit, technologically-sensitive sectoral structures and creative environments is of critical importance for a successful technological transformation process. Since new technology is an important weapon in a competitive market, firms will consider

---

[5]The periods 1840-1880, 1920-1940 and 1970-1980 are period of interregional divergence in the USA.

[6]See for example the discussion in Armstrong and Taylor (1985:118-121).

a favourable geographical location as an important dimension of their entrepreneurial strategy. Consequently, locational aspects have become an important focus of current technology research. Besides, even when technological innovations have materialized, this does not mean that all firms or regions are able to 'reap the fruits' of a new technology. Apparently, there are many bottlenecks to be overcome. This leads for instance to the question of which transfer mechanisms (e.g. networks) are favourable for ensuring a smooth diffusion and adoption of new technologies.

The reason for which technological change can create spillovers and increasing returns is that technological inputs are non-rival goods. New inventions are produced at a high cost for the first unit but subsequent units (e.g. photocopies) can be produced at virtually zero cost. This generates nonconvexities in production (e.g. Romer 1990) even if such goods are partially excludable (i.e. appropriable) through patents for example. The technological spillover phenomenon is better captured by human capital accumulation or the introduction of new goods rather than by physical capital accumulation. Internal economies of scale may generate some increasing returns to capital, but the scope is likely to be limited. Thus, the new models of growth emphasize formal education (Lucas 1988), public knowledge (Romer, 1986), entrepreneurial imitation (Schmitz 1989) and the introduction of new goods (Stokey 1988).

In each of these cases, the characteristics of dynamic competitive equilibrium can be traced by setting up an optimal control problem similar to the one discussed in the previous section. Where a steady-state exists, the presence of an externality, e.g. through R&D, creates a divergence between the private and social rates of return and the competitive equilibrium may not be Pareto-optimal. While the transitional dynamics of the neoclassical growth model have been researched (see King and Rebelo 1993), they are not easily tractable in the more complex models.

The properties of the steady-state are informative however, about the growth process and the role of policy. To highlight a common feature of most of the endogenous growth models, we will now formulate a simple model of endogenous technological change in which the existence and properties of the steady-state are readily established without having to explicitly solve the underlying dynamic optimization problem.[7] The model forms then the basic framework for discussion of the impact of spatial interaction in the next section.

As in the previous section, technological change is considered a labour-augmenting process. Thus, if N measures the effective labour input, $N = L\,T$, where L is the quantity of workers and T is an index of the average quality of labour input, which depends on the stock of knowledge and practices. The model of the previous section is a special case in which T grows at the exogenous rate g. However, here we relax this assumption. Central to the current view about the process of technological innovation is that a change in T requires a production process with real resource inputs, a multi-product output, its own technology, market structure, spatial differentiation and, indeed, its own changing technology

---

[7]Solow (1994:49) has in fact argued that the intertemporally-optimising representative agent formulation in itself has had little to add to the insights of the New Growth literature.

(e.g. Dosi 1988). Hence we shall assume that a change in T is generated by the following process of knowledge creation:

$$T = H(\frac{R}{L}, T) \qquad (11.12)$$

where R/L is expenditure per worker on activities such as education, training, R&D etc.[8] Thus, the change in T is positively related to the intensity of the effort devoted to the enhancement of labour quality as well as the current level of labour quality. This function is assumed to be homogeneous, of degree one, twice differentiable and concave.

Both the public sector and private sector in the economy carry out knowledge creating activities, funded through taxes and retained profits respectively. For simplicity, we lump these activities together and assume that a fraction m of national income is allocated to the process of technical change. Hence,

$$R = m \, Y \qquad (11.13)$$

As in the case of the accumulation of physical capital, a trade-off arises in that a large value of m reduces current consumption, but yields a higher level of output in the future. If the production function relating effective per capita output to effective per capita capital is the same as in the previous section, i.e. $\hat{y} = f(\hat{R})$, then it is easy to see that

$$\tilde{T} = H(\frac{R}{LT}, 1) = H(m\hat{y},1) \equiv h(mf(\hat{k})) \qquad (11.14)$$

Households now maximize lifetime utility according to equation (11.5) as before, but consumption per capita c at any time cannot exceed $f(k) - \dot{k} - m \, f(k) - \delta \, k$. There are now two decision variables: the propensity to save and the propensity to allocate resources to technical change. It can be shown that in this type of model both propensities will be constant on a steady-state growth path.[9] Thus, as in the standard neoclassical model, income per head will grow in the steady state at a constant rate, but (11.14) shows that this rate is now a function of both m (the proportion of resources devoted to education, innovation etc.), and $\hat{R}$ (the effective capital intensity).

It is straightforward to show how savings behaviour affects $\hat{R}$ and, therefore, the rate of growth of income per capita. The propensity to save for physical capital

---

[8]This equation is a generalisation of a model of endogenous technical change proposed by Conlisk (1967), who assumed that dT/dt would be a linear function of Y/L and T.

[9]See e.g. Lucas (1988:17-27).

accumulation, s, is by definition equal to

$$s \equiv \frac{\dot{K} + \delta K}{Y} \qquad (11.15)$$

Now consider the steady-state growth path on which s and m are constant. If we assume that the labour input L again grows at an exogenous rate, n, we can derive a 'fundamental growth equation' similar to equation (11.4) for the Solow-Swan model. For given s and m, the path of the effective capital intensity $\hat{k}$ is given by

$$\dot{\hat{k}} = s \, f(\hat{k}) - \delta \, \hat{k} - [n + h(m \, f(\hat{k}))] \, \hat{k} \qquad (11.16)$$

The long-run equilibrium level of the effective capital intensity is given by $\hat{k}^*$ for which $\dot{\hat{k}} = 0$. Under the specified conditions, such an equilibrium exists and is stable.[10] Inspection of Figure 11.1, which depicts the components of the rate of growth $\tilde{\hat{k}}$, will make this clear.

It may be surprising that the 'positive feedback loop' from output growth to technological change is still consistent with a steady state. In this model this is due to diminishing marginal returns to physical capital, combined with constant returns to scale. At very high levels of effective capital intensity, capital accumulation does not proceed fast enough to accommodate growing effective labour input. This reduces the capital intensity. However, if we allow increasing returns to scale in production, ever increasing growth rates emerge. An example is Romer (1986). Romer justifies such a model by the observation that in the very long-run (over the last three centuries) worldwide labour productivity growth has been accelerating, although it is equally true that during the last forty years productivity growth has exhibited a downward trend (Romer 1989). Nijkamp and Poot (1993b) formulate a model of increasing returns in which such ever increasing growth is eventually checked by technological, social and economic capacity constraints.

The merit of the simple model we discussed above is that it shows how endogenous technical change generates a link between thriftiness and per capita growth: if, for example, a removal of tax distortions raises the optimal steady-state savings ratio, the per capita growth rate becomes permanently higher. By means of Figure 11.1 it can be easily established that

---

[10]This follows from the Inada assumptions regarding f(k) and the concavity of H(.,.), which implies strict concavity of h(.).

i   If the savings ratio s increases, $\check{R}^*$ and the rate of growth of output per worker h(m f($\check{R}^*$)) increase;

ii   If the rate of depreciation δ increases, $\check{R}^*$ and the rate of growth of output per worker h(m f($\check{R}^*$)) decrease;

iii If the rate of growth of labour supply n increases, aggregate output will grow faster, but $\check{R}^*$ and the rate of growth of output per worker decline;

iv If the optimal proportion of income devoted to the production of technical change m increases, the growth rate of aggregate output increases, $\check{R}^*$ decreases, but here the rate of growth of output per capita increases.

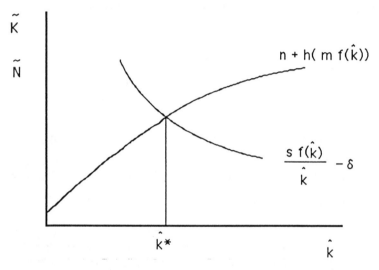

Fig. 11.1.   The effective capital intensity on the steady-state growth path in a model with endogenous technical change

It should be noted that the rate of population growth, n, is assumed here to be exogenous. Theories have been formulated which explain fertility decisions in an intertemporal optimization framework similar to that for the neoclassical growth model. What is required for this is a notion of dynastic utility: parents care about the utility attained by their children (see Becker and Barro 1988). Becker et al. (1990) show that if endogenous fertility is combined with human capital accumulation and the latter process exhibits increasing returns, multiple steady-states emerge: an undeveloped steady state with little human capital and high fertility and a developed state with growing human capital and low fertility. As this is a common feature of these increasing returns models, it suggests that historical endowments and luck are critical determinants of the differentials in growth observed between countries or regions.

11.4. Technological Change and Spatial Interdependency

The analysis has been thus far confined to the case of the closed economy. The importance of trade, capital flows, the diffusion of product and process innovations and net migration at the interregional and international levels, suggest that spatial interactions need to be explicitly considered, both in terms of their direct effects on growth and their effects on technological change. In this section we address these issues briefly by considering, in turn: factor mobility, diffusion and trade.

11.4.1 Factor Mobility

If interregional differences in technological change generate interregional differences in growth rates, a reallocation of production factors may be expected. In the neoclassical model, such a reallocation would generate a convergence in the rate of technological change. This can be easily demonstrated by means of the model of the previous section. If production factors are paid their marginal product and the effective amount of capital per worker is $\hat{k}$, the real rate of return on

capital is $f'(\hat{k})$ and the real wage at time t is $e^{h(mf(\hat{k}))t} [f(\hat{k}) - \hat{k} f'(\hat{k})]$. Thus, net

capital movements would be in the direction of low income regions (with low values of $\hat{k}$) while net labour migration would be in the direction of high income regions. It is well known however, that capital does not always flow from rich to poor countries as the neoclassical model predicts. It can be argued nevertheless that real rates of return differentials are in fact very small because of significant differences between countries in human capital accumulation, the external benefits of human capital and capital market imperfections. (Lucas 1990). Similarly, human capital also migrates from places where it is scarce to places where it is abundant (this is often referred to as the 'brain drain' effect), rather than vice versa (Lucas 1988).

Let us consider labour migration to suggest an explanation for such observations. Separating the effect of 'natural' growth and migration, the change in labour supply is given by

$$\dot{L} = n L + M \qquad (11.17)$$

in which net migration M may be assumed to be given by

$$\frac{M}{L L^f} = q(w - w^f) \qquad (11.18)$$

in which q measures the speed of response of the imperfectly mobile production factor, labour, to a real wage differential. Job search models of migration readily lead to such a specification (Borjas 1989, Poot 1993). Combining (11.17) and (11.18) with the earlier model, the fundamental growth equation now becomes (in growth rate form):

$$\tilde{\hat{k}} = \frac{s\, f\, (\hat{k})}{\hat{k}} - \delta - n - q\, \{\, e^{h(mf(\hat{k}))t}\, [\, f(\hat{k}) - \hat{k}\, f'(\hat{k})]\, w^f\} \, L^f - h(m\, f(\hat{k}))$$

(11.19)

Equation (11.19) shows how migration could act as an equilibrating mechanism: in low income regions the real wage would be low and grow slowly (due to a lower rate of technological change). Outward migration raises the effective amount of capital per worker $\hat{k}$ and therefore raises the level of the real wage as well as its rate of change. The reverse is true for high income regions. The parameter q would measure how fast this process is.

However, the problem with this model is that it ignores the effect of migrants themselves on technological change. If migrants provide new ideas and encourage investment which embodies new technologies, there are dynamic gains from inward migration not captured in (11.19). These effects may lead to the "brain drain" type of migration referred to earlier. Empirical studies suggest that there are indeed dynamic gains from migration. This implies that migration may have a slightly diverging rather than converging effect on real income disparities. Examples from the literature on international migration are Poot et al. (1988), Simon (1989) and Withers (1991). This literature suggests that net immigration in developed countries has raised per capita incomes. Regional studies suggest that migrants on balance move in the 'right' direction, but that this reallocation does not reduce interregional disparities, e.g. Van Dijk et al. (1989). These observations would be consistent with a 'cumulative causation' process rather than neoclassical convergence.

11.4.2 Diffusion

Introducing diffusion in our simple model of endogenous technological change provides additional insights regarding the growth paths. Diffusion analysis has in recent years become an important field of research in industrial economics. At the micro-level, it not only focuses on the distribution and adoption of new technologies (see Brown 1981, Soete and Turner 1984), but also on business services and networks related to technological transformations (Cappellin 1989). In most diffusion studies the S-shaped curve forms a central component (see Davies 1979, Metcalfe 1981, and Morrill et al. 1988). Both the adoption time and the adoption rate can be pictured in this curve. The precise shape of the S-curve can then be explained from firm size, market structure, profitability of innovations etc. (see Kamien and Schwartz, 1982). An important negative role can be played in this context by barriers to information transfer in a multi-region system (see Giaoutzi and Nijkamp, 1988).

At the macro level, we have simply interpreted technology as the 'know how' augmenting the labour input. Equation (11.14) can be modified to explicitly consider the transmission of the accumulation of 'know how' from one region to another (see also Nijkamp et al. 1991):

$$\tilde{T} = h(\ m\ f(\hat{k})) + d\ h(\ m^f\ f(\hat{k}^f)) \qquad (11.20)$$

where d is a diffusion parameter, which for simplicity is assumed constant.[11]

It is straightforward to show that diffusion is compatible with a steady-state in which both regions could grow at different rates. The equilibrium effective capital intensities $\hat{k}^*$ and $\hat{k}^{f*}$ can then be found as the steady-state solution to the simultaneous differential equations:

$$\dot{\hat{k}} = s\ f\ (\hat{k}) - [\delta + n + h(m\ f(\hat{k})) + d\ h(\ m^f\ f(\hat{k}^f))]\hat{k} \qquad (11.21a)$$

$$\dot{\hat{k}}^f = s^f\ f\ (\hat{k}^f) - [\delta^f + n^f + h(\ m^f\ f(\hat{k}^f)\ ) + d\ h(\ m\ f(\hat{k})\ )\ ]\hat{k}^f$$

$$(11.21b)$$

The existence and stability of a solution $(\hat{k}^*, \hat{k}^{f*})$ depends on the properties of the functions f and h and the values of the parameters. One situation is depicted in Figure 11.2. The curve $d\hat{k}/dt = 0$ in the figure represents the locus of points $(\hat{k}, \hat{k}^f)$ at which the first region experiences steady-state growth (given by equation 11.21a), while the curve $d\hat{k}^f/dt = 0$ similarly defines steady-state points for the second region. Given the assumptions for f and h, these curves are both downward sloping and concave. Figure 11.2 shows a situation of global stability. Hence, in this model of technological change and diffusion, there can be persistent differences in regional growth rates.

If the diffusion parameter is very large, 'overshooting' may take place and the trajectories of the effective capital per worker could become unstable. This situation is illustrated in Figure 11.3, where the steady-state is a saddle point. Here, for example, starting from situation $(\hat{k}_0, \hat{k}^f_0)$ the effective capital intensity in both regions begins to grow at a diminishing rate until $d\hat{k}/dt = 0$, but $d\hat{k}^f/dt$ is

---

[11]It is expected that in reality the parameter d varies over time. The literature suggests that this is the result of Schumpeterian swarming effects of new basic technologies and feedback effects from adopted innovations on spatial structures (see Alderman 1989). However, varying d over time would in our simple model introduce an unnecessary complication.

226

then still positive and generates a declining capital intensity in the first region, while $\hat{k}^f$ continues to grow.

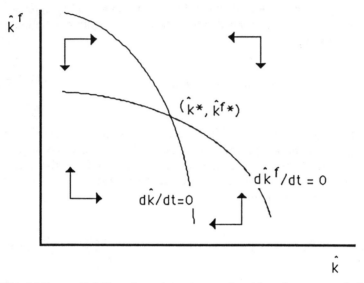

Fig. 11.2. Stability of steady-state growth with endogenous technical change and diffusion

However, if there exists a steady-state in which the growth rates differ because the equilibrium capital intensities differ, there will be a persistent, and constant, difference in the rate of return on capital and an increasing real wage gap unless migration and capital movements (in opposite directions) are significant enough to reduce the factor price gaps.

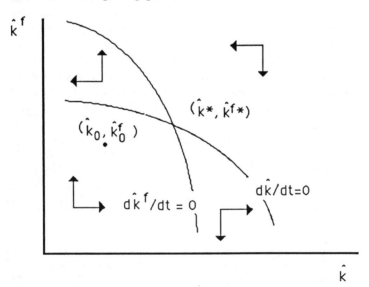

Fig. 11.3. The case of unstable effective capital intensities

In conclusion, factor mobility in this model has the usual equilibrating effect of bringing capital intensities closer, but large values of the diffusion parameter d can have a de-stabilizing influence in terms of generating diverging growth.[12] In the circumstances in which a steady-state exists, it is easy to identify the benefits of diffusion: compared with the situation of autarky, the equilibrium effective capital intensity is lower, the rate of return to capital is higher and income per capita grows at a faster rate.

11.4.3 Trade

In reality, trade, factor mobility and diffusion of technical change occur simultaneously. Freeman (1988) noted that trade theory offers two quite different views of the interrelationship between these flows. On the one hand, the Heckscher-Ohlin theory suggests that trade and factor mobility are substitutes in achieving factor price equalization and a final equilibrium with a static allocation of factors. However, if trade results from differences in technology between regions, the flows are likely to be complementary. Increases in net migration can then increase trade and generate capital inflows as well.

An extensive discussion of innovation and growth in open economies is given by Grossman and Helpman (1991). Here we will focus on one key issue, namely whether such openness will lead to interregional convergence or not. As before, we commence with the standard neoclassical model and then introduce endogenous technological change.

The traditional neoclassical trade-and-growth model, e.g. Oniki and Uzawa (1965), suggests that two trading regions (in which the rate of growth in labour supply is identical) would, under standard conditions, move towards a long-run balanced growth path. The two regions grow on this path at identical rates and the pattern of specialization is determined by the equilibrium factor intensities, i.e. the regions would produce more of the good which uses the abundant production factor more intensively. The extension of this two-good two-factor model to incorporate endogenous technical change along the lines discussed earlier is straightforward. In the trade model, there are two goods: a consumption good and an investment good. The consumption good is chosen as numéraire, whilst the price of the investment good (i.e. the terms of trade) is p. Under standard neoclassical conditions, domestic product per capita is fully determined by the effective capital intensity $\hat{k}$ and p, i.e. $y = y(\hat{k},p)$ in each region. The demand for the investment good is given by

$$p \, I = s \, Y \qquad (11.22)$$

---

[12]Simulations with a three-region model with endogenous technical change and spatial diffusion confirmed the possibility of such a situation (see Nijkamp et al. 1991).

and since again $\tilde{k} = \tilde{K} - \tilde{L} - \tilde{T}$ we get here

$$\dot{k} = s\ y\ (\hat{k},p)\ /\ p\ -[\delta + n + h(\ m\ y(\hat{k},p)\ ) + d\ h(\ m^f\ y(\hat{k}^f,p)\ )]\ \hat{k}$$

(11.23a)

$$\dot{k}^f = s^f\ y^f\ (\hat{k}^f,p)\ /\ p\ -\ [\delta^f + n^f + h(\ m^f\ y^f(\hat{k}^f,p)\ )\ +\ d\ h(\ m\ y(\hat{k},p)\ )\ ]\ \hat{k}^f$$

(11.23b)

At any point in time, for given labour forces L, L^f and average labour qualities T and T^f, the terms of trade are uniquely determined, in Heckscher-Ohlin fashion, by the existing capital intensities ($\hat{k}$, $\hat{k}^f$), through assuming trade equilibrium. This implies that p, the price of the investment good in terms of the consumption good, is given by

$$P = \rho\ (\hat{k},\hat{k}^f\ |\ L,\ L^f,\ T,\ T^f)$$

(11.24)

The equations (11.22) to (11.24) completely specify, from any given factor endowments and initial technologies T_0 and T^f_0, the growth of both regions and the pattern of specialization. However, it is easy to see that steady-state capital intensities $\hat{k}*$ and $\hat{k}^f*$ will exist if and only if both regions grow at the same rate, otherwise p continues to change. This requires that

$$n + h(m\ y(\hat{k}^*,p^*)\ )\ +\ d\ h(\ m^f\ y^f(\hat{k}^{f*},p^*)\ ) =$$
$$= n^f + h(\ m^f\ y^f(\hat{k}^{f*},p^*)\ )\ +\ d\ h(\ m\ y(\hat{k}^*,p^*)\ )$$

(11.25)

This is the generalization of the usual assumption that the natural growth rate in both regions must be identical for the existence of a steady-state. While differences in the growth rates of labour supply n and n^f may indeed be small between regions, the introduction of endogenous technical change is clearly a disequilibrating factor in a trade model, since it is not necessary that the steady-state solution of (11.23) and (11.24) would satisfy (11.25) also. Hence, from any given starting position, the growth rates of effective labour supply in the two regions may not converge.

It is obvious from equation (11.20) that with a positive parameter d, the process of innovation diffusion increases the rate of growth in income per head. There are similarly dynamic gains from trade. These are not captured in the trade model above. However, the dynamic gains from economic integration have been formally modelled by Rivera-Batiz and Romer (1991), who show that integration of regional economies raises the average growth rate. It is nonetheless possible that a specialization based on comparative advantage leads to a sub-optimal investment in R&D activities by resource rich economies (Grossman and Helpman 1994).

The neoclassical trade model assumes that labour-augmenting technical change affects both the consumption and the investment goods sectors equally. It would be more realistic to assume that labour productivity improvements could vary between sectors, or that a trade advantage is generated by product innovation described in the product cycle theory (Vernon, 1966, Krugman, 1979). Alternatively, it may be the level of activity in specific sectors which provides a 'learning by doing' spillover benefit for the whole economy. In this case it is straightforward to show that an increase in the supply of the resource used intensively in the knowledge generating sector speeds up growth (Grossman and Helpman 1990a). Similarly, the market allocation of resources to this sector is suboptimal because firms do not take a spillover benefit into account. Not surprisingly, the presence of a positive externality implies that subsidizing the R&D sector improves welfare.

In an open economy context, the capture of spillover benefits from other regions increases growth, but what matters from the policy perspective is which of the regions has a comparative advantage in the R&D sector. If subsidies are given to regions which are better at manufacturing rather than innovating, the overall growth rate may decline.

A more in-depth analysis of comparative advantage and long-run growth is provided by Grossman and Helpman (1990b). In their model, there are three sectors: an R&D sector, which produces blueprints for new products and also generates increases in the stock of knowledge; an intermediate goods sector and a final consumption goods sector. Resources devoted to R&D raise the number of available varieties of differentiated inputs in final production and this in turn raises total factor productivity. If this model is applied to two regions, each with fixed primary resources, a steady-state growth rate can be computed and its sensitivity to policies analysed. For example, a small R&D subsidy in both regions increases the rate of growth, while a national trade policy that switches spending toward the consumer good produced by the region with comparative advantage in R&D will cause long-run growth rates to decline. Grossman and Helpman (1990b) note that, as in our simple model discussed above, diffusion can have a significant effect on the long-run growth rate.

When technical change is positively related to output, the model generates a feedback mechanism in which production exhibits increasing returns to scale. It is intuitively clear that 'uneven development' is a necessary outcome of such a situation: an initial discrepancy in capital-labour ratios between regions will be reinforced over time.

Trade specialization may also generate such uneven development. An example of such a situation is a model formulated by Krugman (1981), which leads to a phase diagram similar to Figure 11.3. Krugman assumes that two products, an agricultural good and a manufactured good, can be produced by means of Ricardian production techniques, with increasing external economies of scale. Such external economies are of course often empirically indistinguishable from technical change. In either case, the technical coefficients representing the input requirements per unit of output decline as the capital stock increases. In this situation the region with the larger initial capital stock has the higher profit rate and, if all profits are saved, generates the fastest capital accumulation. The result is an ever-increasing divergence between the regions, which only ends when a boundary of some kind has been reached. Krugman assumed this to be a limit to labour supply.[13]

Moreover, factor mobility and commodity trade may reinforce each other through technical change. Lucas (1988) suggested that a difference in human capital accumulation is one of the main causes of a difference in growth rates between regions or countries. Different goods have different potentials for human capital growth through on-the-job training or through learning-by-doing. Consequently, the comparative advantage which determines which goods get produced also determines the rate of growth in human capital (and therefore technical change). Lucas' (1988) model of trade and growth has features similar to Krugman's (1981) model, although the increase in the efficiency of the Ricardian production technology in the former is due to human capital accumulation through learning by doing, rather than economies of scale through physical capital accumulation. Nonetheless, if two goods are produced which are "good" substitutes (i.e. they have a substitution elasticity greater than one), there will be a tendency for complete specialization with the direction of specialization determined by the initial conditions.

The immediate implication is that policy should encourage the creation of initial conditions on the growth path which take the possibility of a technological comparative advantage into account. To ensure that more resources are devoted to goods with a high learning-by-doing propensity, an industrial policy of 'picking winners' would appear helpful. The introduction of trade in this framework also generates complete specialization. Over time, the terms of trade change continuously to reinforce the pattern of comparative advantage. Provided that the goods concerned are good substitutes, regions which produce the good which enjoys a faster technical change will continue to have a higher growth rate, resulting in a continuing change in the terms of trade. Thus, this dynamic trade model again suggests a persistent pattern of uneven development.

There is of course, in the regional growth literature, a fairly long tradition of expounding uneven development, for example, in Myrdal's (1957) cumulative causation theory.[14] The current challenge in this type of modelling is to be able to endogenize changes in the position of individual regions in this growth continuum. Possibilities for such growth switches would include - on the demand side - the introduction of different income elasticities for different classes of goods; and on the supply side the continuing introduction of new goods, with

---

[13]In another paper, we point to the eventual emergence of external diseconomies as the long-run constraint (see Nijkamp and Poot 1993b).

[14]Features of cumulative causation such as imperfect competition, increasing returns to scale and product differentiation have emerged as central themes of the 'new international economics' of which the models mentioned in this section are examples (see also Krugman 1988).

learning potentials declining with the amount produced. Such factors could continuously shake up the existing pattern of specialization and explain why, for example, the rapid growth in NICs has been associated with a growth of exports in products initially not produced in these countries. Kaldor's (1961) stylized facts of growth, which included the observation of a large inter-country variance in productivity growth rates, led at the regional level to a search for explanations in terms of export demand and patterns of specialization. Because this literature was influential in traditional thinking about regional policies, it is useful to contrast in the next section this approach with the supply and technology-oriented models discussed earlier.

11.5 Divergence from a Kaldorian Perspective

As noted above, the conventional regional growth literature has tended to proceed along Keynesian lines with a heavy emphasis on demand considerations. Thus, in the well known Kaldor-Dixon-Thirlwall model, output growth in a region is driven by relative competitiveness and income growth outside the region (Dixon and Thirlwall 1975). In such an export-led growth model, the only role played by supply-side factors relates to the effects of cost inflation and productivity on relative competitiveness, with the latter effect being generated by means of Verdoorn's law (Verdoorn 1949). This model explains differences in equilibrium growth rates between regions in terms of differences in price and income elasticities in the demand for exports and differences in rates of autonomous productivity growth.

In more formal terms, output growth is assumed to be export-led:

$$\tilde{Y} = \varphi \, \tilde{X} \qquad (11.26)$$

where $\sim$ refers again to a rate of growth and X to the volume of exports. The export demand function has constant price and income elasticities

$$\tilde{X} = - \, \eta \, \tilde{p} + \xi \, \tilde{p}^f + \tau \, \tilde{Y}^f \qquad (11.27)$$

Price inflation results from fixed mark-up pricing on production costs, which in turn depend on unit wage costs w and labour productivity. Thus, in rate of change terms

$$\tilde{p} = \tilde{w} - \tilde{y} \qquad (11.28)$$

Central to this growth model is that labour productivity is partly dependent on growth of output itself, i.e. Verdoorn's Law:

$$\tilde{y} = \beta + \lambda \ \tilde{Y} \qquad (11.29)$$

An extensive literature exists regarding the empirical evidence for this relationship (reviewed in, for example, Bairam 1987), which suggests that the observed relationship may be the result of simultaneous responses in output and labour markets to changes in demand, combined with the effects of economies of scale and technical progress. Naturally, a simultaneous equation approach is required for empirical estimation of the parameters in (11.29). By and large, the empirical evidence suggests that λ is positive. Hence dynamic increasing returns appear to exist, but what is missing here is an explicit specification of economies of scale and the link between output growth and technological change. Moreover, Skott (1989) noted that the link between productivity and competitiveness implied by the Verdoorn relationship and the export demand function is too strong: nominal wages could react to an increase in productivity, which could partly offset the effect of productivity growth on export growth.

The reduced form of the model (11.26)-(11.29) is readily computed and suggests a constant rate of growth of income per worker:

$$\tilde{y} = \beta + \frac{\lambda \varphi \ [\ -\eta \ (\tilde{w} - \beta) + \xi \ \tilde{p}^f + \tau \ \tilde{Y}^f]}{1 - \lambda \ \varphi \ \eta} \qquad (11.30)$$

This model has unrealistic implications if it is considered in an explicit two region situation in which income growth in either region affects growth in the other region through trade between them. It is fairly straightforward to compute the reduced form for the per capita income growth rates in both regions, but depending on the choice of parameters, these growth rates could obviously differ and would suggest a persisting trade imbalance (Nijkamp and Poot 1993a). Krugman (1989) noted that long-run balance of payments equilibrium in such a regional growth-and-trade framework necessitates a strict relationship between differences in growth rates between regions on the one hand and income elasticities of the demand for exports and imports on the other.[15] The Kaldor-Dixon-Thirlwall model is itself not informative about the processes which would ensure that the growth rates generated by this model would be consistent with long-run balance of payments equilibrium.

If, for example, technical change proceeds at a different pace in two regions, growth in the more innovative region could be hampered by lower demand for its

---

[15]Interestingly, the latter condition appears indeed consistent with international trade data, i.e. countries which grow fast tend to experience a high income elasticity of the demand for their exports, while the income elasticity of their demand for imports is low.

output from the less innovative, and less competitive, region. Indeed, if the Verdoorn effect is strong enough, a situation of 'immiserizing' growth may be generated in which a detrimental shock in the trading partner's economy (e.g. a rapid growth in nominal wages) is more than compensated for by the negative effect on the local economy.

The model discussed above does not explicitly take into account the possibility of migration between the regions, nor the diffusion and adoption of technological innovation. These phenomena cannot be readily introduced here. For example, net migration would respond to the difference in growth rates in per capita incomes, but the latter are again likely to be themselves affected by net migration. Moreover, production capacity limits are assumed here to be unimportant. In essence, the model describes the properties of a demand-driven steady state growth path rather than full dynamics.[16] Yet it does make it explicit that an exogenous shock to trade can have a long-term impact on the equilibrium growth rate, although our discussion suggests that the introduction of simple explicit feedback effects (here, aggregate demand and relative competitiveness) can strongly modify behaviour.[17]

11.6 Conclusions

The conceptual framework discussed in this paper served to identify and explore new departures for the analysis of economic dynamics in an open system, with a specific focus on spatial interdependencies in the form of trade, factor mobility and innovation diffusion. Many new models have been proposed in the literature. Such models capture one or more of the important features of development: sectoral composition, human and physical capital accumulation, natural endowments, economies of scale, trade, technological innovation and diffusion, factor mobility, government policies and market imperfections. However, the design of a coherent and unified framework appears to be far from easy. Moreover, most of the new theories still require extensive empirical scrutiny. In empirical work it will be important to distinguish between transitional dynamics and long-run steady-state tendencies.

Both the export-led growth model and the general equilibrium models considered in this paper have the ability to generate persisting differences in long-run growth rates in the presence of some spatial interdependency, provided there are barriers

---

[16]It is possible to introduce lags in the behavioural equations. Dixon and Thirlwall (1975) showed that the introduction of one period lags in the export demand function still generates convergence to the equilibrium growth rate for plausible values of the elasticities.

[17]This is a general conclusion for models of interdependent regions. See also, for example, the models which have been developed by Frenkel and Razin (1987) to describe the effects of fiscal policies and monetary conditions on equilibrium output in a 'two-region world'.

to other types of flows. However, in the presence of endogenous technical change generating increasing returns there is a tendency for a highly interdependent system to be unstable, with a likelihood of 'uneven development'. While the new growth models offer interesting and appropriate foundation stones for a thorough analysis of the evolutionary patterns of a multi-regional system, it is obvious that much work in this area remains to be done, both at the micro and macro levels.

For example, the locational aspects of R&D creation, diffusion and adoption deserve much closer attention, as is highlighted in various recent OECD reports on technology policy. To some extent, this issue is comparable to the infrastructure debate presented, among others, by Biehl et al. (1986) and Nijkamp (1986). Production theories may be used to assess the implications of a favourable infrastructure in particular regions with respect to differential competitiveness. In our context, a regional dynamization of a production function, accompanied by a technological diffusion function with parameters dependent on information barriers on the one hand and competitive behaviour on the other, would provide a promising starting-point. Changing trade patterns, factor flows and public policies could then be incorporated to identify the long-run growth tendencies of the regions in the system.

References

Aghion P and P Howitt (1992) A Model of Growth through Creative Destruction, Econometrica 60(2):323-351

Alderman N (1989) Models of Innovation Diffusion. In: Cappellin R and P Nijkamp (Eds) Theories and Policies of Technological Development at the Local Level. Gower Aldershot

Armstrong H and J Taylor (1985) Regional Economics and Policy. Philip Allen Oxford

Arthur W B (1990) Positive Feedbacks in the Economy. Scientific American. February 80-85

Bairam E I (1987) The Verdoorn Law, Returns to Scale and Industrial Growth. A Review of the Literature. Australian Economic Papers. 26:20-42

Barro R J and X Sala-i-Martin (1992) Convergence. Journal of Political Economy 100 2:223-251.

Barro R J and X Sala-i-Martin (1995) Economic Growth. McGraw-Hill New York

Baumol W J (1986) Productivity Growth, Convergence and Welfare. What the Long Run Data Show. American Economic Review. 76 5:1072-1085

Becker G S and R J Barro (1988) A Reformulation of the Economic Theory of Fertility. Quarterly Journal of Economics. 103 1:1-25

Becker G S, Murphy K M and R F Tamura (1990) Human Capital, Fertility, and Economic Growth. Journal of Political Economy. 98 5.2:S12-S37

Biehl D, et al (1986) The Contribution of Infrastructure to Regional Development. Brussels. Commission of European Communities

Borjas G J (1989) Economic Theory and International Migration. International Migration Review. 23 3:457-485

Brown L A (1981) Innovation Diffusion. Methuen London

Cappellin R (1989) The Diffusion of Producer Services in the Urban System. In: Cappellin R and P Nijkamp (Eds) Theories and Policies of Technological Development at the Local Level. Gower Aldershot

Cass D (1965) Optimum Growth in an Aggregative Model of Capital Accumulation, Review of Economic Studies. 32 3:233-240

Conlisk J (1967) A Modified Neo-classical Growth Model with Endogenous Technical Change. Southern Economic Journal. 34:199-208

Davelaar E J and P Nijkamp (1989) Spatial Dispersion of Technological Innovation. Journal of Regional Science. 29 3:325-346

Davies S (1979) The Diffusion of Process Innovations. Cambridge University Press Cambridge

De Long J B (1988) Productivity Growth, Convergence and Welfare. Comment. American Economic Review. 78 5:1138-1154

Dixon R and A P Thirlwall (1975) A Model of Regional Growth-Rate Differences on Kaldorian Lines. Oxford Economic Papers. 27:201-214

Dosi G (1988) Sources, Procedures, and Microeconomic Effects of Innovation. Journal of Economic Literature 88 3:1120-1171

Dowrick S and D Nguyen (1989) OECD Comparative Economic Growth 1950-85: Catch-Up and Convergence. American Economic Review. 79 5:1010-1030

Easterlin R A (1960) Interregional Differences in Per Capita Income, Population and Total Income. 1840-1950. In: Conference on Research in Income and Wealth. NBER Studies in Income and Wealth 24

Ehrlich I (1990) The Problem of Development: Introduction. Journal of Political Economy 98 5.2:S1-S11

Freeman R (1988) Immigration, Trade and Capital Flows in the American Economy. In: Baker L and P Miller (Eds) The Economics of Immigration. Australian Government Publishing Service. Canberra

Frenkel J A and A Razin (1987) Fiscal Policies and the World Economy. An Intertemporal Approach. MIT Press Cambridge, Mass

Giaoutzi M and P Nijkamp (Eds) (1988) Informatics and Regional Development. Gower Aldershot

Grossman G M and E Helpman (1990a) Trade, Innovation and Growth. American Economic Review. 80 2:86-91

Grossman G M and E Helpman (1990b) Comparative Advantage and Long-Run Growth. American Economic Review. 80 4:796-815

Grossman G M and E Helpman (1991) Innovation and Growth in the Global Economy. MIT Press Cambridge, Mass

Grossman G M and E Helpman (1994) Endogenous Innovation in the Theory of Growth. Journal of Economic Perspectives. 8 1:23-44

Inada K (1963) On a Two-Sector Model of Economic Growth: Comments and A Generalization. Review of Economic Studies. pp 119-127

International Monetary Fund (various years) International Financial Statistics. Washington DC.

Kaldor N (1961) Capital Accumulation and Economic Growth. In: Lutz F A and D C Hague (Eds) The Theory of Capital. St Martin's Press New York

Kamann D J F (1988) Spatial Differentiation in the Social Impact of Technology. Gower Aldershot

Kamien M I and N L Schwartz (1982) Market Structure and Innovation. Cambridge University Press Cambridge

King R G and S T Rebelo (1993) Transitional Dynamics and Economic Growth in the Neoclassical Model. American Economic Review. 83 4:908-931

Krugman P (1979) A Model of Innovation, Technology Transfer, and the World Distribution of Income. Journal of Political Economy. 87:253-266

Krugman P (1981) Trade, Accumulation and Uneven Development. Journal of Development Economics. 8:149-161

Krugman P (Ed) (1988) Strategic Trade Policy and the New International Economics. MIT Press Cambridge Mass

Krugman P (1989) Income Elasticities and Real Exchange Rates. European Economic Review. 33:1031-1054

Krugman P (1991) Geography and Trade. MIT Press Cambridge, Mass

Levine R and D Renelt (1992) A Sensitivity Analysis of Cross-Country Growth Regressions. American Economic Review. 82 4:942-963

Lucas R E (1988) On the Mechanics of Economic Development. Journal of Monetary Economics. 22 1:3-42

Lucas R E (1990) Why Doesn't Capital Flow from Rich to Poor Countries? American Economic Review. 80 2:92-96

Malecki E J and P Nijkamp (1988) Technology and Regional Development. Some Thoughts on Policy. Environment and Planning. C6:383-399

Mankiw N G, Romer D and D N Weil (1992) A Contribution to the Empirics of Economic Growth. Quarterly Journal of Economics. 107 May:407-438

McCombie J S L (1988a) A Synoptic View of Regional Growth and Unemployment. I - The Neoclassical Theory. Urban Studies. 25 4:267-281

McCombie J S L (1988b) A Synoptic View of Regional Growth and Unemployment. II - The Post-Keynesian Theory. Urban Studies. 25 5:399-417

Metcalfe J S (1981) Impulse and Diffusion in the Study of Technical Change. Futures. 13:347-359

Morrill R, Gaile G L and G I Thrall (1988) Spatial Diffusion. Sage Beverly Hills

Myrdal G (1957) Economic Theory and Underdeveloped Regions. Duckworth London

Nijkamp P (1986) Infrastructure and Regional Development. A Multidimensional Policy Analysis. Empirical Economics. 11:1-21

Nijkamp P (1991) Regional Economic Growth and Regional Policy. A European Perspective. In: Evans L, Poot J and N Quigley (Eds) Long-Run Perspectives on the New Zealand Economy. New Zealand Association of Economists Wellington

Nijkamp P and J Poot (1993a) Technological Progress and Spatial Dynamics. A Theoretical Reflection. In: Kohno H and P Nijkamp (Eds) Potentials and Bottlenecks of Spatial Economic Development. Springer Verlag Berlin

Nijkamp P and J Poot (1993b) Endogenous Technological Change, Innovation Diffusion and Transitional Dynamics in a Nonlinear Growth Model. Australian Economic Papers. 32 4:191-213

Nijkamp P, Poot J and J Rouwendal (1991) A Nonlinear Dynamic Model of Spatial Development and R&D Policy. Annals of Regional Science 25:287-302

Oniki H and H Uzawa (1965) Patterns of Trade and Investment in a Dynamic Model of International Trade. Review of Economic Studies. 32:15-38

Passinetti L and R Solow (Eds) (1994) Economic Growth and the Structure of Long-Term Development. Macmillan London

Poot J (1993) Trans-Tasman Migration and Economic Growth in Australasia. In: Carmichael G (Ed) Trans-Tasman Migration: Trends, Causes and Consequences. Australian Government Publishing Service. Canberra

Poot J, Nana G and B Philpott (1988) International Migration and the New Zealand Economy. A Long-Run Perspective. Victoria University Press Wellington

Porter M (1990) The Competitive Advantage of Nations. Macmillan London

Rebelo S (1991) Long-Run Policy Analysis and Long-Run Growth. Journal of Political Economy. 99 3:500-521

Rivera-Batiz L A and P M Romer (1991) Economic Integration and Endogenous Growth. Quarterly Journal of Economics. 106 2:531-556

Romer P M (1986) Increasing Returns and Long-Run Growth. Journal of Political Economy. 94 5:1002-1037

Romer P M (1989) Capital Accumulation in the Theory of Long-Run Growth. In: R J Barro (Ed) Modern Business Cycle Theory. Harvard University Press

Romer P M (1990) Are Nonconvexities Important for Understanding Growth? American Economic Review. 80 2:97-103

Romer P M (1994) The Origins of Endogenous Growth. Journal of Economic Perspectives. 8 1:3-22

Schmitz J A (1989) Imitation, Entrepreneurship, and Long-Run Growth. Journal of Political Economy, 97 3:721-739.

Scott M (1989) A New View of Economic Growth. Oxford University Press Oxford

Simon J L (1989) The Economic Consequences of Immigration. Basil Blackwell Oxford

Skott P (1989) Kaldor's Laws, Cumulative Causation and Regional Development. Paper presented at the 29th European Congress of the Regional Science Association. Cambridge UK

Soete L and R Turner (1984) Technology Diffusion and the Rate of Technical Change. Economic Journal. 94:612-623

Solow R M (1956) A Contribution to the Theory of Economic Growth. Quarterly Journal of Economics. 70:65-94

Solow R M (1988) Growth Theory and After. American Economic Review. 78 3:307-317

Solow R M (1994) Perspectives on Growth Theory. Journal of Economic Perspectives. 8 1:45-54

Stern N (1991) The Determinants of Growth. Economic Journal. 101 January: 122-133

Stokey N L (1988) Learning by Doing and the Introduction of New Goods. Journal of Political Economy. 96:701-717

Summers R and A Heston (1991) The Penn World Table (Mark 5). An Expanded Set of International Comparisons, 1950-1988. Quarterly Journal of Economics. 106 2: 327-368.

Swan T W (1956) Economic Growth and Capital Accumulation. Economic Record 32. Reprinted In: Peter Newman (Ed) Readings in Mathematical Economics. Vol 2. Capital and Growth. Baltimore. Johns Hopkins Press 1968

Van Dijk J, Folmer H, Herzog Jr H W and A M Schlottmann (1989) Migration and Labour Market Adjustment. Kluwer Dordrecht

Verdoorn P J (1949) Fattori che regolano lo Sviluppo della Produttività del Lavoro. Factors Governing the Growth of Labour Productivity L'Industria. 1:3-10. English translation by AP Thirlwall and G Thirlwall. In: Research in Population and Economics 1979

Vernon R (1966) International Investment and International Trade in the Product Cycle. Quarterly Journal of Economics. 80:190-207

Withers G (1991) Economics, Immigration and Interdependence. In: Evans L, Poot J and N Quigley (Eds) Long-Run Perspectives on the New Zealand Economy. New Zealand Association of Economists Wellington

World Bank (various years) World Bank National Accounts. Washington DC

12 Innovation and Strategy in Space: Towards a New Location Theory of the Firm

Maryann P. Feldman and Aydan S. Kutay

531

R32

12.1 Introduction

D21

Traditional location theory suggests that individual firms freely scan the environment and select a location which minimizes production costs (see Garafola and Fogerty 1988). According to this theory, locational advantage reflects conventional natural advantages associated with land, labour and capital. Emphasis is given to production and specifically to the exploitation of economies of scale, but the theory does not accommodate the increased importance of innovation, or the ability to produce higher quality products. In order to engage in innovation, firms must coordinate a variety of activities both within their functional boundaries and with their external environments. An important dimension of innovation is product variety, which is defined as the adaptation of new products to specific market segments. Product variety becomes critical to competitive strategy in a global economy, as it offers firms a way to increase market share through improved time-to-market acceptance. Business strategists have focused on questions of coordination in an increasingly borderless world. Less attention however, has been paid to developing a locational theory which would consider the geographic organization of the firm's activity. Location may offer firms an effective way to organize resources in light of the increased importance of product variety in innovation. In addition, such a theory should also consider the potential of new technologies in communication and production for altering locational dynamics and requirements.

The accelerated rate of growth in information and communication technologies in recent years and the increasing impact of information technology on the way business strategies are formed has redirected attention to the relationship between innovation, technological change and the spatial structure of the industrial landscape. The inadequacy of traditional location theory in explaining the location decisions of firms in the current context has been demonstrated in previous research (see Clark et al. 1988; Kutay 1988). Recent research on the geographic dimensions of innovation focuses on the ways in which particular places have acquired a comparative advantage for innovation and economic development (see Storper and Walker 1988; Scott 1993; Feldman 1994). The theory put forward in this paper suggests that locational advantage reflects the cumulative investments in human and technological capability in specific places. According to this perspective, locational advantage in the capacity to innovate is ever more dependent on the agglomeration of specialized skills, knowledge, institutions, and resources that make up an underlying technological infrastructure.

While the infrastructural perspective on locational advantage differs sharply with traditional location theory, it has not considered the ways in which new communication and information technologies affect and alter the locational requirements of firms or the ways in which firms can best accommodate global markets. Specifically, although these new technologies make physical distance less important, there is evidence that innovative activity tends to concentrate in specific locations. This fact suggests that our conceptualization of location should be redefined from a focus on distance and transportation costs to one which views location as a spatially defined collection of people and resources. Geography can be re-conceptualized as a stage upon which firms may organize resources and develop accumulations of expertise. The attributes and characteristics of places may provide the most relevant conceptualization of location given decreasing transportation costs and increasing access to information.

In this chapter, we suggest that location decisions are an important part of the global strategy of the firm and require a more complex decision-making process than previously considered by location theory. Integral to such a theory is the consideration that today's firms, whether large or small, have to consider competition from not only domestic businesses but also from international businesses. In this chapter, we examine how recent trends in technology and in the world economy affect location decisions and describe how the dynamics of globalization, technological change and management philosophies change location requirements. We propose a new framework for location decisions and suggest a model of the global firm.

Global strategy can be formulated as the optimal configuration of the firm's resources in terms of intellectual capital, manufacturing labour, operational assets, and technology. Today's firms select particular locations in accordance with the requirements dictated by the portfolio of the activities in which the firm is involved. Location provides a stage upon which the firm can organize its varying activities and is a key component of global strategy. Different locations provide unique comparative advantages which can be used in order to satisfy the objectives of the firm. This comparative advantage includes the conventional advantages of land, labour, and capital as well as the elements of the technological infrastructure and depending on the objectives and capabilities of the firm. In addition, each country will have varying consumer tastes and preferences, differing business practices and distribution channels, and unique government policies. A firm may have to customise its product and reconfigure its strategy to compete effectively in different locations. Through this process, the firm develops an understanding of each market in which it sells products. Such knowledge is tacit and specific and cannot be easily transferred from one location to another. In the global economy, the ability to gain from locational advantage is ever more dependent on the strategy of the firm.

The remainder of this chapter is organized as follows. The next section provides a series of considerations which should be incorporated in a new theory of firm locational decision-making. These include the emergence of global markets, new patterns of trade and regional economic integration, technological change in communication and information technologies and technological change in production systems. We then consider how these factors affect firm strategy and how location under these conditions becomes an element of strategy. We then present a framework for location decisions by the firm.

12.2   Considerations for a New Location Theory

Three fundamental developments with implications for location theory of the firm have been observed. First, we are witnessing the globalization of the world markets in which we are moving away from an economic system characterized by national markets as distinct entities, isolated from each other by barriers of trade, distance, time, and culture. As a result, individual firms are increasingly dispersing parts of their production processes to different locations around the globe. This globalization of production is not limited to large firms like Ford or General Motors, but extends to small and medium-sized firms which recognize the importance of global operation as an element of competitive strategy (see Trager 1989). Two major factors seem to underlie the trend toward globalization of the world economy. The first is the decline in barriers to the free flow of goods, services and capital that has occurred since the end of WWII. The second factor is the increasing rate of innovation in communication, information and transportation technologies over the same period. While the lowering of trade barriers made globalization a theoretical possibility, technological change has made it a tangible reality. Each of these factors has strong implications for location theory.

12.2.1   The Emergence of Global Markets

During the past twenty years, almost all major industries have seen the development of a global marketplace for their products. In 1993, more than 70 percent of goods were traded in an international marketplace (see The Economist Yearbook 1993). The size of the world economy in 1994 was $26 trillion. It is projected to grow by 85% by the year 2010, reaching $49 trillion (see Business Week 1994 special issue).

Following the trends of the last decade, the growth rate of world trade is expected to surpass the growth rate of the world economy. World trade is projected to grow by more than 300 percent during the next decade and reach the value of $16.6 trillion in the year 2010. New markets are developing in countries of south East Asia and Eastern Europe. Countries which were once regarded as less developed or developing are now the fastest growing economies in the world. These countries now constitute large markets for leading edge, quality products.

The established industrial powers such as the U.S. or Western Europe no longer possess the fastest growing economies of the world. Japan now has the largest per capital income among all industrialized nations.

Globalization of the world economy would not have been possible without a declining trend in trade and investment barriers. During the 1920s and 1930s many of the nation-states of the world erected formidable barriers to international trade and investment. Many of the barriers took the form of high tariffs on imports of manufactured goods. The typical aim of such tariffs was to protect domestic industries from foreign competition.

Ultimately this depressed the world demand and contributed to the Great Depression of the 1930s. Having learned from this experience, the advanced industrial nations of the west, under U.S. leadership, committed themselves after WWII to removing the barriers to the free flow of goods, services and capital between nations. Under the General Agreement on Tariff and Trade (TAGG), there has been a significant lowering of barriers to the free flow of goods since the 1950s. More recently, attempts have been made to expand GATT to include trade in services. At the same time many countries have been progressively removing restrictions on capital inflows and outflows. In 1991, for example, 34 countries (both industrialized and developing) made changes to their laws governing direct foreign investment to encourage both inward investment by foreign firms and outward domestic firms (see The Economist World Economy Survey 1992).

The development of large and sophisticated markets dictates a global perspective for all firms. Thus, in many industries it is no longer meaningful to talk about the 'American market', the 'French market' or the 'Japanese market'. There is only the 'global market' in which goods from 'Coca-Cola' to 'Perrier water', from 'McDonald's hamburgers' to 'Honda Accords' get 'global' acceptance. As a consequence, it is increasingly difficult to identity a product based on nationality.

A firm might design a product in one country, produce component parts in other countries, assemble the product in yet another country, and export the finished good around the world. Consider, for example, Ford's Mercury Capri. Based upon the comparison of production factor costs and quality in various countries, Ford has dispersed many of the Capri production activities to other countries. The car's body and interior were designed in Italy, its engine and drive train are manufactured in Japan, it is assembled in Australia and the final product is marketed and distributed world-wide. As a result of the emergence of global markets, firms are able to distribute their activities in order to find the optimal location for each activity. In this way, firms can tap into the most appropriate expertise for a specific task regardless of national borders. This dictates a strategy of assessing the requirements of each component activity, finding the most appropriate location and then linking the activities together to produce a product.

12.2.2 New Trade Patterns and Regional Economic Integration

While trade barriers have been declining, protectionist measures are once more on the rise in the U.S. and elsewhere. Trading limitations such as non-tariff barriers (NTBs) and voluntary export restraints (VERs) have found favour as a trade-restricting tools. By placing restraints on local content, sales volume and market share, these measures interfere with a firm's ability to enter a market through export-based strategy and force a shift toward foreign direct investment (FDI). During the 1980s, for example, the percentage of U.S. imports subject to non-tariff barriers increased to 25 percent (Japan Trade Organization 1990). The trend has been even more severe in the European community where the value of trade affected by VERS increased by 60 percent during the 1980s.

During the 1980s, the world economy has shown signs of developing into a tripolar global economy with hubs in Europe, North America and Southeast Asia. Each area composed of member states linking their economies by relatively free internal movements of goods and resources, and the establishment of common standards and coordinated macroeconomic policies. The European community (EC) encompasses fifteen countries and the North American Free Trade Agreement (NAFTA) links the economies of the U.S., Canada, and Mexico. The Pacific Rim countries, centred principally around Japan, with the Association of South-East Asian Nations (ASEAN) has become the fastest growing trading region in the world. Three other trade blocs are currently operating in Latin America. The largest is MEROCOSUR which first originated in 1988 as a free trade pact between Brazil and Argentina and which now also includes Paraguay and Uruguay.

While the lowering trade barriers enables firms to see the world as their market, the regionalization of trading economies forces a shift toward direct foreign investment so that firms can benefit from a manufacturing presence in each region of significant demand. The establishment of regional manufacturing centres allows firms to better understand varying consumer preferences and hence tailor products in order to increase customer satisfaction and product acceptance. In this way, modifications in products can create greater variety and may ultimately lead to new product extensions and innovations (see Maruca). Expertise may develop independently at each regional site.

12.2.3   Technological   Change   in   Communication   and   Information Technology

While the lowering of trade barriers made globalization of world markets a theoretical possibility, it was technological change which made it a tangible reality. Major advances in communication and information technologies, particularly the development of the microprocessor, enabled the explosive growth of high-power, low cost computing, vastly increasing the amount of information that could be processed by firms while simultaneously decreasing the cost of this information. Development in satellites, communication and fibre optics technologies has revolutionized global communications enabling the flow of vast amounts of information along pathways that are so densely used and rapid that they are known as information highways.

As a result of these developments, the real costs of information processing and communication have fallen dramatically in the past two decades. This has made it possible for a firms to manage globally dispersed production systems more easily. For example, Texas Instruments (TI), U.S. electronics firm, has approximately 50 plants in 19 countries. A satellite-based communications system allows TI to coordinate production planning, accounting, financial planning, marketing, customer service and human resource management. This system enables managers of its worldwide operations to exchange vast amounts of information instantaneously and to effect tight coordination between the firm's different plants and activities (see Dicken 1992). Hewlett and Packard, another U.S. electronics firm, uses satellite communications and information processing technologies to link its new-product development teams based in Japan, the U.S., Great Britain and Germany. When developing new products, teams based in different countries use teleconferencing technologies to meet on a weekly basis, and electronic mail and fax to communicate daily. Communications technologies have enabled Hewlett-Packard to increase the integration of its globally dispersed operations and to reduce the time needed to develop new products. By taking advantage of time zone differences, companies are able to compress product development time by shifting projects to follow the course of the working day around the globe.

12.2.4   Technological Change in Production Systems

In the new global economy, firms increasingly use technological innovation and novelty as a source of competitive advantage. The pace of technological change has accelerated since the start of the industrial revolution and continues to do so today. The result has been a dramatic shortening of product life cycles. New technological breakthroughs can make established products obsolete virtually overnight, or alternatively, can make a myriad of new products possible. This process of 'creative destruction' unleashed by technological change makes it absolutely critical for a firm to stay at the leading of edge of technology, or risk the possibility that their products will become obsolete and non-competitive. Whereas these products, in the past, these products may have been sold in developing nations, this strategy has become less viable because of the greater trading opportunities and the increased sophistication of these markets today.

   The decline in product life cycles forces manufacturers to achieve greater product variety within shorter lead times. The emergence of new manufacturing technologies such as computer-integrated manufacturing systems brings flexibility to the production process, making it possible to cope with the numerous different parts which need to be manufactured in relatively small batches. The new manufacturing technologies that integrate computer-controlled tools and material handling systems with a centralized monitoring and scheduling function offer significant advantages when the nature of product demand requires differentiation. This type of product variety is not the product of the R&D lab, but is a multi-phase process which integrates various types of expertise within the organization and also incorporates feedback between producers and their suppliers and product users.

   One of the greatest impacts of the new manufacturing technologies has been the reduction in plant scale (see Jaikumar 1986; Luria 1990). The new manufacturing strategy is now toward a range of scale-reducing technologies, which allow smaller plants to be cost efficient despite lower production volumes. This strategy, accompanied by new production methodologies such as Total Quality Management (TQM) and Just-in-Time (JIT) manufacturing, can lead to productivity increases which mean that a plant is able to produce to the same demand with fewer overheads (see Mody et al. 1992; Inman and Mebra 1990). Reduction in plant scale and the integration of technology to manufacturing operations decrease labour costs to a lower percentage of overall product costs. Product costs are primarily determined by material, equipment, capital and overheads, none of which is located specific. All are embedded in the product through technology and the processes used to manufacture it. Depending on the level of integration of technology in manufacturing operations, direct labour costs become a less significant consideration in location decisions.

The development of markets over global space, on the other hand, forces firms to differentiate product versions so that it can satisfy different market segments. To succeed in global markets, it is often necessary to introduce new products in all major industrialized markets simultaneously. Location in a market provides a means of facilitating rapid customer feedback and hence giving the firm a product development edge in tailoring the product to local tastes and preferences.

To summarize, there are two important trends that should be incorporated in a new locational theory of the firm. The first concerns the configuration of the new global economy, the second concerns the effect of new technologies. The implications of the former globalization for location theory are the following:

- The increased integration of the world economy into a single marketplace is increasing competition and forcing firms, whether large or small, to acquire a global business strategy. The scale of location decision-making is therefore on a global rather than a local basis.
- Regional economic integration and increasing levels of non-tariff barriers are forcing firms to consider the location of production resources. This factor is increasing the benefits of decentralized locations on a global scale. This decentralization is based on the functional requirements of the firm's activities and is part of the firm's global business strategy.

The implications of the latter are that:
- Developments in communications and information technologies enable the flow of vast amounts of information to manage and coordinate globally dispersed operations, therefore reducing the need for centralized operations.
- New production technologies make smaller plants cost efficient without penalizing operation at lower volumes, further enabling the decentralization of production operations.
- The need for customer feedback to tailor products to market demand has increased the benefits of being close to customers and has further encouraged decentralization.

The next section provides a location decision-making model which incorporates these considerations.

12.3   A Framework for Location Decisions

Firstly, we summarize observations previously made about current location times. Globalization of the world economy is forcing firms of all sizes to adopt a global approach to business strategy; location is an important dimension of this strategy. Regionalization of trading economies and the need to tailor products to different customer tastes oblige firms to establish a manufacturing presence in each region of significant market demand. New production and communication technologies facilitate the establishment of decentralized, smaller and more flexible plants through reduced economies of scale and increased ease of information transfer. Consumer tastes and preferences, business practices, technical expertise, distribution channels and government policies differ from place to place. In order to succeed in each country, a firm may have to customise its product and redefine its strategy. Through this process, the firm develops an understanding of each market in which it sells products. Such knowledge is tacit and specific and cannot be transferred wholesale from one location to another. These factors can be synthesized into a new framework for location theory.

12.3.1   Using Location to Reduce the Uncertainty of Increasing Market Share

Increasing market share is the critical objective of competitiveness in today's global markets. Given the rapid obsolescence of products, the best strategy to increase market share is to produce new high quality products with the lowest possible costs. The conceptual justification for integrating locational decision to business strategy rests on the assertion that by siting in a given location, a firm can reduce its uncertainty about obtaining market share through the timely introduction of new products, exploiting the advantages of reaching the market rapidly. To express the analytical foundation of the integration of location into the business strategy, we use the Lancasterian method of viewing a product as an n-element vector

$$P = [p_1, p_2, p_3, \ldots, p_n]$$

where P is the product characteristics vector and each of the n dimensions is a different characteristic sought after in the market. Similarly, a given user of the product has a user characteristics vector:

$$U = [u_1, u_2, u_3, \ldots, u_n]$$

in the same n-dimensional space.

If P=U for the case of a given user, the probability of the product being sold to a user would be 1; i.e. the characteristic of the product would exactly match the desires of the consumer. The probability of the product being sold decreases as U and P diverge from each other. Thus the probability of sale to a potential user can be expressed as a stochastic function:

$$Pr(S) = \Psi$$

where $\psi = p_i - u_i$ for every ith characteristic.

In a world of complete certainty about the users and zero cost to offer variety, the firm could capture the entire market by reducing Ψ to zero. In the absence of this situation, the firm must minimize Ψ subject to the constraint that the marginal revenue from an additional variety is greater than the marginal cost of offering that variety.

The Ψ factor can be influenced by finding the average user characteristic vector. This vector can be obtained by dividing a population of potential users into subpopulations whose preferences are relatively close to each other. In this way, the firm tries to find the average user characteristics vector of each subpopulation and tries to offer a product to fit that average. If population j has N_j users, then the average user characteristics vector would be:

$$\mu_i = \Sigma_{i=1...Nj} \ U_{ij}/N_j, \text{ for every characteristic i.}$$

Given the above equation, and considering that user characteristics u_i are not mutually independent, a firm can maximize the certainty of sales to a subpopulation j by offering a variety such that $P = \mu$.

However, even when $P = \mu$, a firm still faces uncertainty in the magnitude of demand for its product. The larger the subpopulation of j, the greater will be uncertainty, expressed in terms of the standard deviation factor σ:

$$\alpha = \sqrt{\Sigma_{j=1...N} \frac{(n_{ij} - \mu_i)}{N_j}}$$

Thus, increasing the product variety offered by the firm through decentralized manufacturing plants increases the number of subpopulations and reduces the number of users in each subpopulation. Thus, the uncertainty of the firm in determining the magnitude of demand for its products is reduced.

Location decisions at a particular locale must be set in the context of the overall business environment considering both internal and external constraints. Locational strategies of competitors and the internal constraints such as availability of capital, the resources available to implement the corporate strategy, the existing technological ability to manage the command and control of additional sites must be considered in the decision making process.

12.4   Framework for Establishing Location Policy

Our results, and the key considerations we have identified, suggest a new framework for establishing location policy on a global scale. Together, these factors can be synthesized onto a four step procedure which is given below.

The first step in formulating a location strategy is to identify the characteristics of the market for each country in which the firm produces. This allows performance measures of the firm's competitive position in the market to be identified and integrated into the location decision. Three measures can be used to characterize the firm's market. These are the magnitude of the market, the strength of the market and the degree of variability in the market. The size of market demand and the rate of growth in the market are standard considerations. The degree of variability, as measured by the frequency of new product introduction provides a measure of market sophistication and the need for the company to seek additional variety.

The second step is to for the firm to determine its strategy for the market. Several considerations may be used to define the firm's goal for its competitive position in the market. Most important are the firm's technological capability and the desired share of the market. These will dictate the frequency of new product introductions required for the firm to capture the market and depend on new growth in the market and the response time required for the firm to respond to changing market conditions by producing new products for the market. Response time is exogenously determined and represents the time necessary for the firm to introduce a new product from the time of its conceptualization to actual manufacturing. This value is constrained or facilitated by the firm's technological capability both in terms of manufacturing and communication technologies. The firm's technological capability will determine the ability to undertake product differentiation in order to achieve the stated goal.

The third step is an assessment of the degree to which the targeted market share can be achieved by giving weights to cost, quality, time-to-market, and flexibility. The differences in weight assigned to each of these factors determines the aspects of location strategy. If cost carries high weight, location would affect transportation costs, labour, and inventory costs. Flexibility, as dictated by time-to-market, would require customer proximity, coordination of manufacturing with marketing, R&D, and product design through communication networks, and a highly educated labour force who can exploit the benefits of new manufacturing technologies through implementation of management methodologies such as JIT or TQM.

In the fourth step, the impact of all the above factors would be synthesized to decide on the type of production process and technologies to employ in each plant or whether to serve the market from one consolidated plant or several decentralized facilities that use flexible manufacturing technology to stay cost efficient.

12.5 Conclusion

Location decisions are an important part of the global strategy of the firm which requires a more complex decision-making process than that suggested by existing perspectives on location theory. Today's firms, whether large are small, have to consider competition from not only domestic businesses but also from international businesses. Every organization needs to formulate strategies within a global context to acquire a competitive edge in today's markets. The growth in information and communication technologies in recent years has redirected attention to the relationship between innovation, technological change and the spatial structure of the industrial landscape.

In this chapter, we have proposed a framework for integrating location decisions within the global business strategy of the firm. We suggest that a firm can reduce its uncertainty about obtaining market share through the timely introduction of new products into the market, by exploiting the locational advantages of reaching the market from a particular location. Locational advantage can change in accordance with the functional requirements dictated by the global strategy. In this new framework, the conventional advantages of land, labour and capital may be as important as the infrastructural elements.

References

Business Week 1994. 21st Century Capitalism. Special Issue

Clark G L, Gertler M and J Whiteman (1988) Regional Dynamics: Studies in Adjustment Theory. Allen and Unwin Boston

Dicken P (1992) Global Shift. Guillord Press New York

Feldman M P (1994) The Geography of Innovation. Kluwer Academic Press Boston

Garafola G A and M S Fogerty (1988) The Role of Labor Costs in Regional Capital Formation. Review of Economics and Statistics. November 1988 70: 593-599.

Inman A R and S Mebra (1990) The Transferability of JIT Concepts to American Small Businesses. Interfaces March-April: 30-37

Jaikumar R (1986) Postindustrial Manufacturing. Harvard Business Review. November-December: 69-76

Japan Trade Organization (1990). Research Report

Kutay A (1988) Technological Change and Spatial Transformation in an Information. Economy I. Environment and Planning A. 20: 569-593

Kutay A (1988) Technological Change and Spatial Transformation in an Information. Economy 2. Environment and Planning A. 20: 707-718

Luria P (1990) Automation, Markets and Scale: Can Flexible Niching Modernize U.S. Manufacturing? International Review of Applied Economics. June: 127-165

Maruca R F The Right Way to Go Global, An Interview with Whirlpool CEO David Whitman. Harvard Business Review 72: 135-145

Mody A, Suri R and J Sanders (1992) Keeping Pace with Change: Organizational and Technological Imperatives. World Development 20: 1797-1816

Scott A J (1993) Technoplis: High Technology Industry and Regional Development in Southern California. University of California Press Berkeley

Storper M and R Walker (1988) The Capitalist Imperative: Territory, Technology and Industrial Growth. Basil Blackwell Oxford

The Economist Yearbook 1993. Economist Books Ltd. London

The Economist World Economy Survey (1992) Another World. September 19

Trager C S (1989) Enter the Mini Multinational. Northeast International Business. March

PART C

Empirical Studies and Policies

13 Investment Policy and Innovation Management: An Exploratory Analysis

Kingsley E. Haynes, Fred Y. Phillips,
Nitin S. Pandit and Carlos R. Arieira

13.1   Introduction

An innovation cannot be recognized, adopted or spatially diffused until it is made 'public'. The most common way an innovation is made 'public' is by bringing it to the market. This process of innovation is increasingly made possible within a context of continuous research and development (R&D). A new idea or invention is designed, tested, costed out and prototyped, benchmarked, examined for scaled up production and evaluated in terms of market acceptability before it is finally offered to the public. Such a procedure requires financial investment at an early stage of the product development cycle. This investment is made in an atmosphere of significant uncertainty and it is the purpose of each stage of the development process to systematically reduce this uncertainty as much as possible.

Traditionally, scholars writing on innovations and the innovation process have focused on either the early research and invention aspect of this R&D activity or on the much later individual adoption and market aspect of product innovation diffusion. These are the crucial dimensions of creative innovativeness and marketability and are of great importance. However, as Kash (1989) has pointed out, the perpetual aspect of innovation is driven by a process of continuous incremental product advancement not by discrete and intermittent leap frog technological jumps. He notes that, historically this process of continuous innovation led to world dominance of the German chemical industry, the US agriculture and defense sectors, and Japan's consumer electronic industries. Japanese industry was at first a pupil of this process but is now the teacher in terms of how this continuous innovation process revolutionizes product innovation. As demonstrated by the speed of product modification and development in the Honda-Yamaha motorcycle war this can be a very rapid innovation process when it is driven by fierce competition.

The issue to be emphasized here is that such a continuous process of innovation requires an investment policy or strategy, since at each stage of the development cycle two choices must be made. One choice is whether to continue to invest. The other choice is which developing product should be invested in. Both of these choices are conditioned by the principle of uncertainty reduction in terms of product success in the market.

Each stage of product development requires investment first in the form of cost commitments and then in terms of specific expenditures. These investment decisions underlie the risk of discontinuing the project (Erlenkotteret al. 1989). Changes are made throughout the development process and are evaluated at each stage of product evolution. However the irreversibility issue is of central concern at each stage of analysis (Henry 1974; Arrow and Fisher 1974; Baldwin 1982; Bernanke 1983). It has been observed that in most cases costs incurred in new project development (from initiation to operation) are committed ahead of actual spending. Further, cost commitments rise most rapidly in early stages of product selection and design as alternatives are discarded and options reduced. Although expenditures usually lag cost commitments in the early phases, they catch up at the final phase where project implementation or production commitment is undertaken. Cost commitments and expenditures converge with project completion, hopefully balancing each other (Figure 13.1).

The purpose of this chapter is to examine the investment pattern that supports this innovation development process and to identify evaluation points in product evolution that will allow uncertainty to be more effectively managed. We propose that at each phase of the development cycle, management should attempt to reduce uncertainty by obtaining cost estimates and measures of desired outcomes in increasing accuracy (McDonald and Siegel 1986; Majd and Pindyek 1987). A prudent manager should not commit to new investments at a rate that exceeds the rate of reduction in uncertainty. This is precisely why information is important and why each new commitment of investment is a Go/No Go decision. Further, we know from cost/benefit analysis that large past expenditures should not condition future commitments especially when it is becoming increasingly certain that the future product innovation will be unsuccessful.

13.2   Uncertainty Measurement for Product Development

In each phase of the product development cycle, the decision maker attempts to reduce uncertainty by pursuing increasingly accurate estimates of production costs and the likelihood of success in the marketplace. It is achieved by means of consumer focus groups, project pilot tests, assessing alternative designs, market tests, cross comparison, willingness-to-pay studies and so on. Wolter Fabrycky (1990) at the National Science Foundation International Workshop on Concurrent Engineering Design and Blanchard and Fabrycky (1990) observed that a firm should not commit funds at a rate that exceeds the rate of uncertainty reduction. This is apparent from the following argument: a given return on investment has been targeted, the subsequent information has confirmed the expected Return on Investment (ROI). What is the incentive for committing additional funds? For a risk-averse manager, the only answer that confirms such commitments is, 'a reduction in the uncertainty surrounding the ROI estimate'.

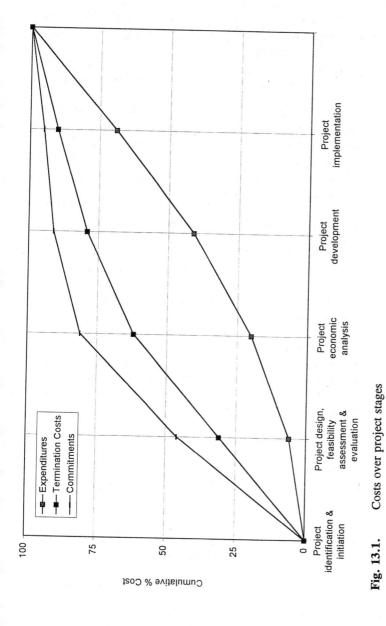

Fig. 13.1. Costs over project stages

To make such an investment evaluation at any stage of the innovation development, the decision maker has to deal with significant problems especially in areas where product innovations are driven by emerging technologies. The task is difficult because data availability is poor and by definition experience is limited. For example, in the case of emerging technologies in micro-electromechanical devices, it was shown (Benson et al. 1993) that it is useful to evaluate these technologies using a triple-gateway. This triple-gateway methodology contends that to be successful a new technology product must pass through a market gateway, a systems management gateway, and a technology gateway. Although this approach does recognize some indicators of information content that may be useful in the characterization of uncertainties in each gateway, it is naive in the sense that it does not consider information gain as an explicit measure of uncertainty.

The triple gateway methodology mentioned above is only partially applicable to our area of interest because it provides a tool for making only a one time decision for investing in a technology. Although not originally designed for incremental investment evaluation, this methodology may be applicable if a temporal or evolutionary framework can be incorporated. Clearly, a systems approach (Sage 1989) could provide an overall multi-phase framework for the triple gateway methodology. Our approach complements this in that we propose a formal information theoretic framework for multiple evaluations. With its origins in the work of Shannon (1948), we propose using an entropy measure to quantify information gain and to evaluate each stage of uncertainty in the product innovation-development cycle.

It is evident that the measures of information gain are not easily quantifiable for all sources of uncertainty. Some measures are qualitative and can potentially be organized only on nominal or ordinal scales. In this paper, we demonstrate the use of information gain mainly for the simple case of quantifiable economic sources of uncertainty (Table 13.1).

Table 13.1. Uncertainty and information gain measures

Source of Uncertainty	Sources of Information Gain
Technology Reliability	Field Test Results
Public Acceptance	Questionnaires & Surveys
Institutional Issues	Analysis of Legal Issues & Liabilities
Fiscal	Alternative Returns on Investment
Economic	Production Costs & Willingness-to-pay

In practice, the uncertainty often centres around a specific goal or criterion and its reduction is usually operationalized through an economic assessment such as rate of return on investment (ROI) analysis. In ROI terms, the return must exceed a hurdle rate which is usually set at a level at least as high as a return on an alternative investment. Whenever projects are first conceived and even after feasibility studies are conducted there is usually considerable uncertainty (say, around 10%) related to some elements in the analysis. When uncertainty in various elements of an ROI analysis are chained together the resulting aggregate uncertainty can be very great (around 30%).

Over time as product estimation becomes increasingly precise, these uncertainties are reduced by such steps as limiting goals, settling on a fixed design and fully assessing levels of utilization. In each phase of project development these uncertainties should decrease until the ROI is relatively certain. Such certainty generates a decision point (Go/No Go) which relates committed costs to future returns on investment. However, even if the uncertainty around the expected ROI drops to zero, it does not mean that the project should go forward. This is especially true if the uncertainty collapses to a mean ROI below that of alternative investments (i.e. below the hurdle rate). However, it is this reduction in uncertainty around a sufficiently high ROI that allows us to move forward in creating new cost commitments and authorizing continued expenditures at each stage of product development. As noted earlier, no risk-averse manager should commit funds faster than the rate of uncertainty reduction.

Below, we formalize this observation by elucidating the nature of committed costs and clarifying associated terminology. First, we develop a framework in order to understand the value of information. Subsequently, we develop a measure of uncertainty that is pertinent to the Go/No Go decision for product development.

13.2.1   The Value of Information

Some elements of the first category of the committed costs (including leases, etc.) are non-recoverable if the project is stopped. In the second category (materials, etc.), although some costs are recoverable, they are irreducible if the project proceeds. The collection of both categories might be called termination costs. If we trace the sources of committed costs and expenditure in Figure 13.1 we will find this distinction meaningful.

Freezing designs and costs makes 'downstream' tasks more efficient, because engineers and managers know what must be designed, built and linked to an existing system, and under what conditions these tasks must be executed. Contingency plans dealing with alternative designs can be discarded. On the other hand, committed costs reduce flexibility. They make less cash available for analyses of alternative investments that may come to light. If market estimates or consumer requirements change, costs sunk into the wrong design are not recoverable. Managers may ask, 'What would it cost to shut down this project?' The answer to this question is: the present value of all committed and nonrecoverable costs. This amount, together with speed of response, are components of a project's flexibility.

Under sequential engineering, a prototype can be designed and tested before deciding on the best project production method. The costs of prototyping include only the costs of designing, testing and economic analysis of the prototype. Under concurrent engineering - there is a pressure for the rapid building of the project - the prototyping activity includes looking ahead to implementation. The cost of investigating and deciding on the mode of production is now part of the cost of prototyping. This cost is now committed - and possibly expended - earlier. If the project is terminated, these costs cannot be recovered. At all additional stages of the product development process, termination or bailout costs are increased.

Clearly, maximum flexibility would result from holding design alternatives open for as long as possible, and operating on a cash-and-carry basis, making the three curves of Figure 13.1 converge. But this strategy would eliminate the benefits of early cost commitments noted above and is unrealistic in almost all cases of investment for development of innovative products.

13.2.2   Cost Commitments

What is the relationship between cost commitment and uncertainty reduction? We argue that the prospect of reduced uncertainty is the only justification for postponing cost commitments. In other words, if no further uncertainty reduction is expected, all project cost could be committed at the earliest possible moment and all investment flexibility discarded.

In many stages of the new product development process, data are collected and estimates made regarding product demand, production costs, product performance, production management, investment alternatives, project schedule, and user response at different pricing levels. Such intelligence is critical to the management of the product development process. At each decision point the project manager may specify a value for a 'fraction of the remaining project costs to be committed at this point'. Committed costs defined above are of a different quantity from the expenditures the management authorizes at each decision juncture. In the event of a subsequent project shutdown, some authorized funds that have not been committed or expended may be recovered.

With respect to project continuance, four choices are made at each decision point: (i) 'Go' i.e. to proceed directly to the roll-out of the project with no further analysis; (ii) 'On; to the next study stage; (iii) 'Skip' the next study stage and proceed directly to the final analysis but one; conditions are identified under which a test may be bypassed (Urban and Katz 1975) or (iv) 'No Go' i.e., stop the project (See Charnes et al. 1966; Charnes et al. 1978). The first, second and third options imply a continuation with all scheduled production and other 'non-test' activities. For the fourth option, of course, no further costs should be committed. For the purpose of analysis, we assume that no unscheduled tests or decisions are involved.

These conditions, plausibly reflecting management practice, justify the assertion that all remaining project costs may be committed immediately if no further intelligence is anticipated. Moreover, if the aforementioned benefits of cost commitment are significant and success is assumed, all remaining costs can be committed immediately.

Figure 13.2 summarizes the possible trajectories of committed costs from a decision point 'i'. In the case of the 'On' decision, an additional decision on committed costs must be made at stage $i+1$. Other decision options determine the path of committed costs for longer time periods.

Having established that all costs may be committed at the moment when no further information is expected, we move to the question, 'How much cost may be committed when such information is anticipated?'

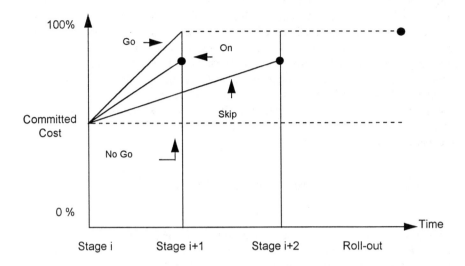

Fig. 13.2. Impact of Go/No Go decision on committed costs
(Following Zirger and Maidique 1990)

13.2.3   Measurement of Uncertainty

Rigorous and unambiguous measures of information and uncertainty have been available since the 1940s for dealing with events that are characterized by one or more probability distributions. What may be at issue in the analysis of a particular problem is the definition of the appropriate random variables and their distribution functions.

The problem addressed here is that of describing the rate at which an organization commits funds, relative to the rate at which it reduces uncertainty. The proceeding section informally established that these two rates should be somehow linked. We note that the proposition is implicit in Akaike's (1973) construction of a decision-theoretic loss function from Kullback's (1959) information measure. Below, we discuss a number of random processes and uncertainty measures in the context of the innovation development process, in order to clarify issues and terminology, and to lead to a measure that will be useful in guiding investment decisions.

The Entropy Measure
'Committed cost' as portrayed in Figure 13.1, refers to money that is expected to be expended at a scheduled (future) time even if the project is cancelled today.

In the language of project management or decision analysis, the level of committed cost at time t or stage i is the 'termination or cancellation penalty', otherwise referred to as bailout cost at time t or stage i. In reality 'committed cost' is the maximum estimate of termination or bailout costs with real cancellation penalties being located between the two curves.

Very few organizations maintain complete product development histories, but from such a history the probability of bailout at stage i can be computed. It is the proportion of products that survive the ith stage review. Table 13.2 displays an attrition pattern for a set of 116 new product innovations from a Booz, Allen study cited by Cunningham and Cunningham (1981). It can be seen that after (economic) demand analysis 14 of the original 116 products will have survived. In the third column, the quantity q represents the updated probability of success. The rightmost column of the table gives the entropy of the probability law, as defined in equation (13.1).

$$H(q) = -q \ln q - (1-q)\ln(1-q) \tag{13.1}$$

where q is the probability of success and $p = (1-q)$ is the probability of failure. We assume that, of the two projects that survive the final analyses and go to full design and cost accounting, only one will later be built.

The entropy measure given in equation (13.1) is often taken as an estimate of uncertainty inherent in such a probability law (see Shannon 1948). A simple illustration will demonstrate the use of H(q). Suppose a completely prescient and truthful being tells us, at the early product design (2nd) stage, whether each given product will succeed or fail to be built. The average information content of this message is $H(q) = .173$. This small number carries the message that will almost always be, 'the project will fail and hence should not be built'. It is usually unsurprising and hence relatively uninformative. This is in contrast to the higher information content of a perfect forecast delivered at the later stage of new product production or testing, when prior success and failure probabilities are more equal.

Although we might intuitively regard uncertainty to decrease at each stage of the development process, Table 13.2 shows that, on the contrary, uncertainty increases. Prior experience tells us that 114 of every 116 product proposals will fail or will be rejected prior to building. Therefore for an arbitrary project, we are almost certain that 'this won't work in terms of being brought to market'. As we work to increase the odds of success for a particular project, uncertainty increases.

The above is a consequence of the fact that the graph of $H = p \ln p$ is concave over the range $0 \le p \le 1$, reaching a maximum when $q = 0.5$. We must look further for an information function that increases monotonically as the odds of project success progress from very small to very high. The reader should note that we are not considering the cyclical aspect of product development cycles which can increase uncertainties non-monotonically when information gain leads to discoveries which in turn affect the uncertainties associated with the set of choices.

Table 13.2. Information based survival probabilities

Development Stage	number of surviving projects	q = {Prob of success of surviving this stage}	H(q)=qlnq -(1-q)ln(1-q)
start (invention)	116	0.009	0.050
product design	24	0.042	0.173
demand analysis	14	0.071	0.257
production	6	0.167	0.451
marketing & testing (roll-out)	3	0.500	0.693

ROI Variance Measure

Information will be gathered at each stage shown in Table 13.2, making the product development process a multi-stage decision problem under uncertainty with recourse. That is, the Go/No Go decision is made repeatedly, using the latest updated information. Where the historical data of Table 13.2 refer to the aggregate of past projects, the updated information pertains to a particular project. It is important to note that whereas the (q, 1-q) law describes the binomial random variable 'success/failure', the project-specific information refers to a different random variable, i.e. a measure of success that for the purpose of this illustration we will assume is return on investment (ROI). The criterion could as well be a cost-benefit ratio, a payback rate or another traditional measure. Although advances in measurement such as data envelopment analysis (Charnes et al. 1978) are now available for the multicriterion evaluation of projects, here we retain a single-criterion measure for simplicity of exposition and to clarify the relationship between cost commitment and uncertainty reduction.

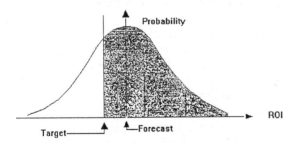

Fig. 13.3. ROI probability estimate

The Go/No Go decision is made according to whether the ROI forecast exceeds a target value, possibly with a confidence margin. Each forecast may be stated as a point estimate with a variance, so that the Go/No Go decision can be conceptualized as in Figure 13.3 under a suitable assumption on the distribution of the ROI estimate. The probability of project survival, immediately prior to a test producing data like that of Figure 13.3, is the expected area of the distribution representing the shaded portion to the right of the ROI target probability. This is taken as the updated probability of success.

Each successive item of information should reduce the variance of the expected ROI distribution. If this distribution is normal (Gaussian), its entropy is uniquely determined by its variance (Hastings and Peacock 1975):

$$H(ROI) = -\ln\sqrt{(2\sigma^2 \pi e)} \qquad (13.2)$$

As $\sigma 2$ decreases, H decreases, indicating a reduction in uncertainty. This uncertainty, however, is uncertainty about ROI, not about the project success/failure question per se. The equation (13.2) measure is therefore not suitable as a guide for scheduling the commitment of costs. To continue the illustration, suppose the ROI variance collapses to zero around a mean that is less than the target ROI. Uncertainty has been reduced to the minimum, yet it is clear no further funds should be committed, i.e. the decision maker is certain that the rate of return will be less than that which is desired.

Directed Divergence
Using q from Table 13.2 as a prior distribution, and p as the updated probability of project failure, the reduction in uncertainty resulted from information at stage i may be written as

$$I(p:q) = p\ln(p/q) + (1-p)\ln[(1-p)/(1-q)] \qquad (13.3)$$

where p and q refer to stage i. This measure, I(p:q), is due to Kullback (1959), and is a generalization of the entropy measure (1) which represents the directed divergence of p against q.

Equation (13.3) may be interpreted as the information gain from an updated sample relative to a baseline sample. Akaike's principle, however, holds that the loss function should equal the directed divergence of the sample against the true distribution. For a single product development, there is no 'true' distribution; or, we might say, the distribution collapses to Prob{success}=0 or Prob{success}=1 when the decision response is known. This situation violates the 'absolute continuity' assumption of the directed divergence measure (see Kullback 1959). More simply and operationally, for a discrete distribution, this means there will be zeros in the denominator.

Equation (13.3) highlights the fact that uncertainty is not reduced in an absolute way; it is reduced relative to some prior state of knowledge. Without bringing in additional considerations, we have no guidance as to how to specify the prior state of knowledge. For example, to measure a reduction in total project uncertainty, we could use as a baseline the 1-in- 100 odds given at the top of Table 13.2. There should, however, be a way to use all the information given in Table 13.2

as that the odds of project success are not 1-in-100 does not tell us directly to what the odds of success are. Thus, equation (13.3) alone cannot be used to guide cost commitment.

Discriminant Functions

Balachandra (1984) uses discriminant analysis in his retrospective study of 100 new development projects and translates the result of the analysis into qualitative guidelines ('red light, yellow light and green light' signals) for the Go/No Go decision. Zirger and Maidique (1990) also use discriminant analysis to find the organizational factors conducive to the success of the development process. In this section, we use the discriminant technique for a different purpose, using increased information as an independent variable. This purpose is to arrive at an information function that will yield quantitative guidelines for the commitment of product development investments.

Suppose that successful projects of the past have returned ROIs distributed as a density function g(.) and failed projects have had a distribution of ROIs as h(.). The latter distribution is constructed using the best estimated ROIs of products that have been terminated prior to execution, and the actual ROI performance of products that have failed after being brought to the market (i.e. did not meet target returns). The specific functional forms of these distributions are immaterial at present. According to Morrison (1976), the log likelihood ratio

$$\lambda = \ln \left[g(.) \, / \, h(.) \right] \qquad (13.4)$$

is a sufficient function for discriminating between the two distributions, based on an observation vector x of study results. In the Kullback (1959) theory, is equal to the information provided by x favouring the hypothesis $H_1 : x \sim g(x)$ over the hypothesis $H_2 : x \sim h(x)$. That is, the discriminant function (4) is the optimal way of classifying a project either as a member of the population of 'products that will succeed' or as a member of the population of 'products that will fail'. The decision rule is: x was generated by a member of the successful population g, if $\lambda > 1$ and by a member of the unsuccessful population h, if $\lambda \leq 1$.

The usual assumption that g and h are continuous density functions over the interval $(-\infty, \infty)$ implies, for example, that a member of the 'successful' population may have an ROI less than the target value. Indeed, this situation is pictured in our Figure 13.2 (following Zirger and Maidique 1990). In reality, projects may succeed or fail for reasons other than ROI performance. Thus this representation is not unreasonable.

A sequence of information gathering activities at each Go/No Go decision stage i (Figure 13.2) where $i = 1, 2, \ldots, n$, occurs throughout the duration of the product development process, resulting in the data vector $x = (x_j)$. Each x_j is a random variable with density $g_i(x_j)$. From this we build a stepwise discriminant function using the cumulatively added information:

If the added information results x_i are not collinear, each successive λ_i will be

$$\lambda_1 = \lambda_1(x_1);$$
$$\lambda_2 = \lambda_2(x_1, x_2); \quad \ldots \tag{13.5}$$
$$\lambda_n = \lambda_n(x_1, x_2, \ldots, x_n).$$

a better discriminator in that the probabilities of misclassification will be reduced.

The ith information stage or test yields discrimination information λ_i. Following this added information, if the project is not terminated, a decision must be made concerning the amount of further cost commitment. We desire to commit costs at a rate less than or equal to the rate of change of information increase (uncertainty decrease). The value of λ_{i+1} is not yet known, so we must estimate it using its expected value under H_1:

$$I_{i+1} = \int g_{i+1}(X_{i+1}) \, \lambda_{i+1}(X_1, X_2, \ldots, x_{i+1}) dx_{i+1} \tag{13.6}$$

The additive property of the logarithmic information measure (see Kullback 1959) allows us to propose the operational rule at stage i (following an information increment) to commit the additional fraction of total project costs equal to

$$(I_{i+1}\lambda_j)/(I_n I_l) \tag{13.7}$$

Table 13.3. Average log odds of correctly identifying a successful product. (From the data of Table 13.2)

Development Stage	q=Prob {market success | survived this stage}	Odds of success	Log odds = q/(1-q)
start (invention)	0.009	0.009	-4.745
product design	0.042	0.043	-3.135
demand analysis	0.071	0.077	-2.565
production	0.176	0.200	-1.609
marketing & testing (roll-out)	0.500	0.000	0.000

However, as in this initial development primarily concerned with descriptive rather than normative measures, we now return to the attrition probabilities of Table 13.2, reinterpreting them in Table 13.3 as the (average) odds of correctly classifying a successful project. The rightmost column of Table 13.3 may be identified with equation (13.6). It is the expected value of the discriminant function at each stage.

It should be noted that perfect information is obtained when x_n, the actual market performance, is observed. At that time, λ will have an infinite value. However at that time all project costs will have been committed, so the relevant span of time for this analysis of cost and uncertainty ends at the moment of the project roll-out decision.

The rightmost column of Table 13.3 is graphed in Figure 13.4. It is, as desired, monotonically increasing, implying that uncertainty is monotonically decreasing. If the rate of cost commitment and uncertainty reduction were parallel, however, and if Figure 13.1 is accurate, the graph of Figure 13.3 would have the concave shape of the upper curve in Figure 13.1. In the next section, we reconcile Figure 13.1 and Figure 13.4.

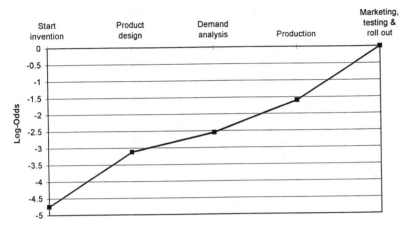

Fig. 13.4. Average log odds of correctly identifying a successful product, by development stage.

13.5   Product Investment Risk: Empirical Results

The Booz-Allen data of Table 13.2 cover a variety of industries, focuses particularly on consumer goods, while other data on committed costs tends to be dominated by avionics and aircraft parts manufactured by aerospace firms (e.g. Shield's data in Appendix 1). In spite of the obvious relevance of both situations to emerging technologies and product innovations, inferences drawn by comparing the costs of the former with the risks of the latter would be suspect. Instead, in this section we present and compare results drawn from two individual firms, one on the basis of secondary data and the other on the basis of primary (original) data.

13.3.1   Comparison of Innovation Investment Decisions

McGrath et al. (1992, p.49) offer the needed cost and risk data drawn from a single organization and from 'other companies considered to be the best' product developers in Company A's industry. The consulting firm with which these sources are affiliated works largely with firms in electronics-related industries. McGrath et al. (1992) refer to the committed costs associated with cancelled

projects as 'lost investment'. The attrition and cost data for Company A appear
in Table 13.4.

Table 13.4. Attrition and Committed Cost Data for Organization A's Failed New Product
Development Project

Development stage	% of cancelled projects failing in this stage	Lost investment due to cancelled projects*
Concept evaluation	19%	3%
Planning & specification	26%	9%
Development	37%	55%
Test and evaluation	14%	24%
Product release	5%	24%

* Expressed as percent of all lost investment on cancelled projects.

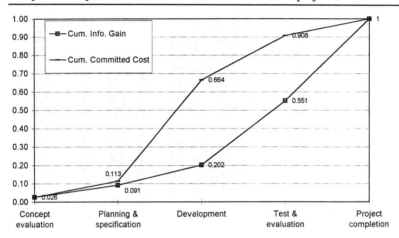

Fig. 13.5. Average uncertainty reduction vs. average cumulatively committed cost, by
development stage: from published data on a single firm.

Figure 13.5 displays the cumulative information gain and cumulatively committed
costs for this firm, normalized to the same scale. The names used on the x-axis
for the development stages are those used by the consulting firm. Figure 13.5
differs from Figure 13.2 in that the lower line is information gain, not
expenditures.

This firm's committed cost (the upper curve) shows the same concave shape as
the earlier data (Figure 13.1). The fact that this curve lies consistently above the
risk-reduction curve shows that the firm is intentionally or not, risk-inclined. In
figure 13.5 the comparison of rates of cost commitment and uncertainty reduction
lends itself to a very simple index of risk behaviour. To construct the index,
simply sum the differences between the latter two rates at each development stage,
then divide by the number of free points of comparison. For the company
represented by the McGrath et al. (1992) data, the risk index is
$(.024 + .462 + .357)/3 = .281$. The completely risk-neutral firm would of course

have an index of zero, and risk-averse firms a negative index.

Figure 13.6 displays the same analysis for the McGrath et al. (1992) data on the industry leaders. However, McGrath et al. do not reveal how many companies are summarized in this aggregate. The closer convergence of the two curves in Figure 6, and the risk index value of -0.04, show that the industry leaders in Organization A's industry were slightly risk-averse but generally matched risk reduction and cost commitment more skilfully than did Organization A. This analysis supports McGrath et al.'s contention that the leaders in Organization A's industry are superior managers of the product innovation development process, at least relative to A itself.

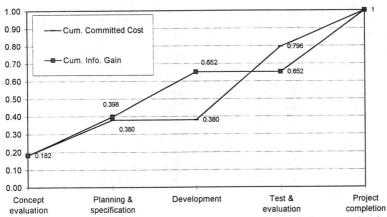

Fig. 13.6. Average uncertainty reduction vs. average cumulative committed cost, before development stage: from published data on an aggregate of industry leaders.

McGrath et al. (1992) note that the middle stage is usually where most of the development investment is made. This is the stage where their industry leaders are most risk- averse. The leaders appear to become more risk-inclined in the test and evaluation stage. It is also where Organization A itself is most inclined to take risk. They note further that 'the best practice companies (shown in Figure 13.7) developed 48 products at a cost of $60 million, while it cost the case-example company $75 million or 25% more (to develop the same number of products)'. McGrath et al. fault Organization A for scuttling an insufficient number of unworthy projects in their early stages. With equal justice, one might fault them for having too many bad product ideas in the first place, or for committing too much investment too early. The new product development battle can be fought in any of these arenas, and the risk profile developed in this paper does not unduly emphasize any specific arena.

For the case of organization B, from a large, diversified manufacturer of industrial and office products, we obtained data on twenty new product development projects randomly chosen from a file representing all the business units. For project planning and evaluation purposes, this organization divides such projects into four stages: Concept, Development, Manufacturing Scale-up and Field Test. In each stage, a project manager must estimate the updated internal rate of return, market risk and technology risk similar to Benson, Sage and Cook's

(1993) triple gateway for emergent technologies. This company uses only its own factories, and prefers line extensions to new businesses. Its committed costs are summarized by avoiding leases and unfamiliar new equipment, committed costs are minimized. Nonetheless, it is remarkable that the two curves of Figure 13.7 match so closely. Indeed in the manufacturing scale-up phase, the curve representing rate of cost commitment falls below the rate of information increase.

The raw data show that a few of the failed projects incurred expenditures in excess of the forecasted life cycle cost. Indeed, the overrun may have contributed to the No Go decision. The upper curve of Figure 13.7 was derived using only the failed projects, for which the termination or bail-out costs were known with certainty. The termination decision may have been based on cost overruns or on the perception of relatively low 'bail-out' costs, so the extent and direction of bias in the placement of the upper curve is a matter for conjecture. (These comments apply also to Figures 13.4 and 13.5.) Also, the location of the left end of the line at zero is arbitrary, as the data contained no projects that failed at the 'start' stage. However, the fact that the lines nearly coincide does not depend on this arbitrary choice. The 50% rate of conversion of concepts into market successes, and the close convergence of the two curves in Figure 13.7, demonstrate that Organization B's reputation as a superior developer of new products is well deserved.

For B, the risk index takes a value of $(.10+.05-.09)/3 = .02$. This index, although a useful summary, obscures interesting diagnostic features like the crossover in Figure 13.7. At that point, the scale-up stage, the firm in question becomes more risk-averse than it was in earlier stages.

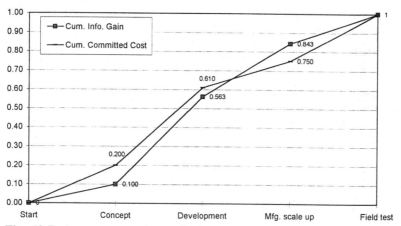

Fig. 13.7. Average uncertainty reduction vs. average cumulative committed cost, by development stage: data from an innovative diversified company.

If we are to draw valid conclusions from these studies, we must be able to answer the following questions. Are the development stages named by the different sources truly comparable, and is terminology consistent? What were the time periods spanned by the studies? As 'committed costs' and 'determined costs' are not conventional cost categories in corporate reporting, how were these quantities defined and how were they culled from financial records? Uncertainties as to the replies suggest a preliminary investigation of the robustness of the discriminant function- based risk profile is in order.

13.3.2 Sensitivity Analysis

A 1990 SAMI report (Wall Street Journal 1990) revealed that of 6,900 new brands introduced in the two previous years, only 240 reached the $1 million annual sales mark. This is slightly less than 3.5%, quite a different percentage from the 50% post-introduction 'success rate' actually achieved by Company B. The low figure may be peculiar to the consumer package goods industry. But how sensitive are the results of analyses like those of Figures 13.4, 13.5 and 13.6 to the post-launch success rate? Table 13.5, which gives the data for the 'information curve' of Table 13.2 for several product success rates suggests they are quite insensitive. (The numbers for 'start' and 'test market' remain the same, of course, due to the normalization).

Table 13.5. Normalized information at each Development stage as a function of product success rate.

Post-launch success	Start	Product evaluation	Economic analysis	Product development	Test market
3%	.65	.79	.83	.90	1
10%	.65	.78	.83	.90	1
17%	.65	.78	.83	.90	1
50%	.65	.77	.81	.88	1
90%	.65	.74	.77	.83	1

13.4   Conclusions

A joint consideration of the three curves 'committed cost', 'termination cost' and 'expenditures' can illuminate the relationship between flexibility and risk in managing product innovation. The relationships drawn in this paper can elucidate the risk behaviour of organizations, aid the comparison of new product development procedures, and provide operational (normative) guidelines for innovation managers on how to commit investments.

The above information model addresses committed costs but does not deal explicitly with termination costs. By including both technological scanning within the scope of product innovation and technology assessment in the ROI estimate for a current design, the mathematical model offered in this paper can address innovation risk without modification.

The overt problem in new product development is to increase the proportion of innovations that become successful in meeting their ROI goals. A collateral problem, of no less importance, is reducing the cost of failures. It is the latter problem to which the present research is most applicable. This paper has discussed the nature of committed and determined costs, and quantified their relationship to the reduction of product development uncertainty. We have emphasized the role of information gain by considering both marketing and production factors in innovation investment evaluation. We have also indicated that the ideas developed can show the relationship between cost commitment and risk behaviour of the organization and provide guidance for operational decisions within a product development framework.

We close with a call for careful collection of product innovation specific data for further examination of these ideas. It may be worthwhile undertaking systematic studies on empirical investment decision behaviour in relation to information gain for specific types of product innovations. Such studies should include both the sensitivity of investment commitments to uncertainty reduction and the consequence of each Go/No Go decision relative to the level of uncertainty. This chapter provides some conceptual groundwork as well as certain limited empirical observations that could provide a basis for further studies.

We should like to add a final remark concerning the practical application of such as approach. It would seem useful to formalize the decision options relating to a certain level of uncertainty for specific types of innovations so that lessons learned in the past can be immediately made available for reference to future innovations of similar types. These options and consequences could be compiled in table form so that users can easily identify their problems with past experiences. The approach is somewhat similar to weather forecasting. When predicting the level of rainfall in one week's time, we need to plug into the empirical formulas derived from past experience many of today's parameters, such as temperature, humidity, wind speed and direction, satellite picture of clouds that are coming in our direction, etc. The uncertainty is high at the beginning but diminishes with time and approximates zero as it approaches the point of forecast. Likewise, uncertainty reduction in new product development is continuous and could be simulated. A well managed product innovation database will provide useful parameters for forecasting uncertainty and managing investment for product development.

Acknowledgement

The authors express their appreciation for support from DOT/FTA Grant #VA-26-0001 entitled 'Assessing Public Policy Issues in the Implementation of IVHS' and DOT/FHWA Cooperative Agreement #DTFH61-93-00027 entitled 'Institutional Issues Research'. The authors are fully responsible for all analysis and interpretation. Dr. Haynes is Director of The Institute of Public Policy and University Professor; Dr. Phillips is Professor and Head, Department of Management, Oregon Graduate Institute of Science and Technology, Beaverton, Oregon; Messrs. Pandit and Arieira are doctoral students and Research Fellows in The Institute at George Mason University.

References

Akaike H (1973) Information Theory and an Extension of the Maximum Likelihood Principle. In Petron B N and F Csaki (Eds). 2nd Annual International Symposium on Information Theory. Akademiai Kiado Budapest 267-281

Balachandra R (1984) Critical Signals for Making Go/No Go Decisions in New Product Development. The Journal of Product Innovation Management 1 2

Blanchard B S and W J Fabrycky (1990) Systems Engineering and Analysis. Prentice Hall Englewood Cliffs NJ (2nd edition)

Baldwin C Y (1982) Optimal Sequential Investment when Capital is Not Readily Reversible. Journal of Finance 37:763-782

Bean J D, Higle J L and R L Smith (1992) Capacity Expansion Under Stochastic Demand. Operations Research 40:5210-216

Benson B, Sage A P and G Cook (1993) Emerging Technology-Evaluation Methodology: with Application to Micro-electromechanical Systems. IEEE Transactions on Engineering Management. 40 2 May 114-123

Bernanke B (1983) Irreversibility, Uncertainty and Cyclical Investment. Quarterly Journal of Economics 98:85-106

Blanchard B S and W J Fabrycky (1990) Systems Engineering and Analysis. Prentice Hall. Englewood Cliffs NJ (2nd edition)

Burwell D G (1993) Is Anybody Listening to the Customer? IVHS Review. Summer 17-26

Charnes A, Cooper W W, DeVoe J K, and D B Learner, (1966) DEMON: Decision Mapping Via Optimum Go/No Go Networks - A Model for Marketing New Products. Management Science XII 11

Charnes A, Cooper W W and E Rhodes (1978) Measuring the Efficiency of Decision Making Units. European Journal of Operations Research 2:429-444

Chryssolouris G, Graves S and K Ulrich (1991) Decision Making in Manufacturing Systems: Product Design, Product Planning, and Process Control. Proceedings of the 1991 NSF Design and Manufacturing Systems Conference Society of Manufacturing Engineers. Dearborn Michigan

Corbett J (1986) Design for Economic Manufacture. Annals of C.I.R.P. 35:1

Cunningham W H and I C M Cunningham (1981) Marketing: A Managerial Approach. South-Western Publishing Company Cincinnati

Fabrycky W (1990) Paper presented at the National Science Foundation International Workshop on Concurrent Engineering Design at the IC2 Institute October

Guba E G and Y S Lincoln (1981) Effective Evaluation: Improving the Usefulness of Evaluation Results through Responsive and Naturalistic Approaches. Jossey-Bass San Francisco

Hastings N A J and J B Peacock (1975) Statistical Distributions. Butterworth London

Haynes K E and R G Stubbings (1987) Planning and Philosophy: The Anthropology of Action. Journal of Planning Education and Research 6: 74-85

Haynes K E, Phillips F Y and R Srivastava (1993) Uncertainty in Infrastructure Management in Infrastructure Planning and Management. Gifford J, Uzarski D and S McNeil (Eds). Published by American Society of Civil Engineers New York 452-461

Haynes K E and Li Qiangsheng (1993) Policy Analysis and Uncertainty: Lessons from the IVHS Transportation Development Process. Journal of Computers, Environment and Urban Systems 17:28-35

Jakeila M (1989) Approaches to Product Optimization Proceedings: Product Optimization Interest Group Meeting CAM-I Computer-Aided Manufacturing - International Inc. (May 16-17) P-89-PO-3. Arlington TX

Kash Don E (1989) Perpetual Innovation: The New World of Competition New York. Basic Books

Kullback S (1959) Information Theory and Statistics. Wiley New York 196-207

Kydland F E and E G Prescatt (1982) Time to Build and Aggregate Fluctuations. Econometrika 50:1345-1370

Lazarsfeld P F and M Rosenberg (Eds) (1955) The Language of Social Research. Free Press Glencoe IL

Lewin K (1948) Resolving Social Conflicts: Selected Papers on Group Dynamics. Harper and Brothers New York

Lewis R M (1989) Product Optimization (PROPT) Program CAM-I Computer-Aided Manufacturing - International Inc. (Oct 23). Arlington TX

Luce L, Richard H and Wesley S C Lum (1992). The Influence of Human Factors and Public/consumer Issues on IVHS Programs. In Gifford J L, Horan T A and D Sperling (Eds) Transportation, Information Technology, and Public Policy, Fairfax, VA: George Mason University and Davis C A University of California Davis Institute of Transportation Studies

Majd S and R S Pindyck (1987) Time to Build, Option Value, and Investment Decisions. Journal of Financial Economics 18:7-27

Manne A S (1961) Capacity Expansion and Probabilistic Growth. Econometrika 29:632-649

Manne A S (1967) Investment for Capacity Expansion: Size, Location, and Time-Phasing. MIT Press Cambridge Massachusetts

McDonald R and D Siegel (1986) The Value of Waiting to Invest. Quarterly Journal of Economics 101:707-727

McGrath M E, Anthony M T and A R Shapiro (1992) Product Development: Success through Product and Cycle-Time Excellence. Butterworth and Heinemann Boston

Meier P M (1977) Game Theory Approach to Design Uncertainty. Journal of the Environmental Engineering Div ASCE 103:99-111

Morrison Donald F (1976) Multivariate Statistical Methods. McGraw-Hill New York (2nd edition)

Munnel A H (1990) How Does Public Infrastructure Affect Regional Economic Performance. New England Economic Review Sept 11-23

National Science Foundation (1990) International Workshop on Concurrent Engineering. Design IC2 Institute. University of Texas at Austin

Port O, Schiller Z and R W King (1990) A Smarter Way to Manufacture. Business Week (April 30) 110-117

Rasmussen A (1990) Producibility and Product Optimization in Concurrent Life Cycle Management. In Phillips, F Y. (Ed). Manufacturing, MIS and Marketing Perspectives IC² Institute. University of Texas at Austin

Rényi A Diary on Information Theory. John Wiley & Sons New York

Riddle F R (1989) Product Optimization Issues Proceedings: Product Optimization Interest Group Meeting CAM-I Computer-Aided Manufacturing - International Inc. (May 16-17) P-89-PO-3. Arlington TX

Robert K and M L Weitzman (1981) Funding Criteria for Research, Development and Exploration Projects. Econometrika 49:1261-1288

Sage A P (1989) Systems Management of Emerging Technologies. Information Decision Technology 15 4:307-326

Shannon C E (1948) A Mathematical Theory of Communication. Bell System Tech. Journal 37:10-21

Shadish W R Jr, Cook T D and L C Leviton (1991) Foundations of Program Evaluation: Theories and Practice. Sage Publications Newbury Park

Shields M D (1989) Product Life Cycle Cost Management Strategies for Product Optimization Proceedings: Product Optimization Core Group Meeting CAM-I Computer-Aided Manufacturing - International Inc. (Apr. 20) P-89-PO-02. Arlington TX

Stough R R and K E Haynes (1988) The Nature and Evaluation of Mega projects. In Roborgh L, Stough R R and T A Toonen Groen (Eds) Public Infrastructure Redefined, Netherlands 125-130

Srinivasan T N (1967) Geometric Rate of Growth of Demand. In Mannel (Ed) Investments for Capacity Expansion: Size, Location, and Time-Phasing. MIT Press Cambridge Massachusetts 151-159

Thore S and W R Touma (1990) The Management of Emerging Computer Technologies. A Portfolio Network with O-I Variables Working Paper #90-10-06. IC² Institute University of Texas at Austin

Urban G L and G M Katz (1983) Pre-Test-Market Models: Validation and Managerial Implications. Journal of Marketing Research 20:221-234

Wall Street Journal (1990) Odds and Ends second section - front page (April 4)

Warfield J N, Christakis A N, Keever D B and T LaBerge (1990) Interactive Management and Defense Systems Acquisition Management. Institute for Advanced Study in the Integrative Sciences. George Mason University, Fairfax Va

Washington Post (1992) Vital Statistics May 29

Weitzman M, Newey W and M Robin (1981) Sequential R&D Strategy for Synfuels Bell. Journal of Economics 12:574-590

Whitney D E (1989) Manufacturing by Design in Managing Projects and Programs Harvard Business Review Book Boston MA

Zirger B J and M A Maidique (1990) A Model of New Product Development: An Empirical Test Management Science 36(7):867-883

Appendix: Committed Costs Vs. Expenditures

It is the rare conference or publication on new project (product) development in which a variant of Figure 13.1 (in the text of this chapter) is not discussed, but Figure 13.2 is also often outlined without identifying sources. In this appendix, we trace the genealogy of the several versions of this graph that have been circulated throughout the project research and management community.

In each portion of Table 13.A1, the data series (columns) are given the names used by the original source. All numbers are cumulative.

Jakeila (1989) cites Rasmussen's 1989 presentation to CAM-I as the source of his figures, and Rasmussen in turn (1990) cites the Kodak corporation. Shields (1989) does not cite a source. Port, Schiller and King (1990) cite CAM-I as the source of the graphic in their Business Week article; the figures therein seem to be an amalgam of those presented by Riddell (1989) and Lewis (1989) at various CAM-I symposia. Riddell's original slide cites the Boeing corporation as the source of data. Lewis' slides cite the January, 1989 issue of Mechanical Engineering (denoted in Table A.1 above as ME) and the August, 1980 issue of Military Electronics/Countermeasures (MECM in Table A.1 below) respectively.

In addition, Whitney (1989) notes that 70% of General Motors's cost of manufacturing truck transmissions is determined in the design stage. Whitney also cites a study at Rolls Royce (Corbett, 1986) showing that 80% of the final production cost of various components is determined in design.

Whitney, and Riddell and MECM, use the terms determined cost and impact on cost, respectively. That their numbers are slightly higher than the corresponding numbers in the committed cost columns lends support to the distinction between committed and determined that we advanced in the text of this paper. Of course, there is no assurance that the product development stages named by these several sources are strictly or even approximately comparable, but these are the sources of the best and most commonly used (only) data sets on this topic.

Table 13.A1. Patterns of Committed/Determined Costs

I. Rasmussen (Kodak)

stage	'committed cost'	'spent'
concept	65%	5%
validation	85	8
development	95	20
production	100	100
operation	--	

II. Shields

stage	'committed cost'	'cash flow'
conception	65%	8%
design	85	14
testing	92	15
process planning	100	25
production	---	100

III. Riddell

stage	'determined'	'incurred'
concept development	70%	1%
advanced development	85	7
full-scale development	95	18
production	100	50
operation & support	---	100

IV. CAM-I (MECM)

stage	'impact on cost'	'cost'
concept	70%	3%
full development	85	15
production	95	50
operation & support	100	100

V. CAM-I (ME)

stage	'committed'
system planning & conceptual design	55%
preliminary system design	85
detail design and development	
production, construction & evaluation	100
system/product use & logistics support	

14 Diffusion and Acceptance of New Products and Processes by Individual Firms

Marina van Geenhuizen and Peter Nijkamp

14.1 Prologue

In recent years, the completion of the internal European market has exerted far reaching impacts on ways of articulating the region. In addition, the internationalization of the national economies has caused a greater specialization of each regional economy in its comparative advantage. This has led to an increased emphasis on the intrinsic regional profile and potentials. In practice, this may mean a strong appeal to the self-organizing capacity and perhaps even also the self-financing capacity of the region. There has always been rivalry between regions in Europe, but it has now become more intense than ever.

It is important to note that the European countries are showing a development towards an integrated (or at least open) network economy in which trade barriers are progressively being removed and spatial interactions are increasing. In such a borderless network economy the position of former border regions is likely to change dramatically as a result of a profound shift in their relative position, especially when they have a high indigenous capacity for innovation (Cappelin and Batey 1993; Nijkamp 1993, Suarez-Villa and Cuadrado-Roura 1993). Thus, both the European integration and the emerging network-economy make newcomer regions potential challengers of established centres and render each of these a potential loser.

The conditions for the diffusion and adoption of new technologies and innovations are unequally dispersed over space: patterns of innovation exhibit a clear geographic component, not only because of sectoral variation among firms in different areas, but also because of differences in locational requirements, available local and regional assets, and spatially discriminating urban and regional policies. Consequently, in the recent past many attempts have been made to analyze the geographic aspects of innovative behaviour.

Despite major advances, in the identification of regional key factors for innovation (cf. Aydalot and Keeble 1988), such as skilled labour and access to venture capital, the relationship between regional development and technological change still deserves much attention. While new technology clearly affects the fortunes of every region and city, it is much less clear under what influences and through what mechanisms individual economic actors adapt and innovate themselves and how the local environment affects this process. In fact, the process of technological change within companies has essentially remained a black box. Accordingly, there is a need for a more behavioural oriented *micro-perspective* on both regional and industrial dynamics, supported by empirical evidence.

The approach to diffusion and adoption of new technology in this article presupposes an *evolutionary* pattern of change (Nelson and Winter 1982). Accordingly, it will be taken for granted that whilst firms are making many incremental and sometimes also substantial changes, the pattern of change will to a large extent be subject to its historical development.

This chapter explores first the phenomena of diffusion and adoption of technology theoretically from an adoption process perspective. The aim of the article is to uncover - in a spatial context - the strategies which enabled individual companies to adopt new product and process technology. To this purpose, the textile industry in the Netherlands will serve as an example of a sector that could survive through both a strong contraction and major technical innovation. The empirical approach used in this article is company life-history analysis which links key decisions of companies to their past and current strategic context. The results will also serve to depict various conceptual issues encountered in the current analysis. These issues stem from the fact that acceptance of new technology is neither a simple nor an isolated phenomenon within the framework of a firm.

14.2   Diffusion and Adoption within the Context of the Firm

In an analysis of diffusion of innovation, it is necessary to distinguish between the subprocesses of generation, adoption and diffusion. Once inventions have been made innovations, knowledge about them is diffused to actors in other regions and countries. Some of the technical knowledge is recorded in manuals, papers and patents whereas other knowledge is embodied in hardware such as equipment and devices. But most of it is tacit and resides with individuals as employees of firms (Dosi et al. 1988). Depending on the kind of knowledge and the kind of innovation, the diffusion process follows different communication mechanisms and channels (Geenhuizen 1994, Charles and Howells 1992). Innovations are of interest to enterprises mainly when they affect, directly or indirectly, their competitive position. In decisions on adoption, therefore, two elements are relevant, i.e. the direction of the decision (positive or negative) and if positive, the time of adoption (early or late) in comparison with competitors.

Most research on innovation relies on the *adoption perspective*, in which the focus is on the process in which adoption occurs (see e.g. Fischer 1989). This demand-side approach views the diffusion of innovation in terms of adoption behaviour of the firms deciding to use the innovation in question. There is general agreement that four types of factors influence the decision on adoption, i.e. (1) characteristics of the innovation such as its cost and complexity, (2) firm's characteristics, particularly size, (3) industry characteristics, such as the competitive structure within the sector, and (4) institutional factors such as laws and regulations (e.g. Brown 1981). Furthermore, the focus of attention has recently shifted toward the role of networks and the interaction within these networks, such as between customers, suppliers and technical institutes (Alderman 1994, Gertler 1993, Grabher 1993, Hagedoorn 1993, Lundvall 1988, Williams and Gibson 1990).

However, the present analysis will elaborate on the adoption perspective, particularly the impact of the nature of the innovation and the firm on adoption. The remaining part of this section will therefore focus on these two influences on the process of adoption.

The *innovation* characteristics identified in the literature as being important are the capital cost of adoption, the technical complexity of the innovation, and the relative profitability or economic advantage of the innovation. Capital costs include the initial investment in implementing the innovation, operating expenses and the risk of adopting an unproved innovation. Shortage of financial resources may delay the adoption of an innovation at least in an early stage, or it may lead to an adoption in a limited manner, particularly in the case of a complex technology. In general, low cost innovations tend to diffuse more rapidly than high cost innovations. Moreover, it is expected that economically more advantageous innovations will diffuse more rapidly. Empirical research into this matter is, however, fraught with difficulties because the economic advantage (or profitability) of new product and process technology compared with alternative (conventional) technologies changes over time (Fischer 1989).

Various *firm* characteristics are also considered important influences on adoption behaviour. In this respect, *firm size* has received a great deal of attention. The influence of firm size is based on the rationale that larger firms can more easily bear the adoption risks, have easier access to capital and expert advice in order to evaluate the benefits (or disadvantages) of the new technology and to implement it within the factory. The adoption of a more complex innovation often needs a further in-situ modification before the innovation can be introduced successfully. Many studies point to early adoption amongst larger firms, although other research reveals that in particular fields (such as biotechnology) smaller firms may take a lead. In the latter fields, smaller firms benefit from their more flexible and adaptive nature. The adoption curve in these cases may, therefore, take an U-shaped form.

A perspective which seems to be somewhat overlooked in the context of adoption behaviour is the *position* of the firm in the various segments of the production system (Taylor and Thrift 1982, Morphet 1987). In the segmented economy approach it is recognized that production organizations fall into a number of company types, based upon their control over essential resources for production. For example, 'leader' companies are permanently engaged in the development of new products and markets, while 'laggard' companies focus strongly on manufacturing products which are nearing the end of their lifecycle. This type of specialization occurs within large corporations as well as within the small firms segment based upon subcontracting relations.

Close attention has been paid to the management style or *attitude* towards new technology (Fischer 1989, Freeman 1982, Porter 1985). In this respect, three major types of strategy can be distinguished, i.e. (1) offensive strategies, (2) defensive strategies and (3) imitation.

Firms adopting *offensive* strategies aim to maintain a lead over competitors by being the first to innovate. These strategies are very costly in terms of overhead on research laboratories and equipment, and staff of top researchers required. In research intensive sectors, a common strategy is to attempt to gain a lead in innovation, but this leadership is often limited to narrow fields such as sectors of the pharmaceutical industry. Apart from the costs, leader firms (pioneers) take various risks connected with both the market and the technology (simple failure or being overtaken by a group of innovators selling a superior and newer technology). The market may be opened up too late, for example, due to a shortage of market channels, decreasing the payback period needed for major breakthrough innovations. Leader firms can easily burn out and be taken over by rival firms (Olleros 1986). Solutions for reducing risks in offensive strategies include subcontracting in order to keep capital costs low and joint ventures with established firms in order to speed up sales. The latter is a common strategy in biotechnology (cf. Daly 1985).

Firms with a *defensive* strategy aim to react rapidly when a competitor introduces a new technology. They may introduce a lower-cost version or a slightly superior version of the new product. Firms that tend toward this technological follower-tactic must also maintain a sizable research and engineering presence.

A third strategy that can be adopted towards new technology is simple *imitation*. An imitative strategy will usually involve an emphasis on engineering expertise focusing on low-cost manufacturing processes.

One aspect which has largely been neglected is the *evolutionary* path of the firm (e.g. Alderman 1994). A firm's strategic changes are always a product of the recent past, such as the inherited technological direction and skills base, established output markets, contact networks and the organizational format. When there is a need for adaptation to new circumstances, the firm's search behaviour is to a large extent based upon accumulated knowledge and routines developed in the past, leading to solutions which are close to the previous situation. Consequently, an evolutionary view may significantly contribute to an explanation of adoption behaviour.

14.3   Diffusion and Adoption in Space

An inherent feature of regions is their 'struggle for life', in the sense that their final aim will be to survive and achieve the highest possible level of wealth. Their survival is, however, not a random phenomenon. It is based on their behaviour in competitive national, as well as increasingly international markets. Total demand in these markets is more or less given, and hence the only possibility of a regional system to attract a maximum market share is to be as competitive as possible. In many cases this requires continuous restructuring of the economic, environmental, industrial and technological base of the region. Thus, spatial-economic competition is a basic feature of regional and urban dynamics: the more competitive a region, the higher its survival chances. This competitive behaviour of regions needs to be seen as a rational decision-making process, including decision-makers from the business sector, the public sector and the public at large.

A dynamic regional system striving for a winning position in competition is facing specific characteristics influencing its behaviour (cf. Ewers and Nijkamp 1990, Nijkamp 1990), particularly the potential creation and adoption of new technology. These characteristics can be summarized as follows:

- A *limited carrying capacity*. This concerns land and resources (physical and human).
- *Multifunctionality*. This leads to the benefit various activities gain within the regional territory from each other (symbiosis). It strongly influences the incubation potential and underlying local learning processes.
- *Interaction and communication networks*, both internal and with other regions. The latter networks affect the region's potentials for adoption of new technology from elsewhere.

Spatial (regional and urban) dynamics are thus the result of internal and external responses by various actors (institutions). In this respect the self-organizing capacity of the region is of crucial importance. It provides, for example, consensus among the actors in the system, as well as a certain level of coordination between them. The self-organizing capacity of regional systems is also closely related to permanent and dynamic learning processes based upon various local amenities (cf. Camagni 1991). Local learning processes are, however, not an obvious attribute of regions. It is increasingly realized that under particular conditions (strong local and personal links) regional networks may blind and limit the receptiveness of firms for the need for innovation. Consequently, the competitiveness of regional firms may be strongly reduced (cf. Grabher 1993).

With regard to the spatial pattern of innovation diffusion, many theoretical analyses have made a distinction between two kinds of effects, i.e. a neighbour-hood effect and a hierarchical effect (cf. Hägerstrand 1967). In this 'classical' theoretical view, the diffusion pattern of a new technology is considered to be dependent upon the communication and information channels of a specific area. The neighbourhood effect reflects the crucial importance of physical distance in the diffusion process, i.e. the diffusion tends to concentrate in the vicinity of the originating source and decays strongly with distance. The hierarchical effect means that diffusion is first found in large metropolitan areas, then gradually trickles through the urban hierarchy to smaller centres.

What the above approach to spatial diffusion implicitly assumes, however, is that every potential adopter has an equal opportunity and need to adopt and that spatial variation in adoption rates depends solely on spatially biased information flows. Regarding the previous discussion on adoption within a firm's context (Section 14.2), the spatial factor clearly needs to incorporate a notion of different levels of receptiveness and capability of potential adopters (cf. Brown 1981; Capello 1993). This is the more relevant in small countries such as the Netherlands, where a spatially biased information availability seems no longer to exist. Consequently, large parts of such countries can be viewed as one 'urban field' (Wever 1987) in which spatial variation in innovation adoption is only based on a different receptiveness and capability of firms.

This chapter will explore in detail variation in adoption behaviour in the Dutch textile industry and will examine two types of innovation: (1) new product technology (i.e. technical textiles) in spinning and weaving companies, and (2) new process technology (i.e. tufting) in carpet factories. The spatial factor will also be considered. We compare firms located in different regions, noting their distance to the economic core of the Netherlands and their level of urbanization (Table 14.1).

Table 14.1 Spatial characteristics of the study-regions in the Netherlands

Region	Location relative to economic core (a) (the Randstad)	Level of urbanization
SE North-Brabant	Intermediate Zone	Medium
NE North-Brabant	Intermediate Zone	Medium
Twente	Outer Zone	Medium
North-Overijssel	Outer Zone	Low

(a)   The Intermediate Zone is adjacent to the economic core and the Outer Zone is at a further distance from this core.

14.4   A Longitudinal Company Perspective

Much of the empirical research on adoption and innovation has been conducted on a cross-section basis. The use of cross-section data may, however, have led to the neglect of the impact of accumulated knowledge in the firm and the impact of strategic choices made in the recent past.

An alternative approach to industrial dynamics involves an actor-oriented longitudinal analysis, meaning that the focus is on development trajectories of individual companies. For understanding corporate development trajectories it is useful to make a distinction between four components:

- Competitive strategy, which refers to the basis on which competitive advantage is built and sustained.
- Growth direction, which refers to the product-market combination.
- Growth method, which includes the way in which new competitive strategies and growth directions are undertaken, for example, through acquisition or internally.

- Organisational format, which includes a large variation in production organisation, from Fordist production to a complete flexible production.

Adoption of technical innovation is an integral part of two different competitive strategies, although in different ways (cf. Malecki 1986, Porter 1985). The first of these strategies, product differentiation, may be achieved by means of new products or improvement in product characteristics. The second competitive strategy, cost leadership, may be gained through the use of new cheap materials and labour saving processes. It makes often use of standardized production processes in which process improvement is of crucial importance.

In order to achieve in-depth insights into industrial dynamics, the present study makes use of Company Life History Analysis (Geenhuizen et al. 1992, Geenhuizen 1993). This approach can be described as: micro-analytical, longitudinal, qualitative and spatially oriented. The unit of analysis is the company and preferably the decision unit. In a retrospective analysis the company is observed for a period of usually 20 to 40 years in order to depict long to medium-term changes from an evolutionary perspective. The results achieved in company life-history analysis are largely qualitative, in-depth insights. Particular attention is given to key decisions which have led to significant corporate change, and also the strategic background of these decisions. Thus, company life-history analysis has the clear advantage that it enables us to identify explanations for technical innovation at the level where the decisions are taken and to consider an evolutionary context. In view of regional competition, the present analysis will put an emphasis on the variation in *offensive* and *defensive* corporate strategies.

In the first stages of the research, company life-history analysis normally makes use of a limited number of case studies, in order to reach sufficient in-depth detail and explanatory background (Schoenberger 1991). The analysis may, however, be extended to the whole population by a sample survey. In the initial data collection process, a multiple source approach is often used primarily because the data demand cannot be covered with one single source, but also because it increases accuracy and the validity of case study results (Yin 1991). These advantages were particularly helpful in the present analysis since they made it possible to avoid the effects of memory gap and personal bias on the part of respondents. The data sources used in the present case study research included: annual reports, in-depth corporate interviews, and external sources such as branch journals and patent registers.

We will turn now to our empirical results on innovation adoption based on a 'microscopic' analysis of the development path of various firms.

14.5 New Product Technology

The product technology to be discussed in this section is concerned with technical textiles (Werner International Inc. 1991). It embraces transport textiles (tyre cord, furnishing and end trim fabrics for automobiles and airplanes), geotextiles (for road-bed stabilisation, bankings and bridges), technical apparel to protect against specific hazard (heat, chemicals), medical textiles, and leisure textiles based on high-tech fibres (such as carbon and aramid). This section will explore development trajectories of large to medium-sized companies in the textile industry (for confidentiality reasons denoted by T1 to T3), located in different regions in the Netherlands. It puts an emphasis on the strategic context in which acceptance of the technology on technical textiles has taken place.

In the late 1950s and early 1960s, merging was a very common strategy in the Dutch textile industry (Table 14.2). The major aim of these mergers was to achieve economies of scale and further market access. Subsequent acquisitions, however, were based upon divergent growth directions. Since the early 1970s, certain companies (including T1 and T3) were successfully producing upgraded interior textiles and printed fabrics, whereas others were less successfully spinning and weaving for the highly competitive market of garment cloth (T2). This different orientation on output markets *within* the sector explains why companies like T1 and T3 aimed at sustained growth in traditional textiles and why T2 aimed at an early structural reorientation (in the late 1960s) towards new technical textiles and advanced materials (Geenhuizen and Van der Knaap 1994).

The development trajectory of T2 indicates that the first steps in the introduction of the new product technology were taken quite early. The joint ventures aimed at producing new technology (new fibres) were established already in the early 1960s and were largely abroad (United States, France). In addition, within a regional network of companies in the late 1950s, a basis was laid for the development of the so-called geotextiles. The first geotextiles were synthetic fabrics used for the construction of sea-defence works in the Netherlands (Delta works). This offensive strategy of creation and adoption of new product technology was clearly evident in the region of Twente, where the textile industry strongly contracted, but also partly shifted towards advanced materials and synthetic products. In the same region, however, some companies have retained or returned to the manufacturing of traditional textiles. Company T3 illustrates a follower strategy in view of its late adoption of advanced material technology in the early 1980s and a subsequent failure of this adoption followed by a return to traditional textiles.

In Southeast North-Brabant, textile companies were less innovative in terms of new products, but they introduced new process technology (such as in printing) and raised the product quality (design) of traditional textiles. This is exemplified by T1.

The above results provide some support for a spatially differentiated adoption of technical textiles and advanced material technology. It does not, however, conform to 'classical' spatial patterns because the first adoption took place in the Outer Zone at a relatively large distance from the economic core. An important factor in this respect has been the historical focus of companies on differently growing textile segments causing a different demand and receptiveness to new product technology. The results confirm some other empirical findings in the Netherlands which question innovation theory that articulates neighbourhood and hierarchical patterns in space (cf. Kleinknecht and Poot 1990).

Table 14.2 Key decisions in the development of textile companies (T1-T3)

T1 (SE North-Brabant)

From 1957	Increased market penetration of interior textiles.
1964	Merger
1969	Merger, expansion with cotton prints.
1975-84	Increased market penetration of interior textiles (carpets, curtain, wall paper).
1984	Little emphasis on technical fabrics. Forward integration in interior textiles (retailing).
1985	Persistent growth in traditional textiles (increasingly abroad).
1990	Increased penetration of foreign (European) markets by means of takeovers.

T2 (Twente)

1957	Merger and development of *new technical textiles*.
1960	Adoption/generation of *new fibre technology*, partially by means of joint ventures (abroad).
1969	Product development in advanced technical textiles.
1976/8	Withdrawal from spinning, stepwise withdrawal from weaving garment cloth.
1980	Persistent growth in advanced technical textiles. Also, diversification beyond textiles.
1991	Increased growth in advanced technical textiles. Also, withdrawal (divestment) from consumer textiles.

T3 (Twente)

1960	Merger
1970	Growth by means of related activities (e.g. textile services). Also, reduction of spinning and weaving.
From 1972	Product development (advanced interior textiles, i.e. sun blinds) and advanced process of finishing.
1975/6	Withdrawal from spinning.
1984	Adoption of *advanced technical textiles*, (e.g. through a takeover in 1986).
1987	Withdrawal from advanced technical textiles and renewed focus on interior textiles.

14.6 New Process Technology

In this section, we explore the acceptance of a major process innovation in the Dutch carpet industry, i.e. tufting. The analysis draws on various case studies of individual firms in the region of North-Overijssel (Outer Zone) and in Northeast North-Brabant (Intermediate Zone), denoted by T4 to T8 (Table 14.3).

Table 14.3 Strategic shifts in the carpet industry (T4-T8)

T4 (North-Overijssel) (1970: 25 workers)

1950 Small scale manufacturing of mats (carpets) of coir.
1970 Major product improvement (PVC backing of coir mats).
1976 Adoption of *tufting* of soft floor covering (in 1973 in cooperation with another local factory).
1980 Upgrading of the product quality of tufted floor covering.

T5   (North-Overijssel) (1970: 150 workers)

1950 Manufacturing of mats (carpets) of coir and sisal.
1969 Adoption of *tufting* of soft floor covering.
1974 Additional specialization in carpet backing and dying on demand of other carpet manufacturers.
1986 Adoption of tufting of synthetic grass.

T6   (North-Overijssel) (1970: 50 workers)

1950 Small scale manufacturing of mats (carpets) of coir.
1967 Adoption of product technology of needle-felt.
1972 Adoption of *tufting* of soft floor covering.
1985 Product differentiation of needle-felt.
1989 Upgrading of the product quality of tufted floor covering.

T7   (NE North-Brabant) (1970: 700 workers)

1960 Large scale manufacturing (weaving) of high-quality carpets and furniture coverings.
1965 Adoption of *tufting* of carpets and floor covering.
1977 Closing down of weaving department and major contraction.
1982 Liquidation.

T8   (NE North-Brabant) (1970: 1,000 workers)

1950 Large scale manufacturing (weaving) of carpets, also needle-felt.
1963 Adoption of *tufting* of carpets and floor covering.
1965 Expansion by means of various takeovers (in Belgium).
1975 Adoption of new dying technology (within a joint venture).
1981 Generation of new product technology of synthetic grass covering (for sports grounds).

Tufting of carpets is different from weaving in that the carpet piles are stitched into a base and then fixed with a latex backing. Due to the far greater labour productivity compared to weaving, tufted carpeting is much cheaper than producing similar woven qualities and as a consequence, comes within easy reach of mass consumption. Tufting therefore, is considered as an important process innovation (Toyne 1984, Rothwell and Zegveld 1985). Together with the growing market demand due to the increased prosperity of the Dutch population (after the late 1960s) and a massive construction of public housing, tufting caused a tremendous growth in the carpet industry. This growth was at odds with the Dutch textile industry at large, which was clearly in a stage of stagnation (Geenhuizen and Van der Knaap 1994).

Regarding the time of adoption, a clear variation can be found between carpet factories in North-Overijssel and those in Northeast North-Brabant. Compared with the average adoption time for the total country (1968), adoption was relatively early in Northeast North-Brabant and late in North-Overijssel (Table 14.3). Our analysis points to two factors that contributed to a late adoption in the latter region. The shift to tufting here involved not only a new process technology but also new product types and particularly a new production organization. The adoption included a shift from hard floor covering (made from sisal and coir) to soft floor covering (wool, synthetic) and, concomitantly, a move from small scale (home) industry to large scale factory work. This comprehensive change, which clearly went beyond the technology itself resulted in a wait-and-see policy (follower strategy). In other regions, however, tufting technology was added to an already existing large-scale weaving of carpets.

The results also indicate an influence of firm size. In terms of the number of employees, the companies in North-Overijssel were significantly smaller than those in North-Brabant (Table 14.3). The modest financial capacity of the former advanced a wait-and-see policy, particularly in the early 1960s when large market risks were still perceived. The results also indicate the influence of the strategic context. Adoption may be delayed or absent when the technology in question does not fit the core of the firm's activities and products. In both T4 and T6 late adoption was partly caused by the preceding product improvement and new product technology in a product area slightly outside the technology in question.

In conclusion, the spatial variation in adoption time confirms 'classical' theories that emphasize the significance of the spatial information availability factor, as adoption came relatively late in a region in the Outer Zone with a low level of urbanization. However, a corporate receptiveness factor seems to have reinforced this influence.

14.7 Conceptual Issues

By using a micro-focus and an evolutionary perspective the analysis developed in this study leads to some significant conceptual conclusions concerning the phenomenon of adoption of new technology by firms. It highlights the importance of two issues have been underestimated in past research.

The first issue concerns the *definition of adoption*, put forward by Alderman (1990). The theoretical concept seems not to be entirely consistent with empirical reality in two ways. Adoption in corporate reality is more a process than an isolated event in time. The question then arises on the time of completion of the adoption, for example, when the new machinery (equipment) has been implemented or when the new technology contributes significantly to production? There may be a gap between these points of several years, especially when a complex technology is involved.

A further question of definition matter concerns the adoption of technology within the framework of interfirm collaboration. As Dicken (1990) has underlined, the confines of a company in relation to other organizations has become progressively 'fluid' in the past decade. New technologies tend to be increasingly developed in joint ventures, jointly owned research companies and various temporary networks (Gibbons et al. 1994; Hagedoorn 1993). It may happen that the parent firm is involved in the adoption in terms of finance, organization and the input of know-how, but that the technology is not actually implemented in its own factory, but somewhere else, even in another region. The question then arises as to how the adoption should be ascribed to the various partners in the technical collaboration.

The second issue concerns the *embedding* of the acceptance of technical innovations. The acceptance of new product and new process technology is never an isolated action. The embedding involves the organizational infrastructure of the firm, innovation in other product fields and changes in non-technological dimensions of the firm in question. Certain technical innovations cannot be adopted when an appropriate organizational infrastructure for them is absent. In this respect, our case studies indicated the relevance of organizational requirements enabling the shift from small scale work towards large scale manufacturing. Accordingly, the time of adoption may be affected by the organizational state of the companies at the time the technology is diffused into a region. Furthermore, while certain generic innovations are important for each company in an industry sector (such as the introduction of a fax machine), various other innovations are only crucial for companies in particular *segments* in a sector (particular product-market combinations). In addition, the use of new technology is only one dimension of change on which firms can sustain their competitiveness. In the same time period firms may change their market strategy (e.g. a shift to international markets), their marketing, etc. The latter strategies often need a significant amount of resources, which may cause a delay of new technology acceptance.

The above arguments do have some implications for the definition and identification of the 'potential adopter', a common unit in spatial diffusion and adoption research. In any case, potential adopters are certainly not all firms within one industry sector. They have to be defined in a more precise way.

14.8   Epilogue

Spatial-industrial dynamics is considered a key question in modern regional research. Regional economies are currently going through a new, often revolutionary phase in their economic transformation process, based upon greater competition.

The recent revival of *Schumpeterian* views on current spatial economic restructuring phenomena has stimulated scientific interest in innovation and economic transformation. Both the behavioural stimuli and the selection environment for the creation and adoption of technological and organizational change in firms have become subject to intensive research. The analysis presented here, clearly belongs to this stream of research.

In this chapter we have questioned the purely spatial (information availability) factor in the process of innovation acceptance. To this purpose we have adopted a microscopic view of technology acceptance and a longitudinal perspective, based on company life-history analysis. Our findings provide support for spatially heterogeneous acceptance. However, they only partially confirm diffusion theory which articulates a spatial variation in information availability. The spatial variation we have found seemed to be related to a spatial differentiation in receptiveness of firms for new technologies and in the capability of firms to make use of the information concerned. According to the above argument, what matters more in adoption than simple physical distance and urbanization level, is what may be called *organizational* and *strategic* distance.

It can be concluded that company life-history analysis gives a detailed insight into new product and process acceptance on the micro-level. It can clearly contribute to an explanation of spatial variation in innovation adoption because it uses a differentiated approach, justified by the complexity of corporate reality.

References

Alderman N (1990) Methodological Issues in the Development of Predictive Models of Innovation Diffusion. In: Ciciotti E, Alderman N and A Thwaites (Eds) Technological Change in a Spatial Context. Springer Berlin 148-166

Alderman N (1994) Technological Trajectories at the Local Scale: the Role of Contact Networks. Paper presented at the 34th European congress of the RSA Groningen (NL) August 23-26

Aydalot and D Keeble (Eds) (1988) High Technology Industry and Innovative Environments. Routledge London

Brown L (1981) Innovation Diffusion. A New Perspective. Methuen London

Camagni R (1991) Innovation Networks: Spatial Perspectives. Belhaven Press London

Cappelin R and P W J Batey (Eds) (1993) Regional Networks, Border Regions and European Integration. Pion London

Capello R (1993) Regional Economic Analysis of Telecommunications Network Externalities. PhD Thesis. Free University Amsterdam

Charles D and J Howells (1992) Technology Transfer in Europe, Public and Private Networks. Belhaven Press London

Daly P (1985) The Biotechnology Business; A Strategic Analysis. Francis Pinter London

Dicken P (1990) The Geography of Enterprise. Elements of a Research Agenda. In: Smidt M de and E Wever (Eds) The Corporate Firm in a Changing World Economy. Routledge London 234-244

Dosi G, Freeman C, Nelson R, Silverberg G and L Soete (Eds) (1988) Technical Change and Economic Theory. Pinter Publishers London

Ewers H J and P Nijkamp (1990) Sustainability as a Key Force for Urban Dynamics. In: Nijkamp P (Ed) Sustainability of Urban Systems. Avebury Aldershot 3-16

Fischer M M (1989) Innovation, Diffusion and Regions. In: Andersson A, Batten D and C Karlsson (Eds) Knowledge and Industrial Organization. Springer Berlin 46-61

Freeman C (1982) The Economics of Industrial Innovation. Frances Pinter London

Freeman C, Clark J and L Soete (1982) Unemployment and Technical Innovation. Frances Pinter London

Geenhuizen M van, Nijkamp P and P Townroe (1992) Company Life History Analysis and Technogenesis: A Spatial View. Technological Forecasting and Social Change 41:13-28

Geenhuizen M van (1993) A Longitudinal Analysis of the Growth of Firms. The Case of The Netherlands. PhD Thesis. Erasmus University Rotterdam

Geenhuizen M van (1994) Barriers to Technology Transfer; the Role of Intermediary Organisations. In: Cuadrado-Roura J R, Nijkamp P and P Salva (Eds) Moving Frontiers: Economic Restructuring, Regional Development and Emerging Networks, Avebury Aldershot 247-276

Geenhuizen M van and B van der Knaap (1994) Dutch Textile Industry in a Global Economy. Regional Studies 28:695-711

Gertler M (1993) Being there: Proximity, Organization, and Culture in the Development and Adoption of Advanced Manufacturing Technologies. Paper presented at the Annual Meeting of the Association of American Geographers. April 6-10 Atlanta

Gibbons M et al (1994) The New Production of Knowledge: The Dynamics of Science and Research in Contemporary Societies. Sage Publications London

Grabher G (1993) The Embedded Firm. On the Socioeconomics of Industrial Networks. Routledge London

Hagedoorn J (1993) Strategic Technology Alliances and Modes of Cooperation in High-technology Industries. In: Grabher G (Ed) The Embedded Firm. On the Socioeconomics of Industrial Networks. Routledge London 166-137

Hägerstrand T (1967) Aspects of the Spatial Structure of Social Communication and the Diffusion of Innovation. Papers of the Regional Science Association 16:27-42

Howells J R L (1983) Filter-down Theory: Location and Technology in the UK Pharmaceutical Industry. Environment and Planning A 15:147-164

Kleinknecht A H and A P Poot (1990) De Regionale Dimensie van Innovatie in de Nederlandse Industrie en Dienstverlening. SEO University of Amsterdam

Lundvall B-A (1988) Innovation as an Interactive Process: From User-producer Interaction to the National System of Innovation. In: Dosi G, Freeman C, Nelson R, Silverberg G and L Soete (Eds) Technical Change and Economic Theory. Pinter Publishers London 349-369

Malecki E J (1986) Technological Imperatives and Modern Corporate Strategy. In: Scott A J and M Storper (Eds) Production, Work and Territory: The Geographical Anatomy of Industrial Capitalism. Allen and Unwin Winchester 69-79

Malecki E (1991) Technology and Economic Development: The Dynamics of Local, Regional and National Change. Longman New York

Morphet C S (1987) Research, Development and Innovation in the Segmented Economy: spatial Implications. In: Van der Knaap G A and E Wever (Eds) New Technology and Regional Development. Croom Helm London 45-62

Nelson R R and S G Winter (1982) An Evolutionary Theory of Economic Changes. Harvard University Press Cambridge

Nijkamp P and U Schubert (1984) Urban Dynamics and Innovation. In: Brotchie J, Newton P, Hall P and P Nijkamp (Eds) Technological Change and Urban Form. Croom Helm London

Nijkamp P (Ed) (1990) Sustainability of Urban Systems. Avebury Aldershot

Nijkamp P (1993) Border Regions and Infrastructure Networks in the European Integration Process. Environment and Planning C 11:431-444

Olleros F J (1986) Emerging Industries and the Burnout of Pioneers. Journal of Product Innovation Management 3:5-18

Porter M E (1983) The Technological Dimension of Competitive Strategy. Research on Technological Innovation, Management and Policy 1:1-33

Pred A (1977) City-Systems in Advanced Economies. Hutchinson London

Rothwell R and W Zegveld (1985) Reindustrialization and Technology. Longman Harlow Essex

Schoenberger E (1991) The Corporate Interview as a Research Method in Economic Geography. Professional Geographer 43:180-189

Suarez-Villa L and J R Cuadrado-Roura (1993) Regional Economic Integration and the Evolution of Disparities. Papers in Regional Science 4:369-387

Taylor M J and N J Thrift (1982) Industrial Linkage and the Segmented Economy. Environment and Planning A:1601-1632

Toyne B (1984) The Global Textile Industry. Allen & Unwin London

Werner International Inc. (1991) Situation and Perspective of Technical Textiles in the European Community. Commission of the European Communities Brussels

Wever E (1987) The Spatial Pattern of High-growth Activities in the Netherlands. In: Van der Knaap G A and E Wever (Eds) New Technology and Regional Development. Croom Helm London 165-185

Williams F and D V Gibson (1990) Technology Transfer, A Communication Perspective. Sage Newbury Park

Yin R K (1991) Case Study Research. Design and Methods. Sage Newbury Park

15 Innovative Capacity, Infrastructure and Regional Inversion: Is there a Long-term Dynamic?

Luis Suarez-Villa

(U.S.)

H 54

Ō31
Ō32

R32

15.1   Introduction

Among the most interesting regional phenomena of the twentieth century is the emergence of previously peripheral or outlying regions as major sources of technology and of economic dynamism. In a few decades, several of these regions have become some of the world's most important repositories of technological knowledge, with impacts that reach well beyond their respective national and continental boundaries. The diffusion of innovations has been greatly affected by the technological emergence of those previously lagging or disadvantaged regions, as knowledge networks, shorter product cycles and rising technical capabilities have allowed greater access to innovations than ever before in human history.

The development of endogenous technological capabilities seems to have been crucial for those processes of *regional inversion*, whereby previously peripheral regions have converged with or even overturned the predominant position of older, industrial heartland areas. The new regional order promoted by such inversion processes has accounted for the emergence of new technologies, productive sectors, institutions and powerful political bases in areas that had little human capital, a sparse infrastructure and few industries only a few decades before. The swiftness of these regional inversion processes has also introduced remarkable changes in economic competitiveness and trade, as integration in continental trading blocs advances and regional economies acquire greater importance than ever before.

This chapter will provide an analysis of what are perhaps the two most important factors underlying the regional inversion process: technological invention and infrastructure. A definition of the concept of regional *innovative capacity* will be provided first, along with its application to the American context. This will be followed by a consideration of major historical trends for innovative capacity and infrastructural investment for 1920-90, to determine their impact on the changing regional order favouring the American Sunbelt regions. Finally, the age cycle dynamics of regional innovative capacity and of national educational infrastructure will be examined, to determine whether any possible causal correspondence may exist between them, and its implications for the regional inversion process.

15.2   The Rise of Regional Inversion: Invention, Innovation and Infrastructural Trends

Over the past five decades, the United States has experienced a regional inversion process of unprecedented proportions, as areas of the Sunbelt became the most important sources of new technology and economic growth (see Suarez-Villa 1993; 1992, 1989, Berry 1991, 1988, Chinitz 1986; Crandall 1993, Norton 1990). The American experience of regional inversion is perhaps the most important among all the advanced nations. New technologies and industries, such as computers, semiconductors, telecommunications, biotechnology and advances in aerospace have emerged from geographical areas where little manufacturing or human capital existed only a few decades ago. Massive interregional population shifts have occurred, changing the balance of national political power and redirecting public resources toward outlying states and regions (see, for example, Schulman 1991; Sale 1975). Over time, these regions have gained a pivotal role in deciding national competitiveness and the fortunes of international trade, especially as the North American trade bloc began to take form. All indications are that, as trade and technological change intensify within the North American bloc and internationally, this on-going inversion process stands to gain even greater dynamism.

A major factor in the regional inversion process is the endogenously-generated technological change which develops through the innovative capabilities of a region's human capital infrastructure. A way to measure a region's innovative capabilities is to consider some indicator of technological and scientific inventive output which is statistically reliable, which for historical time series are available, and which has institutionalized selection criteria that are consistent over time. In the United States, invention patent awards fulfil all of these requirements. Such awards are probably the most reliable and historically consistent data resource available in the United States, with annual reporting that spans over a century and that have consistent evaluation and award criteria (Suarez-Villa 1993, 1990; Griliches 1990; Acs and Audretsch 1989; Basberg 1987).

Invention patents are not the only source of innovation, however, and this must be taken into account whenever this data resource is utilized. Other sources of innovation are, for example, associated with aspects of organization, such as those changes implemented to improve networking or to modify hierarchies, communication structures, or the internal division of labour within a firm. It is also possible that some inventions may never be patented, even if they are actually utilized in some economic activity. In a competitive market economy, where monopolies are outlawed and oligopolistic tendencies are restrained through imports, that possibility can be considered to be virtually insignificant, however. As an appropriation safeguard, therefore, the legal and property value of patenting remains high in the United States, and it is a powerful incentive to ensure any appropriation of rewards, given the uncertain nature of innovation and the potential applications that may eventually be found by a patent's owner or by others, including competitors (Griliches 1990; Griliches et al. 1987; Samson 1990; Hounshell and Smith 1988; Narin et al. 1987; Wyatt 1986; Horstmann et al. 1985;

Bound et al. 1984; Scherer 1983; Allison et al. 1982). It should be remembered that many significant inventions have historically found application well outside the immediate field for which they were created, with unforeseen benefits and rewards. In many cases, such innovative uses were not originally envisioned by their creators, and it remained up to others to discover their potential and to seek licensing for their introduction in economic practice (Kuhn 1962; Davies 1979; Popper 1981; Harre 1988; Hull 1988; Holton 1988).

The measurement of endogenous regional technological capability will be carried out by using a macro-performance indicator: *innovative capacity*. As defined in this article and in Suarez-Villa (1990, 1993), innovative capacity is a measure of inventive output comprising all patent awards that are legally available for use in any given year. The innovative capacity is therefore an aggregate measure of successful inventive outcomes: 'success' being defined exogenously by a positive decision in the evaluation process required for a patent award. Meeting all the requirements regarding originality and distinctiveness is the most important objective of that evaluation process, and it confers a property right which is both legally and economically defensible.

Innovative capacity is also a measure of regional innovative *potential*, since it indicates the aggregate inventive output that is available for innovative use, regardless of whether it is actually applied or not. As an indicator of endogenous capabilities, it reflects a region's human capital resource base, which is to be regarded separately from any consideration of economic use or success. The economic application of any invention may be successful or ill-fated, depending on many factors; however, innovative potential must be considered in its own right, whether it leads to successful economic ventures or not. Nevertheless, it can be assumed that the higher the innovative capacity, the more likely it is that economic applications will result, with a possibly greater probability of economic success. A value of the innovative capacity concept is, therefore, that it provides a measurement of the inventive capability and the innovative potential of any given area, regardless of whether economic applications are successful (measured by the usual indicators of profitability, market share, revenues and others), or whether such applications occur within the region of origin or elsewhere.

In essence, therefore, the innovative capacity measures the output of inventions 'at source'. Tracing the economic application of any invention, and whether it is successful or not (in economic terms) is quite a different matter. While the latter can provide an ultimate measure of economic impact or 'success' for any given invention, its use as a measure of national or regional innovative potential would be rather flawed, since it would not include a broad-based appreciation of inventive creativity and output. Such a measure would also seem to ignore the difference between invention and innovation, where the former relates to discoveries and new ideas, and the latter refers to the actual usage or application of an invention. This important distinction is usually ignored in the economic literature, where the terms tend to be confused or are assumed to have the same meaning (see Suarez-Villa 1993; Suarez- Villa and Hasnath 1993).

In the United States, patent statistics are compiled geographically, recording the location of residence of the first-listed inventor, regardless of whether the patent right is actually held by an individual or by a firm (see U.S. Patent and

Trademark Office 1989; U.S. Department of Commerce 1990; U.S. Bureau of the Census, various years). It is therefore possible to obtain a regional distribution of patent awards that is both very reliable and historically consistent (see Suarez-Villa 1993). The legal term for all patent awards is 17 years, and renewals or extensions are rare since they would require a special request by the U.S. Congress.

Regional innovative capacity will therefore be a moving 17-year measure of all valid patents awarded to inventors in a designated geographical area. It is in the best interest of any region to have a growing innovative capacity, since a larger pool of endogenously-generated inventions is likely to result in a greater probability of innovative applications. A declining innovative capacity is bound to have troublesome implications, not only for the rate of endogenously-generated innovation, but also possibly for a region's future technological and economic competitiveness, and for its human capital infrastructure. While many (or even most) of the innovative applications resulting from a region's inventive output may not occur locally, the endogenous technological capabilities that it implies will nevertheless stand to benefit the region of origin, through any local applications that may occur or through the financial rewards to invention that may be realized. It should be noted that regions that are able to sustain a rising innovative capacity may also benefit from an in-flow of actual and potential inventors attracted by the existing or expanding human capital infrastructure, creating the conditions for further increases in endogenous inventive output.

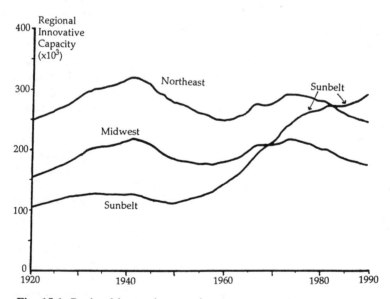

Fig. 15.1. Regional innovative capacity

The innovative capacity trends shown in Figure 15.1 reflect a regional inversion process that has favoured the American Sunbelt throughout the past four decades.

The Sunbelt's rising inventive output after 1950 contrasts sharply with the levelling Northeastern and Midwestern trends of the mid-1970s and with their subsequent decline (for a more detailed treatment of these trends, see Suarez-Villa 1993). The Sunbelt's technological rise was helped by the emergence of corporate research and development (R&D) as a major source of invention. The rise of systematic R&D within corporate organizations was largely aimed at reducing the inherent risk and uncertainty of invention; in time, the location of R&D would also become an integral part of corporate strategy-making, leading to a spatial fragmentation of production and research in many sectors (see, for example, Marschak et al. 1967; Mansfield 1971; Beggs 1984; Donaldson and Lorsch 1986; Rubin and Huber 1986; von Hippel 1988; Chandler 1990; Davelaar and Nijkamp 1990; Kamann and Nijkamp 1990, Suarez-Villa and Fischer 1995, Suarez-Villa and Karlsson 1995). By the early 1940s corporate invention had replaced individual invention as the United States' most important source of patents, leading to a rapid rise in the national innovative capacity (see Suarez-Villa 1990).

Perhaps the most important aspect of American regional inversion and technological change is the role which infrastructural investments have played in the process. It may be assumed that the development of human capital required for invention depends heavily on certain infrastructural investments, particularly those more directly related to education and research. A deeper treatment of the process of regional inversion and the role of technological change therefore needs to examine the relationship between infrastructure and invention from a long-term perspective, to determine their causal potential in the inversion process.

In the United States, highway and educational facility construction are the two largest categories of public infrastructural expenditure (see U.S. Bureau of the Census 1981, American Public Works Association 1976, U.S. Department of Commerce, various years). Figure 15.2 presents both the aggregate and the educational public infrastructure construction trends for the nation (see Suarez-Villa and Hasnath 1993, for a more detailed discussion of these trends). Historically, public construction in the United States has been primarily a state and local government function, although the federal government has provided a significant amount of supplementary funding for state and local projects over the years. Infrastructural construction by all government levels (state, local, federal) has historically accounted for between one-third and one-seventh of all national construction, including both public and private expenditures (see Suarez-Villa and Hasnath 1993; Hasnath and Chatterjee 1990).

A brief overview of the regional innovative capacity and the national infrastructural construction trends in Figure 15.2 reveals some common tendencies. The rises in both aggregate and educational public infrastructural construction of the mid-1940s through the late 1960s appear to correspond with the Sunbelt's rising innovative capacity trend starting in the early 1950s. The situation after 1970 is less clear-cut, however, as the infrastructural trends fluctuate sharply. It should be obvious, therefore, that a better answer to the question of any possible correspondence between infrastructure and endogenous regional invention must be sought by some other means. The next section will provide a close analysis of that relationship by considering the age cycle dynamics of infrastructural construction investment and regional invention patenting.

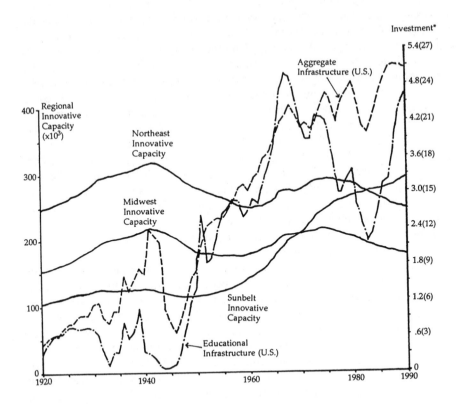

* constant 1958 billion dollars
aggregate infrastructural investment in ()

Fig. 15.2. Public Infrastructural investment and regional innovative capacity

15.3   The Long-term Dynamic: Genesis and Cyclical Change

If infrastructure has been essential to the rise of invention and of technological capabilities in outlying regions, identification of the underlying dynamic linking these processes could contribute much to our understanding of how regional inversion occurs and develops. Invention and innovation are possibly, through their myriad effects, the factors that lend the inversion process its most enduring qualities. It is hard indeed to think of any other endogenous factors that could contribute as much to the self-sustaining, long-term character of the inversion process.

The contribution of infrastructural investment to invention occurs, first, through the provision of facilities that help develop a region's indigenous human capital resources, and accommodate any in-flows of human capital from other regions.

The importance of the latter should not be underestimated, particularly as the inversion process takes off and an outlying region becomes a net recipient of highly skilled labour. Such in-flows have, for example, been very important in the rising technological prowess of several areas in the American Sunbelt, such as Silicon Valley, Orange County and Southern California, and the outer-space exploration enclaves in Texas and the Southeast.

The built infrastructure also provides many operational advantages to innovative firms, particularly to the small ones that rely to a considerable extent on externalities for their technological, human capital and production needs. Such advantages make positive income elasticities a major characteristic of most public infrastructure, particularly for those investments related to education and research, and to the communication of specialized knowledge. Easy access and high quality also tend to enhance the infrastructure's income-elastic characteristics, providing greater opportunities for networking and interfirm cooperation (see Suarez-Villa and Hasnath 1993; Suarez-Villa 1995). For practical purposes, the term 'investment' will be used here synonymously with 'spending'; virtually all infrastructural expenditures can be considered to be investments of a sort, given their positive income-elastic characteristics and their deferred (but nevertheless very important) returns on economic productivity and human capital development.

Analyzing the long-term dynamics of regional inversion will require a consideration of their temporal character. Both infrastructure and invention patents age and can generally be expected to be less effective over time, as obsolescence sets in. Although invention patents have a legal term of 17 years, most usage by the owners of the patent, their firms, or through licensing tends to occur early on. As knowledge of a discovery or idea diffuses, alternatives for which it is unnecessary to seek licensing from the patent's owner, are more likely to be found, thereby reducing its economic potential and technological impact.

Similarly, facilities age and decline over time, becoming less adequate for the purposes for which they were built. State-of-the-art educational facilities, such as research laboratories and learning centres, tend to have a greater impact on invention than older ones. Out-of-date facilities may need to be refurbished substantially, to make them useful again for the purposes for which they were originally built. The refurbishment can be compared to the granting of a new patent in that it provides a significant and distinct improvement on a previously established facility, whose legal life term may already have lapsed. It can be argued that such temporally sensitive characteristics make the comparison of the ageing dynamics of both patenting and knowledge-directed infrastructural spending an interesting exercise.

If a significant correspondence can be found between the cyclical fluctuations of invention patenting and of infrastructural construction, it may provide a positive indication of what is perhaps the most important underlying dynamic favouring the process of regional inversion. This potential temporal association can be measured by grouping patent awards and built infrastructural spending into specific age groups. Utilizing the 17-year life term of patent awards as a guide, three age groups of equal spans can be specified (0-5, 6-11, 12-17 years) for both invention patenting and infrastructural spending. Each age group will then become a moving annual indicator of the valid invention patents and of the monetary value of educational infrastructure spending that they comprise (see Suarez-Villa 1990; 1993; Suarez-Villa and Hasnath 1993).

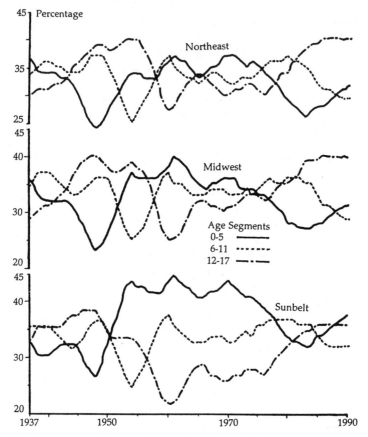

Fig. 15.3. Innovative capacity: regional age cycles for patents

A rising innovative capacity can be predicted whenever a certain rank order of the three age groups for invention patenting can be found. In essence, therefore, interpreting the age cycle dynamics shown in Figure 15.3 make it relatively easy to forecast rises in innovative capacity and in the prospects of self-sustained increases over long periods. A rising innovative capacity trend (in Figure 15.2) can be unequivocally forecast whenever the newer patent age group (0-5 years) has the largest share of all the age categories, followed by the middle (6-11) and older (12-17) age segments. Similarly, a rising infrastructural spending trend can be anticipated whenever the relative shares of total spending favour the younger age group, followed by the middle and older segments. Subsequent rises in the overall innovative capacity (and infrastructural) trends are so predictable whenever these rank-ordered sets occur that they seem to dispel long-held notions about the difficulty of anticipating long-term changes in the performance of inventive activity.

299

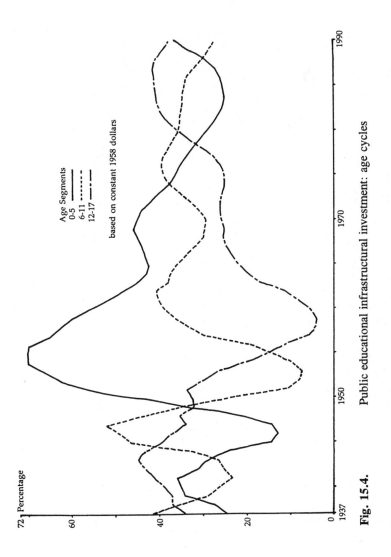

Fig. 15.4. Public educational infrastructural investment: age cycles

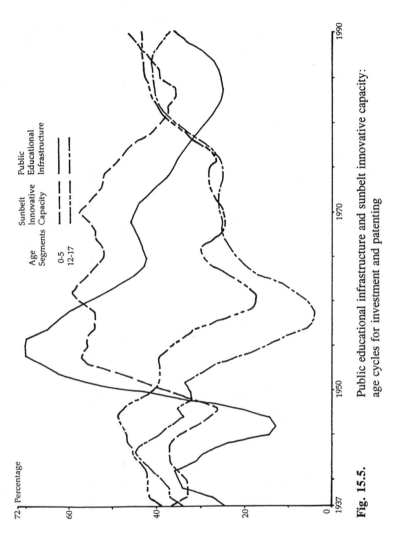

Fig. 15.5. Public educational infrastructure and sunbelt innovative capacity: age cycles for investment and patenting

The regional age cycles shown in Figure 15.3 reveal what could be considered to be the genesis of the American regional inversion process favouring the Sunbelt. By the late 1940s, well before any other socio-economic indicators could show the changes to come, the Sunbelt's patent age cycle dynamics provided the first signs of the changing regional order (see Suarez-Villa 1992, 1993). It should be noted that, of the three regions shown in Figure 15.3, only the Sunbelt shows the prescribed rank order of the age segments, which promoted its rising innovative capacity throughout the 1950s, 1960s and 1970s. Thus, the Sunbelt's newer patent age group (0-5 years) retains the largest share of valid patents, followed by the middle group (6-11 years) and the oldest patent group (12-17 years) throughout those decades.

The age cycles of U.S. educational infrastructure investment, shown in Figure 15.4, provide the desired rank order of age segments from the mid-1950s through the mid-1970s. The newer age segment (0-5 years), however, rose rapidly after the mid-1940s, reaching a 70 percent share of total educational infrastructure investment by the mid-1950s. It is interesting to note that the time span between the bottom points of the newer age segment (0-5 years) in the mid-1940s and in the mid-1980s amounts to approximately 40 years. This period is quite similar to the Sunbelt's 40-year lapse between the two bottoming-out points of its newer patent age group's cycle in the late 1940s and again in the mid-1980s (see Figure 15.3). This brings up the question of a possible correspondence or causal association between the educational infrastructure investment and the Sunbelt's innovative capacity trends.

The age cycles of the Sunbelt's innovative capacity and of U.S. educational infrastructural investment, combined in Figure 15.5, show a remarkable convergence in the timing of the major turning points of their cycles. For the sake of simplicity, Figure 15.5 shows only the cyclical fluctuations of the newer (0-5 years) and the older (12-17 years) age cycles. These two age segments are crucial in determining changes in the overall trends of Figures 15.1 and 15.2, inasmuch as they provide indications of the entry and exit quanta for the 17-year innovative capacity and infrastructural investment terms. It should also be noted that, for all practical purposes, infrastructural spending refers to built (or 'in place') infrastructure, since the data are based on disbursement of construction funds; for most public projects, complete disbursements usually do not occur until a facility either nears completion or is actually completed (see Suarez- Villa and Hasnath 1993).

The convergence of the age cycles shown in Figure 15.5 for the concurrent turnabouts (bottoming points) of educational infrastructure (1945, 1984) and the Sunbelt's innovative capacity (1947, 1984) for their newer age cycle segments (0-5 years), seems to indicate a potential causal correspondence. Similarly, for their older (12-17 years) segments, the turnabouts for educational infrastructure (1958, 1975) and for the Sunbelt's innovative capacity (1960, 1976) cycles show a remarkable convergence in their timing. The situation with the peak points of their cycles also appears to display synchronicity; peaks for the newer age segments (0-5 years) in 1941, 1954 and 1969 for educational infrastructure corresponding respectively to the 1942, 1954 and 1970 turnabouts for the Sunbelt's cycle. For their older age segments (12-17 years) the educational infrastructure's peaks

(1943, 1969-70, 1985) also correspond with the Sunbelt's (1946, 1973, 1986) cyclical turns. It should be noted that neither the Northeast's nor the Midwest's age cycles provided as good a correspondence with educational infrastructure as that found with the Sunbelt's cycles.

On the basis of these observations, it is difficult indeed to ignore or deny the role which infrastructure and regional invention seem to have played in the regional inversion process. One issue that might be raised is whether another factor or variable may have affected both educational infrastructure and innovative capacity, resulting in the correspondence found in the previous analyses. This possibility was, however, thoroughly tested for in Suarez-Villa and Hasnath (1993) with national data and it was found to be unfounded. Another question may be the lagging of inventive output with respect to the educational infrastructure cycles. The short or non-existent lags between the infrastructural and the innovative capacity age cycles may be due to the fact that national data is being used for the educational infrastructure estimates, whereas the innovative capacity age cycles are based on regional data. Unfortunately, the lack of reliable regional data for educational infrastructure spending, spanning the long periods comprised in the analyses, prevented their use in this study. It stands to be seen whether the significant lags which would be expected if a causal relationship exists between infrastructure and invention, would emerge in estimates using regional data for both indicators.

15.4   Conclusion

The preceding analyses have looked into the trends and dynamics of what are perhaps the two most important factors driving the process of regional inversion. The capacity to generate inventions endogenously and to invest in the kinds of infrastructure that promote it, may well be the most crucial prerequisites for the kind of sustained, long-term regional change that promotes interregional convergence. The collapsing of space and time that invention and innovation introduce in a regional economy, increasing productivity and human capital development, is part of an amplifying diffusion process that tends to affect many other geographical areas. Such effects can eventually extend beyond a national boundary, to promote both economic integration in continental markets, and the emergence of certain regions as major international nodes of trade and production.

This contribution leads us to question to what extent any regional development process that does not include the endogenous generation of technology can be sustained over the long-term. The previous analyses may lead to the conclusion that invention, innovation and the infrastructural investments that promote them may be both a prerequisite and a crucial component of the process of regional inversion. The deferred, long-term character of that process should be a source of interest due to the institutional and policy concerns that it raises. Institutional objectives and policies tend to be drawn most of the time on the basis of short-term priorities and results, rather than on an understanding of any long-term benefits and effects. Analyses such as that provided here reveal some of the characteristics of these long-term processes and should help to improve our understanding of how processes of convergence, continental economic integration and technological diffusion occur. Hopefully, this contribution will attract interest to the much neglected analysis of the long-term dynamics of regional change and the roles played by technology and innovation.

References

Acs Z J and D B Audretsch (1989) Patents as a Measure of Inventive Activity. Kyklos 42:171-80

Allison P D, Long J, Scott J and T K Krauze (1982) Cumulative Advantage and Inequality in Science. American Sociological Review 47:615-25

American Public Works Association (1976) History of Public Works in the United States, 1776-1976. Public Works Association Chicago

Basberg B L (1987) Patents and the Measurement of Technological Change: A Survey of the Literature. Research Policy 16:131-41

Beggs J J (1984) Long-run Trends in Patenting. In: Griliches Z (Ed) R&D, Patents and Productivity. University of Chicago Press Chicago

Berry B J L (1988) Migration Reversals in Perspective: The Long-Wave Evidence. International Regional Science Review 11:245-51

Berry B J L (1991) Long-Wave Rhythms in Economic Development and Political Behavior. Johns Hopkins University Press Baltimore

Bound J, Cummins C, Griliches Z, Hall B H and A Jaffe (1984) Who Does R&D and Who Patents? In: Griliches Z (Ed) R&D, Patents and Productivity. University of Chicago Press Chicago

Chandler A D (1990) Scale and Scope: The Dynamics of Industrial Capitalism. Harvard University Press Cambridge

Chinitz B (1986) The Regional Transformation of the American Economy. American Economic Review. Papers and Proceedings 76:300-03

Crandall R W (1993) Manufacturing on the Move. The Brookings Institution. Washington DC

Davelaar E-J and P Nijkamp (1990) Technological Innovation and Spatial Transformation. Technological Forecasting and Social Change 37:181-202

Davies S (1979) The Diffusion of Process Innovations. Cambridge University Press Cambridge

Donaldson G and J W Lorsch (1986) Decision Making at the Top: The Shaping of Strategic Decisions. Basic Books New York

Griliches Z, Pakes A and B H Hall (1987) The Value of Patents as Indicators of Inventive Activity. In: Dasgupta P and P Stoneman (Eds) Economic Policy and Technological Performance. Cambridge University Press New York

Griliches Z (1990) Patent Statistics as Economic Indicators: A Survey. Journal of Economic Literature 28:1661-1707

304

Harre R (1988) Great Scientific Experiments: Twenty Experiments that Changed Our View of the World. Oxford University Press Oxford

Hasnath S A and L Chatterjee (1990) Public Construction in the United States: An Analysis of Expenditure Patterns. Annals of Regional Science 24:133-45

von Hippel E A (1988) The Sources of Innovation. Oxford University Press New York

Holton G (1988) Thematic Origins of Scientific Thought: Kepler to Einstein. University of Chicago Press Chicago

Horstmann I, MacDonald G M and A Slivinski (1985) Patents as Information Transfer Mechanisms: To Patent or (Maybe) Not to Patent. Journal of Political Economy 93:837-58

Hounshell, D A and J K Smith (1988) Science and Corporate Strategy. Cambridge University Press New York

Hull D L (1988) Science as a Process: An Evolutionary Account of the Social and Conceptual Development of Science. University of Chicago Press Chicago

Kamann D J F and P Nijkamp (1990) Technogenesis: Incubation and Diffusion. In: Cappellin R and P Nijkamp (Eds) The Spatial Context of Technological Development. Avebury Aldershot

Kuhn T S (1962) The Structure of Scientific Revolutions. University of Chicago Press Chicago

Mansfield E (1971) Research and Innovation in the Modern Corporation. Norton New York

Marschak T A, Glennan T K and R Summers (1967) Strategy for R and D. Springer Verlag New York

Narin F, Noma E and R Perry (1987) Patents as Indicators of Corporate Technological Change. Research Policy 16:143-55

Norton R D (1990) Population Growth and US Regional Futures. Survey of Regional Literature 16:2-14

Popper K R (1981) The Rationality of Scientific Revolutions. In: Hacking I (Ed) Scientific Revolutions. Oxford University Press Oxford

Rubin M R and M T Huber (1986) The Knowledge Industry in the United States, 1960-80. Princeton University Press Princeton

Sale K (1975) Power Shift: The Rise of the Southern Rim and its Challenge to the Eastern Establishment. Random House New York

Samson K J (1990) Scientists as Entrepreneurs: Organizational Performance in Scientist-Started New Ventures. Kluwer Academic New York

Scherer F M (1983) The Propensity to Patent. International Journal of Industrial Organization 1:221-5

Schulman B (1991) From Cotton Belt to Sunbelt: Federal Policy, Economic Development, and the Transformation of the South, 1938-80. Oxford University Press New York

Suarez-Villa L (1989) The Evolution of Regional Economies. Praeger New York and London

Suarez-Villa L (1990) Invention, Inventive Learning, and Innovative Capacity. Behavioral Science 35:290-310

Suarez-Villa L (1992) Twentieth Century US Regional and Sectoral Change in Perspective. Survey of Regional Literature 20:32-9

Suarez-Villa L (1993) The Dynamics of Regional Invention and Innovation: Innovative Capacity and Regional Change in the Twentieth Century. Geographical Analysis 25:147-64

Suarez-Villa L (1995) Innovative Capacity, Infrastructure, and Regional Policy. In: Batten D and C Karlsson (Eds) Infrastructure, Economic Growth and Regional Development. Springer Verlag Berlin and New York

Suarez-Villa L and M M Fischer (1995) Technology, Organization and Export-driven Research and Development in Austria's Electronics Industry. Regional Studies 29:19-42

Suarez-Villa L and S A Hasnath (1993) The Effect of Infrastructure on Invention: Innovative Capacity and the Dynamics of Public Construction Investment. Technological Forecasting and Social Change 44:333-58

Suarez-Villa L and C Karlsson (1996) The Development of Sweden's R&D-intensive Electronics Industries: Exports, Outsourcing and Territorial Distribution. Environment and Planning A 28

US Bureau of the Census (1981) Construction Reports: Value of New Construction Put in Place in the United States, 1964 to 1980. US Government Printing Office Washington DC

US Bureau of the Census (various years) Statistical Abstract of the United States. US Government Printing Office Washington DC

US Department of Commerce (1990) Technology Assessment and Forecast, 7th Report. US Government Printing Office Washington DC

US Department of Commerce (various years) Construction Review. US Government Printing Office Washington DC

US Patent and Trademark Office (1989) Annual Report. US Government Printing Office Washington DC

Wyatt G (1986) The Economics of Inventions. St Martin's Press New York

16 Telecommunications Network Externalities and Regional Development: Empirical Evidence

Roberta Capello and Peter Nijkamp

16.1   Introduction[1]

In the last two decades economic research has placed much emphasis on the role of advanced technologies, and in particular of the information and telecommunications technology, in processes of economic growth and restructuring. Many regional economic studies have focused on the potential of advanced information and telecommunications technologies for promoting the convergence of regional economies by decreasing socio-economic disparities. In the eighties, the concept of the 'Information Economy' came to the fore. This notion underlies the strategic role played in economic development by information as a strategic resource and, consequently, by telecommunications technologies as vehicles for transmitting information.[2]

The importance of telecommunications technologies as strategic weapons favouring the competitiveness of firms and the comparative advantage of regions has stimulated industrial economists to study the behavioral-economic mechanisms associated with the diffusion of these technologies among potential users. In this vein a new concept, the *network externality*, has been extensively studied in this field. The term stems from the well-known economic concept of externality. In economic theory, an externality is said to exist when a person external to a transaction is directly affected (positively or negatively) by the events of the transaction. The concept of 'network externality' is related to a simple but fundamental observation that the user-value of a network is highly dependent on the number of already existing subscribers or clients. This means that the decision of a potential user to become a member of a network is dependent on the number of current users. In the economic literature, the definition of network externality has up till now mainly been linked to the effects that the number of subscribers

---

[1]This paper is the result of a common research project of the two authors. Sections 2, 3 and 4 have been contributed by R. Capello, while the remaining sections have been written jointly.

[2]We refer among others to Gillespie and Hepworth (1986) and Gillespie and Williams (1988), Gillespie, Goddard, Robinson, Smith and Thwaites (1984); Gillespie, Goddard, Hepworth and Williams (1989); Goddard (1980).

has on the utility function of the new subscriber. In other words, the definition so far concerns *consumption network externalities*.[3]

This chapter addresses a new research issue. The idea presented here, and extensively discussed in other publications,[4] is that telecommunications networks are not only governed by *consumption network externalities*, but also generate *production network externalities*, since the behaviour of a firm linked to a telecommunications network influences the performance of other linked firms and - via multiplier effects - that of relevant regions. The positive effects generated by production network externalities on the performance of firms have little to do with the traditional effects on corporate performance generated by innovation processes or economies of scale. Although the effects of innovative processes and economies of scale are similar, the nature of production network externality effects is rather different, because their advantages stem from the difference between the marginal costs and benefits of being networked. This is not true for positive effects generated by innovative processes, or by economies of scale. The former stem from an increase in productivity of production processes, the latter from a decrease in costs resulting from large-scale production dimensions.

While at a conceptual level the above distinction may seem simple, at an empirical level the separation of the three effects (i.e. innovative process, economies of scale and network externality) in the production function is fraught with difficulties. The present paper provides a contribution in this particular area, and tries to separate the network externality effects from the effects of economies of scale or the innovative process.

The main research questions addressed in this paper from a conceptual and empirical point of view are:

i   what is the relative importance of network externalities for firms in their decision to join a network?

ii   do firms and regions gain from exploiting network externalities?

iii what are the micro- and macro-conditions stimulating or supporting the exploitation of network externalities?

The questions will be treated from both a theoretical and empirical perspective. The empirical analysis is carried out by means of a primary database, viz. small and medium sized firms belonging to different sectors, located in both the North and the South of Italy. For the South of Italy, interviews have been carried out in firms involved in the dedicated EC Programme, STAR, and set up with the aim at decreasing regional disparities through the adoption of advanced and sophisticated telecommunications technologies. In this way, our empirical analysis is therefore able to test the extent to which the STAR Programme generated positive effects on the performance of firms and of regions.[5] This database has

---

[3]See, among others, Allen (1988) and (1990); Antonelli (1989), (1990) and (1992); Bental and Spiegel (1990); Cabral and Leite (1989); David (1985) and (1992); Hayashi (1992); Katz and Shapiro (1985) and (1986); Markus (1989) and (1992).

[4]See for example, Capello (1994); Capello (1995); Capello and Nijkamp (1994).

[5]For a description of the database, see Annex 1.

been used to provide empirical answers to the three questions mentioned above.

The structure of the paper is as follows. Section 16.2 presents a conceptual model for assessing the effects of telecommunications network externalities on the performance of firms and regions. Sections 16.3, 16.4 and 16.5 deal respectively with the three questions from an empirical point of view. Section 16.6 presents some policy implications and some concluding remarks.

16.2   A Conceptual Model of Production Network Externalities

16.2.1   The Economic Symbiosis Concept

Our basic research is related to the linkage between network externalities and industrial and regional performance. Our main aim is the analysis of whether or not network externalities may be measured in terms of industrial (i.e. micro) and regional (i.e. macro) performance. Such a question is fraught with difficulties of a methodological nature, which will be dealt with in the empirical chapters of this study. At the purely conceptual level however, it is plausible to envisage a positive relationship between network externalities and corporate productivity.

The achievement of better economic performance by exploiting benefits derived from network participation is what we call an 'economic symbiosis' effect, an improvement in the economic performance based on non-paid benefits from synergies among firms. The 'economic symbiosis' may be seen clearly in a set of firms strongly interrelated to each other via a physical network. Such firms and their interdependent sectors have a relatively high productivity because of the achievement of strong symbiotic advantages in comparison with the non-networked firms. These advantages may be classified into direct and indirect advantages (Table 16.1). The direct advantages of network participation are those which directly and positively affect the productivity of a firm. These can be classified into:

• static advantages, which may be summarized as synergies among actors operating in different economic environments;
• dynamic advantages, represented by the possible achievement of network-based innovation and access to previously unknown markets.

The indirect advantages which affect the productivity of firms can be divided into:

• static advantages, such as information provision induced via network interconnection;
• dynamic advantages, such as complementary assets.

These advantages are generated by the existence of a physical linkage to the network, leading to what we call 'positive network externalities'. They are expected to generate a positive effect on the performance of firms. As a result of more and better information, more synergies with other sectors operating in different economic environments, a higher degree of innovativeness in bureaucratic procedures and more complementary assets these industries are able to become more efficient (in terms of productivity) than others. The synergetic and 'symbiosis' effect therefore generates economic advantages to those firms which are networked, although they do not pay a marginal price for them. Thus, via a network, both pecuniary and technical externalities may be achieved and exploited. Ceteris paribus, the networked firms are capable of generating dynamic growth in the economy, via spillover and multiplier effects.

Table 16.1 Typology of advantages of network externalities

	STATIC	DYNAMICS
DIRECT	Synergies among firms	Achievement of network-based innovations and access to previously unknown markets
INDIRECT	Information provision	Complementary assets

The 'economic symbiosis' model has many similarities to the 'growth pole' theory of Perroux (1955). In Perroux's approach, a set of firms, strongly interrelated to each other via input-output linkages around a leading industry (*l'industrie môtrice*), is able to generate strong cumulative and multiplier effects on the economy via spillover effects. The effects of a 'growth pole' and of 'economic symbiosis' are in fact very similar. These effects are, however, the outcome of two different phenomena. In the growth pole theory, the determinants of the strong dynamic effects are the existence of advanced technological practices and high innovation rates in the set of industries. In the 'economic symbiosis' approach, the dynamics of these industries is explained by the physical network linkage among firms generating advantages such as greater synergy among economic operators and more information. The access to a network and the non-paid-for advantages a firm gets by joining the network play a crucial role in the performance of firms, primarily via an increase in productivity. This assumption is evident, since a firm may obtain advantages from being networked without paying a marginal price for them. In other words, the (technical and pecuniary) network externalities generated via the network are the intrinsic explanation of the 'economic symbiosis' phenomena.

Although the effects may be similar, the causes of this phenomenon are completely different. Without denying the existence of other major factors which improve the performance of firms (e.g. international market developments, better marketing strategies, etc.), the present analysis focuses on one specific cause: production network externalities. At the conceptual level, the framework therefore rests on the assumption that the main micro-stimulus explaining adoption processes of advanced telecommunications technologies is the existence of network externalities.

310

16.2.2   The Spatial Symbiosis Concept

The existence and exploitation of network externalities are not confined to the single firm: there are also strong geographical implications. The 'economic symbiosis' concept may be translated into a spatial setting by the assuming that the set of networked industries is spatially clustered and, secondly, by focusing on spillover effects likely for immediately adjacent areas.

In other words, a *'spatial symbiosis'* effect takes place when a set of 'networked firms' is present in a specific region (Figure 16.1). This leads to non-zero positive effects on the firms' performance generated by the non-paid for advantages of being networked. A number of spillover effects are generated in the local (regional) economy and can be measured in terms of better regional performance. This phenomenon may give rise to cumulative effects à la Myrdal (1959) and act as the foundation stones for a local sustainable development.

The spatial symbiosis concept is very similar to the 'growth centre' theory of Boudeville (1968), who first provided a 'translation' of Perroux' growth pole theory in geographical rather than economic space. Boudeville's theory emphasizes the importance of geographical agglomeration and concentration of economic activities for local economic growth. The geographical clustering of a set of innovative industries produces strong backwash and spillover effects over space which may lead, in the presence of some local prerequisites (i.e. a highly developed infrastructure, provision of centrally supplied public and social services, a demand for labour) to accelerated local economic growth.

The difference between the 'growth centre' and the 'spatial symbiosis' concept quite clearly emerges. Again, it is reasonable to claim that, although the local economic effects are the same, the nature of these phenomena is quite different. In Boudeville's theory, local economic growth is explained essentially by the geographical agglomeration of innovative firms. In the 'spatial symbiosis' approach, the presence of a set of networked firms is not the only reason for the generation of local economic growth. It is assumed that a regional economy may also benefit from the input obtained by inter-regionally networked firms. In fact, it is likely that the physical connectivity between two firms located in two different regions may generate exchanges of local know-how and information which increase productivity, and consequently regional performance. In other words, when the connectivity takes place at an inter-regional level, the effects it produces may also have import and export implications.

Thus, in summary, the positive effects for a firm are expected to have a beneficial influence at the level of the wider (regional) environment. Thus, in economic terms, we expect a region to be able to gain from network externalities by exploiting advantages stemming from participation in the network. At a macro-level these advantages may be regarded as the achievement of:

* spatially dispersed information;
* new geographic market areas;
* complementary know-how from specialized economic areas; d) additional specialized input from other regions.

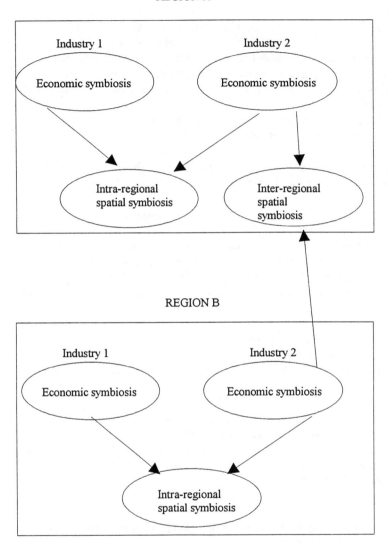

Fig. 16.1. Economic and spatial symbiosis

After this theoretical explanation, a legitimate question is whether the 'economic and spatial symbiosis' effect may be tested empirically. This is the subject of the following sections.

16.3 Telecommunications Adoption: Backgrounds and Regional Variations

As explained above, one of the most important reasons for a firm to join a network stems from the number of firms already linked to the network. Our first investigation concerns this phenomenon of consumption network externalities and was first carried out by means of a survey of information and telecommunications network access and use by firms in Italy. The aim of this empirical exercise was to test the questions:

- what is the relative importance of network externalities for firms in their decision to participate in a network?;

- intensive telecommunications network participation have greater economic significance for some regions than others?

Our empirical investigation was based on two methods of analysis, one explorative and the other interpretative in nature. Firstly, a *descriptive representation* of the results from our database served to discover to what extent network externalities were the incentive to join a network. On the basis of exploratory *contingency table analysis*, we develop a behavioural analysis of the reasons for adoption. In particular, our analysis concentrates on the identification of:

- the main reasons for adoption;
- the main reasons for non-adoption;
- the main conditions for future adoption;
- the main reasons for dissatisfaction with the telecommunications technologies in use.

The second step of the empirical analysis in this section focuses on further confirmation of the results through an *interpretative analysis*. The aim is to identify the reasons for industry and regional variations. The methodology used for our interpretative analysis is based on a *discrete choice modelling* approach using multinomial logit models.

16.3.1 Network Externalities as the Main Reasons for Adoption: Descriptive Results

This section provides contingency table analyses of questions related to the adoption and decision process regarding new networks and services. The contingency tables show the percentage of firms replying positively to each specific choice, based on a subdivision of the sample between the South and the North of Italy. Correlation analyses based on Chi-Square (χ^2) test statistics and P-values[6] are run between each (positive) reply and its associated region. In this way we are able to investigate which replies are correlated with region-specific conditions.

Table 16.2 Main reasons for adoption by macro-areas

Reasons for adoption	South	North	Entire Sample	CHISQ	P-Value
High percentage of subscribers in region already networked	5.7	20.0	12.8	3.18	0.074
High percentage of subscribers in other regions already networked	11.4	17.1	14.3	0.467	0.493
Importance of the network or service for business	37.1	65.7	51.4	0.719	0.017
Suppliers are networked or use the same service	5.7	11.4	8.6	0.729	0.393
Other firms in the same sector are connected	8.6	14.3	11.4	0.565	0.452
No other advanced communication network or service available	11.4	5.7	8.6	0.729	0.393
Low costs of implementation	42.9	22.9	32.8	3.17	0.075
Low costs of use	45.7	17.1	31.4	3.17	0.01
Increasing awareness of these technologies through demonstration centres	40.0	5.7	22.8	11.66	0.001
Image effect	20.0	14.3	17.1	0.402	0.526
Others	31.4	14.3	22.8	2.92	0.088

The results (see Table 16.2) were fairly satisfactory since, at a descriptive level at least, our expectations were fulfilled and the testable hypotheses verified with *strong regional variations*. The importance of the regional dimension in our replies

---

[6]The Chi-square (CHISQ) represents the degree of dependency between two variables. In our case the two variables will be the percentage of positive replies obtained in our interviews and the relevant regional variable. The P-value represents the probability to accept the null hypothesis when this is true (McFadden 1983).

is witnessed by the statistically significant level of both χ^2 and of the P-value between the 'network externality' variables. What is surprising is that a significant regional correlation exists only for the network externality variables, as is represented by the contingency tables.

a *Reasons for adoption.* Comparison between the two macro-areas revealed interesting results on the main reasons for adoption. The most important reasons in our sample were 'the importance of the technologies for the business' and 'low costs of implementation and use'. However, when the analysis is carried out at a regional level, significant variations in the replies appear to emerge. For the *South of Italy* the findings support our first research proposition: that the number of existing subscribers is the main reason for adoption. Reasons put forward for adoption stress very strongly 'the existence of promotion and demonstration centres', and almost as significant 'the importance of the technology for the corporate business' and 'low usage and implementation costs' (Table 16.2). These replies witness the difficulties of adoption processes, since:

- help from the supply side in demonstrating and promoting innovation is of crucial importance. Greater possible connectivity deriving from greater availability is not in itself a sufficient element to give rise to adoption processes;
- another main reason for adoption in the first phases of a diffusion process is 'the importance of innovation for the business'. In other words, if a linkage is shown between the technology and the business areas, there is a higher rate of success for an innovation;
- stimuli on the financial side also become strategic when the risks of failure of the adoption have to be borne by the adopters. Risks are of course higher during the first phases of adoption.

Another remarkable result is the *total lack* of any reference to 'the number of connected users, suppliers, buyers and competitors in the same or other regions'. This result stresses once more the empirical plausibility of our testable hypotheses. In the first phases of the adoption process, such as in the case of the STAR programme, other reasons prevail in the decision to adopt rather than 'the number of existing subscribers'. In this area, in fact, there is still too small a number of subscribers to act as an attracting factor for potential subscribers.

The major reason put forward for adoption by firms located in the *North of Italy* is 'the importance of these technologies for business' (more than 65.7% of replies), followed by 'low costs of implementation' (22.9%) and 'use costs' (17.1%) and by 'the high percentage of already existing subscribers in the region' and 'in other regions' (37%) (see Table 16.2). These replies confirm our expectations, since in an area at an advanced stage of diffusion:

- the total number of existing subscribers is an important motive for previous adoption, provided the technologies themselves are of any interest to the business and their costs are reasonable;
- the number of subscribers plays a more important role in the decision to adopt.

In comparison with the South, where the percentage of responses declaring this motive was around 17%, the number in the North was almost 40%. Given the more advanced stage of the diffusion process, this result can be explained by the 'bandwagon effect', i.e. the higher the number of subscribers (after the critical mass has been achieved), the more likely new subscribers are to adopt;

- in advanced stages of diffusion, help from the supply side, through the establishment of demonstration and promotion centres, no longer play a significant role in the diffusion process (only 5.7% of replies);
- at any stage of diffusion, price incentives (e.g. free technologies in the South) are significant in the adoption process.

b *Reasons for non-adoption.* The importance of the number of adopters in the first phases of diffusion is also confirmed by the reasons given by firms for NON-adoption. The results show that main reasons for NON-adoption are 'the importance of the network or service for corporate business' and 'low percentage of connected subscribers, suppliers and customers'.

Regional variations were once more evident. In the South the two prevailing reasons for non-adoption were: 'the low percentage of connected local users' and 'the irrelevance for corporate business'. Other significant adoption bottlenecks included 'the low percentage of non-local connected users and of connected suppliers and customers', 'the use and adoption costs (including time costs before efficient connection is achieved)' and 'the inability to understand and/or evaluate these technologies' (Table 16.3).

Table 16.3 Main reasons for non-adoption by macro-areas

Reasons for adoption	South	North	Entire Sample	CHISQ	P-Value
Low percentage of connected subscribers	71.4	14.3	42.8	23.33	0
Low percentage of connected subscribers in other regions	45.7	8.6	27.1	12.21	0
No suppliers or customers using them	57.1	31.4	44.3	4.69	0.03
Not likely to be useful for business	51.4	48.6	50.0	0.057	0.81
Not important in area of business	28.6	14.3	21.4	2.12	0.14
High costs of use	22.9	37.1	30.0	1.70	0.19
Others	28.6	14.3	21.4	2.12	0.14

Table 16.4 Main conditions for future adoption by macro-areas

Conditions for future adoption	South	North	Entire Sample	CHISQ	P-Value
NETWORKS					
A higher number of people connected	28.6	11.4	20.0	3.14	0.07
A higher number of suppliers and customers connected	25.7	8.6	17.1	3.62	0.06
A better geographical distribution of the network	11.4	5.7	8.6	0.73	0.39
A reduction in the price of use	11.4	11.4	11.4	0.0	1
A reduction in the price of access	8.6	8.6	8.6	0.0	1
More recent availability of the network	8.6	11.4	10.0	0.16	0.69
Technical progress in the network	25.7	20.0	22.8	0.32	0.57
Increase in the efficiency of the network	20.0	14.3	17.1	0.40	0.53
Good results from the previous adoption	14.2	5.7	10.0	1.43	0.23
SERVICES					
A higher number of people connected	40.0	14.3	27.1	5.85	0.02
A higher number of suppliers and customers connected	31.4	20.0	25.7	1.19	0.27
A better geographical distribution of the service	17.1	2.9	10.0	3.97	0.05
A reduction in the price of use	5.7	20.0	12.9	3.2	0.07
A reduction in the price of access	11.4	17.1	14.3	0.47	0.49
More recent availability of the service	11.4	5.7	8.6	0.73	0.39
Technical progress in the service	20.0	17.1	18.6	0.09	0.76
Increase in the efficiency of the service	25.7	17.1	0.76	21.4	0.38
Good results from the previous adoption	20.0	2.9	5.08	11.4	0.02

The 'bandwagon effect' in the advanced stages of diffusion processes is confirmed by the results in the North. The difference between the North and the South is even greater in the case of 'the reasons for non-adoption' than 'the reasons for adoption'. While in the South the main reason for non-adoption was 'the low percentage of regional subscribers' (71.4%), in the North the main reason for non-adoption has been identified as 'the non-importance of these technologies for business' (48.6%), followed by 'the high costs of use' (37.1%) (see Table 16.3). The reasons for non-adoption in the North have very little to do with numbers of subscribers; they are linked to business interests and financial constraints.

c *Main conditions for future adoption.* Table 16.4 presents the main conditions for future adoptions for both networks and services. From these results too it is clear that 'a higher number of subscribers' does not play an important role in stimulating adoption in the North (11.4% in general, 8.6% in the case of customers and suppliers). Future adoption is related to 'the technical progress of networks' (20%), 'an increase in their efficiency' (14.3%) and 'a reduction in the price of use for services' (11.4%). On the contrary, *in the South* the main condition for future adoption is 'a higher number of subscribers', for networks (28.6%) and services (40%) (see Table 16.4). While 'the reduction in the price of use' still remains crucial in the North (20%), this has the same weight as 'a higher number of suppliers and customers connected' (20%). 'Technical progress' and 'increase in the efficiency of the service' still play both a crucial role (17.1%). *In the South*, 'a higher number of people connected' is stated to be the most important reason, especially in the case of advanced interactive services, such as videotex and electronic mail. For data banks, too, 'a higher number of people connected' is the main expected condition for future adoption, explained by the fact that a higher number of people connected would assure a higher revenue to the service providers, and thus stimulate a better quality service (i.e. a larger variety of information available). For electronic mail and electronic data interchange services, the most important reasons for future adoption were also 'technical progress' and 'increase in network efficiency'.

d *Reasons for dissatisfaction with the telecommunications technologies in use.* The most interesting result is the regional variation in the reasons for the dissatisfaction with the quality of existing technologies. As expected, "the degree of satisfaction with the technologies" among subscribers in the North is higher than in the South (68.6% against 57.1%) (Table 16.5). Moreover, the main reason put forward for a high degree of dissatisfaction is 'the too low number of subscribers' (46%) and 'the too high costs of implementation' (30%).

Table 16.5 Degree of satisfaction of the quality of communication by macro-areas

	South	North	Entire Sample	CHISQ P-Value
Yes	57.1	68.6	62.8	0.98 0.322
No	42.9	31.4	37.1	

It is interesting to see that these two reasons are also very much dependent on the regional dimension. While in the North 'the high costs of their implementation' are the most important reasons (54.5%), this does not apply to the South (13.3%) (see Table 16.6). Dissatisfaction does not stem from 'the low number of subscribers' (only 9.1% of positive replies for this reason), as is the case in the South (73.3%).

The Northern sample in general showed greater interest in these technologies. There was a high percentage of replies recording 'the increased intensity of business relationships after the adoption of telecommunications technologies' (62.9% in the North against 20% in the South). The reasons given for the 37% who claimed 'decreased business relationships after the introduction of these technologies', in the North were not linked to network externality effects. As Table 16.7 shows, none of the choices embodying a network externality effect (such as 'no suppliers or customers using these technologies') are put forward as main reasons for structural business relationships. Again, these results demonstrate that at low penetration levels, as in the South, 'the low number of subscribers' becomes a constraint for the increase in the intensity of use of these technologies. At a more advanced adoption stage, the main reasons explaining the low level of intensity of use is 'the non-importance for business' (17.1% of replies) and lack of staff skills in their use (14.3%).

Table 16.6 Major persistent user problems in the case of dissatisfaction by macro-areas

Reasons for dissatisfaction	South	North	Entire Sample	CHISQ	P-Value
Costs of implementation	13.3	54.5	30.7	5.06	0.024
No longer free services	20.0	0.0	11.5	2.48	0.11
Use of these technologies required profound organizational changes	0.0	9.1	3.8	1.42	0.23
Too few subscribers	73.3	9.1	46.1	10.53	0.001
Suppliers not using these technologies	33.3	9.1	23.1	2.10	0.15
Competitors not using these technologies	0.0	26.7	15.4	3.47	0.063
Other	54.5	20.0	34.6	3.35	0.067

Table 16.7 Main reasons for lack of increase in intensity of use

	South	North	Entire Sample	CHISQ	P-Value
None	20.0	11.4	15.7	0.97	0.32
Costs of initial investment	2.8	0.0	1.4	1.01	0.32
Lack of reliability of the technologies used	2.8	0.0	1.4	1.01	0.34
Level of telecommunication charges	0.0	0.0	0.0		
Lack of staff skills in their use	2.8	14.3	8.6	2.92	0.08
Services not relevant to business	14.3	17.1	15.7	0.11	0.74
Customers do not use them	0.0	31.4	15.7	13.05	0.0
Suppliers do not use them	2.8	25.7	14.3	7.45	0.006
Too few subscribers in general	40.0	0.0	20.0	17.5	0.0
Restricted geographical diffusion of networks and services	8.6	2.8	5.71	1.06	0.30
Other reasons	8.6	5.7	7.14	0.21	0.64

16.3.3 Estimated Logit Model for the Regional and Industry Level of Analysis

The interpretation of our first research proposition is based on a standard discrete choice modelling approach, with economic random utility theory as the underlying theoretical rationale and revealed preferences as the empirical orientation. Discrete choice models such as multinominal logit, nested multinominal logit and multinominal probit models are now well-established modelling approaches applied in a wide range of fields.[7]

The importance of these models for our analysis stems from the fact that in most cases the decision of a firm to join a network is of a discrete nature; in our case, too, the behavioral analysis is based on revealed preferences and the database obtained may only be applied for discrete models.

The logit models in our empirical analysis will be based on the complete

---

[7]On theory and applications of logit models see, among others, Ben-Akiva and Lerman (1985); Bishop et al. (1977); Camagni (1992); Domencich and McFadden (1975); Fischer and Nijkamp (1985); Fischer et al. (1992); Griguolo and Reggiani (1985); Leonardi (1985); Nijkamp et al. (1985).

database. In the next section, the willingness to participate in the network will be analyzed, highlighting industry and regional variations in the decision. The results of the analysis go further than the exercise of testing our first research proposition. They also have policy implications, especially in the case of Southern Italy, where the results will be able to prove whether the financial effort made by the EC to promote the use of these technologies in less developed regions has, in fact, generated a willingness to adopt by local firms.

A number of *explanatory variables* regarded as theoretically plausible reasons to join a network have been selected:

- the *price incentives* for a firm. Here a distinction is made between firms having replied that low implementation and use costs have been a crucial variable in the decision to adopt (PRICE=1) and firms which in previous adoptions have not recognized low financial costs as a basic reason for adoption (PRICE=0);
- the *role of the supply* in supporting the adoption of new networks and services through, for example, demonstration and promotion centres. The question here is whether firms have recognized the efforts made by the supply side as helpful in their decision-making process of previous adoptions (ROLE=1) or whether they have never taken the supply efforts into consideration (ROLE=0);
- the *bandwagon effect* in the decision to adopt. In this respect, a distinction is made between firms which have recognized the number of adopters as a crucial variable in the decision making process for previous adoptions (NET=1) and firms which have assigned no role at all to the number of adopters in decision-making processes for previous adoptions (NET=0).

Table 16.8 Degree of dependency between the dependent variable and the categorical variables

Willingness to adopt*	CHISQ*	P-Values
Price	5.66	0.017
Role	3.81	0.066
Net	3.02	0.082

The choice of these variables was made with reference to a wide range of potential explanatory variables, such as the size of firms, the sector firms belong to, the innovation capacity of firms, their flexibility with respect to change, the importance of the technology for the business, learning processes, etc. Among all plausible categories, we have chosen those having the highest degree of dependency. The results of this analysis are shown in Table 16.8. Among all possible variables, it is interesting to underline that neither the *size of firms* nor the *sector firms belong to* can be shown to affect the willingness to adopt.[8] This

[8]The CHISQ and P-Values between the willingness to adopt and the sector variable are respectively 2.02 and 0.15. CHISQ and P-values between the willingness to adopt and the size of firms (measured as the level of turnover) are 1.11 and 0.77. Even in the case of

means that contact patterns do not significantly differ between small and medium sized firms. Furthermore, the reasons for adoption do not vary between sectors. This second result is in line with the finding of the descriptive analysis at the sectoral level presented above.

The above variables a), b) and c) represent the expected explanatory variable of the willingness to adopt. A measure of the willingness to adopt is given in our database by the revealed interest to adopt advanced (interactive) services in the near future. A 0-1 variable has thus been built on the distinction between firms which have revealed a preference for future adoption of advanced interactive services (i.e. electronic mail) (WILL=1) and those which have not (WILL=0).

From a *statistical point of view*, the estimated results of the logit models are satisfactory at *the regional level*. This model has a 0.65 probability value to explain the willingness to adopt. Moreover, the good fit of the model from a statistical point of view is also demonstrated by the low number of significant categorical variables and by the lack of significant interaction effects between these variables.

Table 16.9 Estimated logit models with respect to the willingness to adopt at the regional level

Variable	Estimated Parameter	CHISQ	P-Value
Constant	0.135	0.09	0.762
Price	0.4913	3.29	0.069
Role	0.3879	1.28	0.257
Net	0.1037	0.07	0.793
Reg	-0.3226	1.25	0.264

CHISQ = 4.18
P-Value = 0.6526

The *economic interpretation* of the model is interesting. A first conclusion from Table 16.9 is that firms having chosen 'low implementation and use costs' as an important reason for adoption are also the most dynamic firms in terms of future adoption, as is shown by the positive sign of the estimated parameter of the PRICE variable. Consequently, *financial incentives represent a very important stimulus* for future adoption. This result has very important policy implications, since it shows that in the early diffusion stages, price variable plays an important role in the decision to adopt. Moreover, it assumes even greater importance when

---

the size of the firm measured as the number of employees, the results are not significant: CHISQ is 3.69 and the P-value 0.29.

linked to the STAR programme which provided these technologies free of charge. Our analysis reveals how strategic this choice has been in stimulating local demand. However, it also represents a useful lesson for future transitions to a commercial phase: the move towards market prices needs to be gradual in order to ensure a high adoption level in the early stages.

As far as the supply role is concerned, it is evident from our results that the existence of promotion and demonstration centres has positive effects on the willingness to adopt. This is indicated by the significant positive estimated parameter value with respect to the variable ROLE. This result confirms once again an established idea in the literature about the necessity for a *bridging mechanism* between demand and supply for the successful adoption of these technologies. The profound technological and organizational changes required in order to adopt and exploit these new technologies can be achieved only if technical and organizational support is provided by the supply side.

Another interesting result is the non-existence of a *'bandwagon effect'* in the willingness to adopt. Contrary to what we expected, the number of existing subscribers does not seem to stimulate future adoption, as is witnessed by the estimated parameter of the variable NET. The parameter of this variable, in fact, has a P-value of 0.79.

As far as the regional dimension is concerned, we introduced into our model a variable reflecting the location of firms (REG), with a value 1 when located in the North and 0 when located in the South. This variable had a negative value, underlining the fact that firms in the North of Italy were less willing to adopt than the ones located in the South. This result is not surprising at all, since the North of Italy has higher adoption rates than the South for networks and services commercially available. Low adoption rates in the North are typical for networks and services which are either in an experimental phase (such as ISDN or video-conferencing) or are still very limited in their geographical extension (such as fibre optic networks).

From the same estimated parameter (with opposite sign) we deduce a clear interest in future adoption for firms located in the South. This is an extremely positive result when viewed in the context of the STAR programme. One of the aims of the EC was to stimulate an interest in these technologies among firms in the South, to show their importance for business activities and for the future of these firms. This aim seems to have been achieved, as the sample in the South has demonstrated the positive attitude of firms towards future adoption. However, to confirm the positive results of the STAR programme, it is also necessary to test whether the other extremely important aim of the programme, i.e. the economic revitalization of backward regions, has been achieved. This is examined in the next section.

At *a sectoral level of analysis*, although interesting from an economic point of view the estimated logit model is less satisfactory. The P-value is equal to 0.405 and there is a statistically significant interaction effect (Table 16.10). In any case, the sectoral results confirm what was previously shown at a regional level, i.e. financial incentives, as well as the supply support, explain quite clearly the willingness to adopt, while the 'bandwagon effect' loses much of its explanatory effect.

As far as the *sectoral component* is concerned, it emerges quite clearly that firms belonging to the service sector are more in favour of future adoption (indicated the statistically significant negative estimated parameter for the industry sector). Moreover, the fact that the variable ROLE * SET is statistically significant demonstrates that those service firms for which supply support is important are especially oriented towards future adoption.

Table 16.10 Estimated logit models with respect to the willingness to adopt the industrial level of analysis

Variable	Estimated Parameter	CHISQ	P-Value
Constant	0.36	0.51	0.473
Price	0.74	6.4	0.011
Role	0.57	2.36	0.124
Net	0.03	0.01	0.939
Set	-0.86	4.91	0.027
Role*Set	-0.53	2.08	0.149

CHISQ = 7.23
P-Value = 0.405

The results obtained so far are in general satisfactory, although they do not seem to support our idea that *consumption network externalities play a critical role in the diffusion process of these technologies.* They do show however, that it is interesting to develop the analysis at a territorial level. In fact, *the regional dimension explains part of the innovative behaviour* of firms (and it is not just an additional variable in an already complex interpretative framework).

The present behavioral analysis has some policy implications, especially in terms of strategies for developing local demand for technological innovation. This is however only the first part of the innovative process and of an innovation policy. To be successful, an innovation policy is expected to generate positive results on the production side, stimulating productivity and economic growth. We present empirical evidence concerning this aspect in the next section.

16.4   Network Externalities and the Performance of Firms and Regions

In this section we move on to the measurement of *production network*

324

externalities. All problems associated with the empirical analysis will show up in this section. In fact, the empirical test on our theoretical framework is fraught with difficulties. One of the main problems is that, in order to determine the impact of network externalities on corporate and regional performance, it is necessary to have a reliable measuring rod of both network externalities and corporate and regional performance. Moreover, production functions are influenced by a large number of elements, such as innovation effects and economies of scale, which are similar to network externality effects. Disentangling specific network externality effects from all these other effects is not so simple.

16.4.1   The Conceptual and Methodological Approaches

Until now study the concept of network externalities has been explained in terms of the positive and increasingly intensive relationship between the number of subscribers and the performance of firms. The higher the number of subscribers, the greater the incentive for a firm to join a network and the better its performance. In reality, this definition is far too broad to explain fully the concept of network externality. In a *static perspective*, the interest of a firm is not to join the highest possible number of other firms connected via the network, but only the highest number of firms directly or indirectly related to its own business activities. Thus, the decision to join the network is not simply related to the total number of firms already networked, but to the number of specific *business-linked firms* already present in the network. The most obvious reason for entering a network is, in fact, the possibility of contacting relevant groups such as suppliers, customers or horizontally related firms in a more efficient and quicker way.

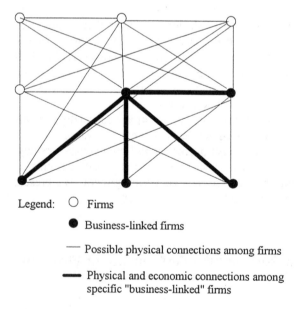

Legend: ○ Firms

        ● Business-linked firms

       — Possible physical connections among firms

       ▬ Physical and economic connections among specific "business-linked" firms

Fig. 16.2. Undirected graph representing connectivity among firms

Connectivity is a measure of linkage between two or more firms in a network. Economic connectivity measures the economic relationships among firms. When these relationships are pursued via a telecommunications network, then we can also speak of physical connectivity. What we argue here is that there is a strong relationship between these two kinds of connectivity, but that *physical connectivity has no reason to exist if economic connectivity does not exist.*

Figure 16.2 is a schematic representation of physical connectivity with the use of graph theory. If we represent firms as 'nodes', or 'vertices', and the physical linkages among them as 'arcs', or 'edges', the outcome is an (undirected) graph of vertices and edges representing all potential physical communication (or contact) that firms can entertain among themselves.

As we have just mentioned, the real interest of a firm, in a static world, is not to be linked to all other possible subscribers, but to achieve full connectivity among those firms related to its specific business. If we represent such firms in our graph with a bold vertex, and their economic relationships with other firms with bold edges, the real matrix of first order relationships will emerge. With this matrix it is possible to measure the proportion of physical connectivity of a given firm with regard to its potential economic connectivity.

It is physical connectivity which generates network externalities. *If the benefits a firm receives from physical connectivity is an increasing function of connectivity itself, then positive network externalities exist.* This is a situation represented by the positive derivative of the benefit function (Figure 16.3). So far we have described a way of measuring network externalities under the assumption of a static world. This is represented in Figure 16.4.

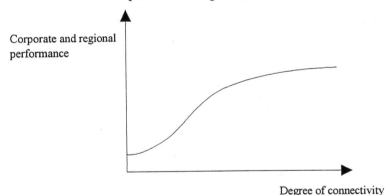

Fig. 16.3. Increasing relationship between the degree of connectivity and corporate and regional performance

From a methodological point of view, various important analytical questions remain. Some limitations of this method are that:
- it measures only direct connections, as second and higher order connectivity linkages are not taken into account. It can easily be argued that the relationships among suppliers or customers of a firm (representing what we call second and higher order connectivity for that firm) do not have the same effect on the performance of the firm as direct connections;

- it does not take into account the *intensity* of information flows. While in the above case we can disregard the effect of indirect connections on a firm's performance, in this case it is more difficult to justify avoiding the problem. The intensity of use of a network inevitably has an impact on corporate performance. Thus, any connectivity index has to be adjusted in order to include a measure of the intensity of use. This will be taken into account in our empirical analysis.
- *in terms of economic importance, the same weight is given to each link*, although it is likely that first order connection's will be of greater strategic importance for the firm.

These limitations apply equally to the individual firm and at the regional level. As discussed above, whereas the first and third are not so crucial, the intensity of use of new technologies is extremely important for our analysis. The empirical analysis was therefore also run with a weighted connectivity index to take this into account.

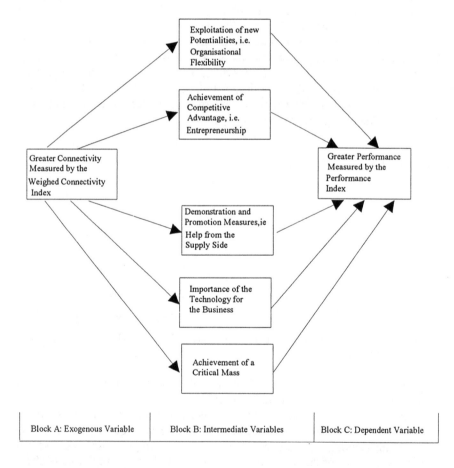

Fig. 16.4. A general model for estimating network externality exploitation

On the basis of the indications obtained by graph theory, a very simple connectivity index was constructed, representing *the ratio between the number of real connections to the total number of potential connections.*[9] Although simple in its formulation, it gives a satisfactory measure of connection for each firm. In the empirical analysis, the results obtained with this index provided evidence of the existence of production network externalities. The effects on regional performance were measured by calculating the sum of the positive effects on all firms located in the region. Thus, we postulate that the higher the number of firms enjoying network externalities in a regions, the higher the regional performance.

The methodology used to test the existence of production network externalities was based on a correlation analysis between the connectivity index and the performance index. The analysis was run in two steps:

• an initial estimation of the correlation coefficient between the 'row' connectivity index,[10] which measures the simple adoption of these technologies by firms, and the performance index, at the national level. We analyze later the extent to which the inclusion of the regional dimension leads to better correlation coefficients as we expect strong regional variations.

• the second step of the empirical analysis consists of the introduction of the 'frequency of use' variable into our framework. Thus, instead of measuring the correlation between the degree of adoption of these technologies and the performance of firms, the analysis is run between the use of these technologies and the performance of firms.

The second index for the empirical analysis is the performance index. A very simple performance index was chosen, representing the labour productivity of each firm, i.e. the ratio between the turnover of firms in 1991 and the number of employees in the same year. This measure may vary according to:

• the sector firm belongs to. In fact, there may be capital-intensive and labour-intensive sectors;

• the region where a firm is located. It might very well be that a sector is more productive in one region than in another because of a different regional penetration of innovation in capital and different labour force skills.

To avoid a biased result with the use of our connectivity index, an analysis was undertaken on the database to see whether there was any consistent relationship between any of these features and productivity. In particular, an analysis was

---

[9]The potential connection of a firm is defined as the total number of existing telecommunication services offered to firms.

[10]In this study 'row' connectivity index means the connectivity index constructed taking into account only the adoption data. This index is different from the so-called 'weighted' connectivity index which is derived by taking into consideration the 'frequency of use' of adopted telecommunication services. This index was in fact constructed by multiplying the adoption data which a weight derived by the data on the frequency of use: a weight of 1 was given to services used every day, 0.7 to services used weekly, 0.3 for services used monthly and, finally, 0.1 for services used annually.

carried out to see whether the most 'labour-intensive' firms belonged to a particular sector, or were located in a specific region; whether the largest firms were located in the same region and the same sector. The results of this analysis showed a completely random relationship among these variables. For this reason we have some confidence that the simple performance index measured as the 'labour productivity' could be used in our analysis. The next sections are devoted to the empirical results regarding the existence of production network externalities.

16.4.2   Empirical Results

In this section we present empirical evidence for our second research issue, i.e. *whether network externalities play a role in the performance of firms and regions.* In particular, we attempt to determine whether there is any correlation between the performance index and the connectivity index.

In light of our conceptual framework, we would expect to find no correlation between the simple adoption of networks and services and the performance of firms. It is in fact not simply being connected to a network which generates benefits to a firm, it is the use of these technologies which creates production network externalities to the networked firms.

In order to test the first hypothesis, we applied the two indices already described above: a connectivity index, i.e. the ratio between the real number of connections to the number of potential connections for each firm of our sample and a performance index i.e. the ratio between the turnover of firms and the number of employees in 1991.

In order to be sure that the results were not biased by sector or size effects, the analysis was also run taking these into account. The 'sector' variable has been introduced by running a multivariate correlation between the performance and connectivity indices and the sector firms belonged to. The size variable was taken into account by running the multivariate correlation in four different size groups. If the size of firms has an impact, we would expect an increasing (decreasing) value of the correlation coefficient when the size of firms increases (decreases). This methodology was also applied at a regional level. Multivariate analysis with sector and size variables allow us to separate out network externality effects from more traditional effects of economies of scale and innovative processes. If there is any relationship between the level of connectivity and the performance of firms and this turns out to be independent from sector or size effects, variations in the performance of firms can be reasonably attributed to the existence of (production) network externalities.

A correlation analysis was run on these two indices at both the national and regional level (Table 16.11). The Pearson correlation coefficient R, having a value of only 0.069 confirmed our first impression and allowed us to go a step further by claiming that almost no correlation exists between these indices. *Our first hypothesis is thus confirmed, since the empirical analysis allows us to conclude that the simple adoption of these technologies has no effects on the performance of firms.*

The results were not significantly different when the size and the sectoral variables were introduced into the analysis. Introducing the sectoral variable leads to a similar result for the correlation coefficient, a value of 0.08. When the analysis is repeated in the four size groups of firms (in terms of both employment and size), the correlation coefficient changes randomly, and does not demonstrate any relation with the size of the firms.

At the regional level our expectations of strong variations appear to be verified. The regional dimension is important in explaining the results of the empirical analysis as the national result was an average of the two regional analyses. The Pearson correlation coefficients were to 0.398 for the North of Italy, and -0.058 for the South of Italy (see Table 16.11 before). For the South the correlation is absent, with a value near zero and a negative sign. For the North of Italy, it is undoubtedly true that the situation improves achieving 0.39 as a correlation value and thus showing a weak correlation between the two indices. This result confirms our hypothesis of the limited effect of adoption on the performance of firms. Results do not vary, when the analysis is run taking into account the sector to which firms belong. In fact, the multivariate correlation analysis shows a similar correlation coefficient value: 0.4 in the North and 0.03 in the South. Moreover, the correlation run separately for the four size groups of firms does not show any clear relation between the dimension of firms and the correlation coefficient values.

Table 16.11 Correlation coefficients between the row connectivity index and the performance index by macro areas

NATIONAL RESULTS		
	Performance Index	Row Connectivity Index
Performance Index	1	0.069
Row Connectivity Index	0.069	1
RESULTS FOR THE NORTH		
	Performance Index	Row Connectivity Index
Performance Index	1	0.398
Row Connectivity Index	0.398	1
RESULTS FOR THE SOUTH		
	Performance Index	Row Connectivity Index
Performance Index	1	-0.0588
Row Connectivity Index	-0.0588	1

If we adjust the connectivity index to the frequency of use of these services, the results of the correlation analysis at the *national level* still show a Pearson correlation coefficient R with very low values, viz. 0.11 (Table 16.12). At the regional level, however it is immediately clear that there is a better fit for a linear correlation in the North than for the South. This impression is confirmed by large differences in the Pearson correlation coefficients, whose value varies from 0.085 for the South, to 0.473 for the North (see Table 16.12). These results show that:

- the regional variation in correlation analyses is even greater in the case of the correlation between the simple adoption and the firms' performance. The national correlation value is nevertheless still very low, because it averages an even lower R in the South and a higher value for the North;
- as expected, the most developed regions are the ones which gain more from network externality effects, while backward regions are not yet able to achieve economic advantage from the adoption;
- the use of these technologies is strategic for the exploitation of production network externalities. In Northern Italy, where these technologies are used more frequently, the economic advantages from their adoption is certainly higher than in Southern Italy.

Table 16.12 Correlation coefficients between the weighted connectivity index and the performance index by macro areas

NATIONAL RESULTS		
	Performance Index	Row Connectivity Index
Performance Index	1	0.11
Row Connectivity Index	0.11	1
RESULTS FOR THE NORTH		
	Performance Index	Row Connectivity Index
Performance Index	1	0.473
Row Connectivity Index	0.473	1
RESULTS FOR THE SOUTH		
	Performance Index	Row Connectivity Index
Performance Index	1	0.0856
Row Connectivity Index	0.0856	1

16.5   Conditions for the Exploitation of Production Network Externalities

We have in previous sections analyzed the reasons for adopting telecommunications technologies and participating in such networks, but there is a need for more comprehensive assessment in which the regional implications are also addressed. In this context, various new problems emerge. The first is to reveal the 'inner working' of the linkage between the connectivity and the performance index identified above, by defining the variables and the elements which characterize the relationship between the connectivity and performance indices. The analysis run in the previous section does not tell us *why and under which conditions the correlation takes place*. The second problem is that the results show the existence of a correlation between the two indices, without showing the *direction of causality*. We have interpreted the results as indicating that the higher the connectivity, the better the performance, but this could easily be interpreted in reverse, by claiming that the better the performance, the greater the likely connectivity. Therefore, we need a more comprehensive analytical model which is able to take into account multidirectional causalities. For this purpose, we have developed a path model.[11]

The ambiguities mentioned above can, in principle, be overcome by imposing and testing a clear causal path, starting from greater connectivity to better performance. Figure 16.4 summarizes our conceptual model for production network externalities. The upper part of the chart represents the conditions needed on the demand side to exploit better production network externalities, while the lower part summarizes the conditions on the supply side. When these conditions are present, we expect firms to be able to exploit the advantages from these technologies.

Our intention in this section is to explain the relationship between the connectivity index and the performance index. Other variables such as the size of the firms or the sector they belong to have been taken into account, as this may have a direct or an indirect effect on either the performance or the connectivity index. However, for the following reasons we have restricted our analysis to some specific variables:

a from the theoretical point of view, the variables used in our model are those most commonly mentioned in the literature on telecommunications diffusion as having strong influence on the decision to adopt these new technologies;

b again from a theoretical point of view, there is no reason why larger firms should have different contact patterns from small firms, as is witnessed by the results obtained above, where the size variable was insignificant with respect to the estimation of consumption and production network externalities. Larger firms find it easier to accept innovation, as is well known in theory, but this aspect is indirectly taken into account in one of the variables included in our model, namely the variable reflecting the innovative behaviour of a firm;

c from a statistical point of view, a model with too many explanatory variables has low explanatory power. Thus, only the variables quoted in the literature as the most important variables explaining adoption processes were tested in the model. Moreover, the good results achieved in the empirical analysis suggest that no important explanatory variable is missing.

The methodology used is the path analysis model, which has an important characteristic: relations between variables must have an evident causal direction. This allows us to disentangle the linkage between the connectivity and the performance index. As mentioned already, the intermediate variables explain the conditions under which the correlation between the connectivity and the performance index takes place. In particular, their presence guarantees that the adoption of new technologies generates better corporate performance. To be sure

---

[11]Another application of path models can be found in Davelaar and Nijkamp 1992.

that the direction of causality between the performance and the connectivity index was the most appropriate one, we also estimated the model imposing the reverse direction of causality, i.e. a model where greater performance was the independent variable explaining greater connectivity. The poor results obtained have demonstrated once more that our hypothesis was the right one.

Concerning the supply side, at least three conditions have to be present in order to allow firms to exploit production network externalities:

a the *achievement of a critical mass* of adopters, especially for interrelated services, e.g. electronic mail. The user value of these technologies is in fact related to the number of existing subscribers. Our idea, tested in another study (Capello, 1994), is that the number of existing subscribers is one of the most important reasons for joining networks and services. The existence of at least a certain level of subscribers is a necessary condition to stimulate a *cumulative self-sustained mechanism*. If a critical mass is not achieved, the risk is that potential adopters have not sufficient incentive to adopt these technologies. In other words, the adoption process of these technologies has to be strongly supported by the supply until a critical mass is achieved;

b another important factor is *constant assistance* in the first phases of development from the supply side, in terms of the technical, managerial and organizational support necessary for an innovative exploitation of these technologies. This process requires a strong organizational effort by the subscriber, who generally needs organizational support from people with expertise and experience. The STAR programme itself has certainly taken this aspect into account. Some specific measures were devoted to the implementation of 'demonstration and promotion actions', through specialized centres whose task was to develop a 'telematics culture' among potential users. However, in some countries such as Italy, these centres have emphasized *technical* rather than *organizational aspects*. Centres in fact promoted these services, mainly by concentrating on their technical features. While they assured technical support to users, they did not provide adequate information and advice about organizational problems or the kind of changes which have to be coped with in order to exploit production network externalities. We therefore expect that the lack of organizational support will act as a bottleneck in the exploitation of network externalities;

c another crucial factor for the exploitation of production network externalities is the *clear identification of the way these technologies may be useful for business purposes*. As we explained before, from a conceptual point of view these technologies are multi-faceted and require a certain degree of organizational change in order to be used effectively. For these reasons, the supply side has to demonstrate their importance for business, in order to stimulate an interest on the demand side.

There is a set of critical success conditions required on the supply side, but it has to be added that on the demand side too there are some necessary requirements in order to exploit production network externalities:

d to begin with, due to their complexity, the capacity of firms to accept new technologies and to exploit them is a very important consideration. Firms have to be highly flexible if they want to exploit production network externalities. The better the capacity of adapting the organizational structure to external changes, the greater the probability of exploiting production network externalities;

e another factor which greatly assists firms to exploit production network externalities is their *innovative behaviour* or their level of entrepreneurship. Business acumen guarantees the achievement of competitive advantage through these technologies. In fact, through previous experience of innovation the firm may have acquired the necessary managerial know-how to be able to implement the organizational changes required to achieve competitive advantage. The higher the number of innovations already adopted by the firm, the greater the probability of exploiting production network externalities.

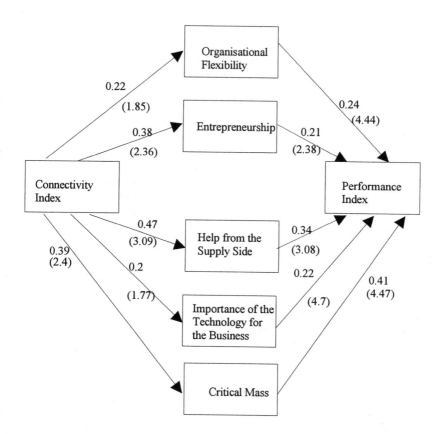

Fig. 16.5. Estimated path analysis model for the North of Italy

334

Figure 16.5 shows the values of the estimated parameters and the T-Student test results (presented in brackets) for the North of Italy. In interpreting the results for both the North and the South of Italy, it should be kept in mind that all variables have been standardized with unit variance and mean zero, in order to be able to compare the relevance of individual parameters. Some conclusions can already be drawn:

a almost all parameters are statistically significant, having T-Student values over 2;
b all estimated relations have the expected sign;
c the model itself fits very well with the given variables (P-value = 0.01).

Our conceptual model thus appears to be confirmed: *micro-conditions allowing firms to exploit network externalities are present in Northern Italy.*

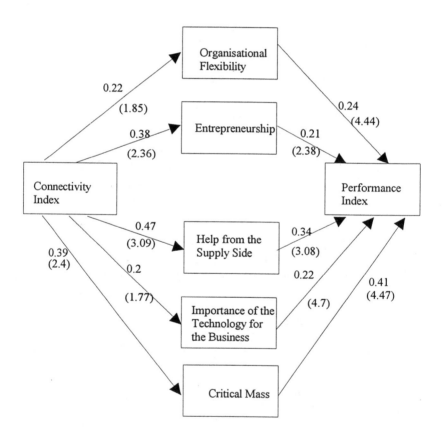

Fig. 16.6. Estimated path analysis model for the South of Italy

A very different result is achieved for the South of Italy. As Figure 16.6 shows, the results of the same conceptual model are quite different and essentially less satisfactory, as expected. At first glance, one can see that the model does not fit the empirical observations very well. Most estimated parameters, are not statistically significant, with T-Student values below the critical value of 2. In any other circumstances this would have been interpreted as a negative result, destroying from an empirical point of view the conceptual framework underpinning the analysis. On the contrary, these results support our expectations: *micro-conditions to exploit production network externalities do not seem to be significantly present in Southern Italy*. We can even argue that with this analysis we have been able to identify the barriers and bottlenecks existing in the exploitation of production network externalities and which hinder the achievement of better economic and spatial performance via the use of advanced telecommunications technologies.

16.6   Conclusions

The chapter has presented a conceptual model of the role played by network externalities in the performance of firms and regions. The basic thesis is that network externalities have a measurable effect on the performance of firms and, via multiplicative effects, on the performance of regions.

An extensive empirical analysis was carried out with the aim of providing answers to three crucial questions:

i what is the relative importance of network externalities for firms in their decision to join a network?
ii do firms and regions gain from exploiting network externalities?
iii what are the micro- and macro-conditions stimulating or supporting the exploitation of network externalities?

The empirical findings confirmed our expectations. The number of existing subscribers is one of the most important reasons for joining a network, especially in the first phases of adoption.

As far as the second question is concerned, a methodology is proposed for measuring production network externalities. The results show that advanced regions are able to exploit network externalities and to take advantage of the use of these technologies, while this opportunity is lacking in backward regions.

These results stimulated a further analysis identifying the micro and macro conditions which explain the capacity to exploit network externalities. Using causal path analysis, a certain number of conditions were identified and from these some policy implications emerged. In particular, it was possible to outline *policy guidelines* for future programmes encouraging the diffusion of advanced technologies.

First of all, it was clear that the promotion of these technologies has to be based on a *bridging mechanism between demand and supply*, i.e. it has to link business needs, or potential business needs, to the existing technological potential. Suppliers should be able to provide not only the physical infrastructure and technical support, but also advice on how to exploit these technologies, on the basis of the specific business needs of potential users. One way of dealing with this problem is to customise the networks and services as much as possible.

A second lesson is related to the 'spatial circumstances' in which these technologies are promoted. A crucial resource for the development of these technologies is *entrepreneurship*. The presence of risk aversion and of non-competitive market structures discourages adoption since, under those conditions, there is insufficient stimulus for firms to bear the organizational, managerial and financial costs necessary. Thus, local entrepreneurship is a strategic element for successful adoption. Innovative policies have to take this aspect into consideration, choosing local areas where *entrepreneurial capabilities* are available. Thus, instead of supplying these technologies to all parts of the less developed regions, efforts should be focused on the most dynamic areas (in terms of both technical and entrepreneurial abilities). These areas, the '*local innovative milieux*', represent the most efficient and dynamic areas, where a technology policy could lead in the long run to effective network externality exploitation. This modern view suggests that technology policy, when implemented without considering local factors, risks resulting either in a cumulative process of spatial concentration of technological development, or an inefficient waste of resources, as it overlooks the adoption problems of small and medium sized enterprises.

Another important feature of technology policy for regional development is its capacity to overcome *local economic constraints*. Telecommunications technologies are a way of acquiring information and know-how which are not present locally, and which are typical of advanced economic areas. For this reason, the STAR programme was seen as a means of shrinking the (physical and economic) distance between backward and more advanced areas in the European Community. However, the way in which the programme has been managed, creating links *within* backward areas, means that its full potential has not been achieved. In the Italian case, the implementation of advanced networks and services in the South, with no direct link to the North of Italy, has acted as a disincentive for adoption. The creation of *a club for poor regions* does not seem to be the right policy for ensuring local sustainable economic development.

Advanced technology programmes must be geared to overcoming the bottlenecks and barriers which hamper the full exploitation of these technologies. The capacity of future intervention policies to solve these problems will determine their impact on regional performance. For this reason, it is of vital importance to assure continuity in the provision of these technologies, via the launch of a second intervention programme. This would have two aims: a) to overcome the present bottlenecks and barriers in the adoption processes, taking into account the 'lessons' learnt from the first phase of the programme, b) to achieve the positive effects on regional performance at present hampered by the low level of adoption and of use.

Annex 1

The database consists of primary data collected via in-depth interviews with users of telecommunication technologies in two different macro-areas[12] in Italy: the North of Italy, in particular the highly industrialized Milan area, and the South of Italy, a typical area of economic underdevelopment.

In the case of the sample in the South, we chose to evaluate the effects of the STAR Programme. The total number of interviews was 70, equally divided between Northern and Southern Italy (35 in each macro-area); they were carried out between September 1992 and February 1993 and were equally distributed between the manufacturing and the service sector. In terms of intra-sectoral distribution, manufacturing firms belong to a varied range of industries, both traditional (food, construction), and modern (engineering, including electronics). In contrast, the service firms had a high presence of business services (often computing and telecommunications) and a small minority of specialized trade services.

The firms were small and medium sized:[13] in 1991 the maximum employment per firm was 78 (with one exception of 160) in the South and 204 in the North. Maximum turnover was 106 billion Italian lire in the South (approximately 58 million ECU) and 48 billion lire (26 million ECU) in the North. The distribution by turnover shows that the sample in the South is constituted predominantly of firms with a maximum turnover of 5 billion lire. In the North, on the contrary, the distribution is in favour of firms having a minimum turnover of 5 billion lire (see Table 16.2). This difference appears also in the case of employees. In the South most firms have a maximum of 20 employees, while in the North most firms have more than 20 employees. This difference mirrors the differences in the economic environment between the two macro-areas: a larger number of small firms in the South, compared to a larger number of medium-sized firms in the North.

The questions put to the sample of firms cover a high range of issues, namely the degree of satisfaction among users of the new services, the intensity of use, the degree of satisfaction with the marketing policies, the interest these services have raised among users, and the relationship between the introduction of these technologies and their business development. The questionnaire was divided into three different parts. The first part served the purpose of testing the first group of

---

[12]By 'macro areas' we mean several regions analyzed together. In the case of the South of Italy, in fact, we analyzed three Southern regions, namely Sicily, Abruzzi and Apulia. The results of the empirical analysis for these three regions are presented together under the heading 'South of Italy'. In the case of the North of Italy, the empirical analysis was run only in one region, i.e. the Lombardy region, and the results are presented under the heading 'North of Italy'.

[13]The STAR Programme was set up only for SMEs. In this study our sample contains only small and medium sized firms: they have no more than 200 employees and a turnover of no more than 100 million Italian lire.

hypotheses related to the first research issue; the questions in fact deal with:

a the most important reasons for the adoption of telecommunication technologies;
b the most important reasons for not having adopted telecommunication technologies;
c the main conditions for future adoption of telecommunication technologies;
d reasons for dissatisfaction with the telecommunication technologies in use.

The second group of questions was related to the second research issue and it focused mainly on:

a level of turnover;
b number of employees;
c telecommunication infrastructure and services used;
d the intensity of use of these technologies;
e number and kinds of innovation achieved via the introduction of telecommunication technologies;
f number and kind of functions having faced organizational changes in the last five years;
g variation in the number of suppliers and customers in the last five years.

Finally, another group of questions related to the third group of hypotheses and concerned:

a the degree of importance of demonstration and promotion centres in the decision-making process to adopt the new technologies;
b the degree of importance of these technologies for the business.

To test the third research issue, other questions included in the questionnaire were also useful, namely the number of functions which had been subject to organizational changes and the amount of innovation achieved.

This broad range of questions was sufficient to create a comprehensive database on network behaviour on which we could build our empirical work.

References

Allen D (1990) Competition, Co-operation and Critical Mass in the Evolution of Networks. Paper presented at the 8th Conference of the International Telecommunications Society, on Telecommunications and the Challenge of Innovation and Global Competition, held in Venice, 18-21 March

339

Allen D (1992) Telecommunication Policy Between Innovation and Standardisation. Paper presented at the IXth International Conference of the International Telecommunication Society held in Sophia Antipolis 14-17 June

Antonelli C (1989) The Diffusion of Information Technology and the Demand for Telecommunication Services. Telecommunications Policy. September:255-264

Antonelli C (1990) Induced Adoption and Externalities in the Regional Diffusion of Information Technology. Regional Studies 24 1:31-40

Antonelli C (Ed) (1992) The Economics of Information Networks. Elsevier Publisher Amsterdam

Ben-Akiva M and SR Lerman (1985) Discrete Choice Analysis: Theory and Application to Travel Demand. MIT Press. Cambridge Massachusetts

Bental B and M Spiegel (1990) Consumption Externalities in Telecommunication Services. In: de Fontenay M and D Sibley (1990) (Eds) Telecommunications Demand Modelling. Elsevier Science Publisher:415-432

Bishop YMM, Fienberg SE and PW Holland (1977) Discrete Multivariate Analysis: Theory and Practice. MIT Press Massachusetts

Boudeville JR (1968) L'Espace et le Pôle de Croissance. Presses Universitaires de Paris France

Cabral L and A Leite (1992) Network Consumption Externalities: the Case of the Portuguese Telex Service. In: Antonelli C (1992) (Ed) The Economics of Information Networks. Elservier Publisher:129-140

Camagni R (1992) Economia Urbana. La Nuova Italia Scientifica Rome

Capello R (1994) Spatial Economic Analysis of Telecommunications Network Externalities. Avebury Aldershot

Capello R (1995) Network Externalities: Towards a Taxonomy of The Concept and a Theory of their Effects on the Performance of Firms and Regions. In: Bertuglia C S, Fischer M M and G Preto (Eds) Technological Change, Economic Development and Space. Springer Verlag Berlin:208-237

Capello R and P Nijkamp P (1995) Corporate and Regional Performance: the Role of Network Externalities. In: Rallet A and C Torre (Eds) Economie Industrielle et Economie Spatiale. Economica Paris:273-296

Davelaar EJ and P Nijkamp P (1992) Operational Models on Innovation and Spatial Development: a Case-Study for the Netherlands. Journal of Scientific and Industrial Research 51 March:273-284

David P (1985) Clio and the Economics of Qwerty. AEA Papers and Proceedings 75 2:332-337

David P (1992) Information Network Economics: Externalities, Innovation and Evolution. In: Antonelli C (Ed) (1992) The Economics of Information Networks 103-106

Domencich TA and D Mc Fadden (1975) Urban Travel Demand: a Behavioural Analysis. North Holland Amsterdam

Fischer M and P Nijkamp (1985) Developments in Explanatory Discrete Spatial Data and Choice Analysis. Progress in Human Geography 9:515-551

Fischer M, Maggi R and C Rammer (1992) Stated Preference Models of Contact Decision Behaviour in Academia. Papers in Regional Science: the Journal of RSAI 71 4:359-371

Gillespie A and M Hepworth (1986) Telecommunications and Regional Development in the Information Economy. Newcastle Studies of the Information Economy. CURDS Newcastle University. 1/October

Gillespie A and H Williams (1988) Telecommunications and the Reconstruction of Regional Comparative Advantage. Environment and Planning A 20:1311-1321

Gillespie A. Goddard J, Robinson F, Smith I and A Thwaites (1984) The Effects of New Technology on the Less-favoured Regions of the Community. Studies Collection. Regional Policy Series EEC 23

Gillespie A, Goddard J, Hepworth M and H Williams H (1989) Information and Communications Technology and Regional Development: an Information Economy Perspective. Science, Technology and Industry Review 5 April:86-111

Goddard J (1980) Technology Forecast in a Spatial Context. Futures April:90-105

Griguolo S and A Reggiani (1985) Modelli di Scelta tra Alternative Discrete: Alcune Note Introduttive. Archivio di Studi Urbani e Regionali 22:47-86

Hayashi K (1992) From Network Externalities to Interconnection: the Changing Nature of Networks and Economy. In: Antonelli C (Ed) (1992) The Economcs of Information Networks. North-Holland Amsterdam 195-216

Katz M and C Shapiro (1985) Network Externalities, Competition and Compatibility. The American Economic Review 75 3:424-440

Katz M and C Shapiro (1986) Technology Adoption in the Presence of Network Externalities. Journal of Political Economy 822-841

Katz M and C Shapiro (1992) Product Introduction with Network Externalities. The Journal of Industrial Economics Vol XL 1:55-83

Leonardi G (1985) Equivalenza Asintotica fra la Teoria delle Utilità Casuali e la Massimizzazione dell'Entropia. In: Reggiani A (Ed) Territorio e Trasporti: Modelli Matematici per l'Analisi e la Pianificazione. Franco Angeli Milano 29-66

Markus M (1987) Towards a Critical Mass Theory of Interactive Media: Universal Access, Interdependence and Diffusion. Communication Research. 14:491-511

Markus M (1992) Critical Mass Contingencies for Telecommunication Consumers. In: Antonelli C (Ed) (1992) The Economics of Information Networks. Elsevier Publisher Amsterdam 431-450

McFadden D (1983) Econometric Analysis of Qualitative Response Models. In: Griliches Z and D Intriligator (Eds) (1983) Handbook of Econometrics. North Holland Amsterdam 1396-1450

Nijkamp P, Leitner H and N Wrigley (Eds) (1985) Measuring the Unmeasurable. Martinus Nijhoff Dordrecht

Perroux F (1955) Notes sur la Notion de Pole de Croissance. Economie Appliquée 7:307-320

17 Strategic Adoption of Advanced Communication Systems: Some Empirical Evidence from the Finnish Metal Industry

Heli Koski

17.1   Introduction

The emerging divergencies in the adoption rates of new technologies - or in the diffusion speed of innovations in general - among firms, regions and countries have stimulated a great number of both theoretical and empirical studies (see e.g., Bertuglia et al. 1995; Mansfield 1968; Nijkamp and Reggiani 1995; Stoneman 1983). In the present study, we focus on entrepreneurial adoption behaviour regarding new technology and examine in particular the determinants of firms' adoption of a new communication technology.

A feature distinguishing advanced communication technology (ACT) from most other innovative technologies is the direct user interdependence - i.e. the presence of network externalities - related to its adoption and use. The different forms of ACT are likely to differ with respect to the business and communication purposes they serve. We assume that these features - along with the management attitude towards technological innovations and willingness to take risks - have a crucial influence on the firm's ACT investment strategies and, thus, on the decision to adopt ACT.

Section 17.2 considers the potential factors determining the firm's adoption of a new communication system. Section 17.3 introduces the data and the econometric models we have used in the empirical analysis. In Section 17.4, we present the results of the empirical estimations regarding the adoption of ACT in the Finnish metal-working industry. First, we explore the differences between the users and the non-users of ACT, then examine the determinants of the diffusion of internal and external communication systems and, finally, we compare the firms' investment strategies related to the adoption of divergent advanced communication systems.

17.2   Determinants of the Firm's Adoption of ACT

Innovations are conventionally divided into two main categories, viz. product innovations and process innovations. The former indicates the emergence of a new product, whereas the latter denotes technological change affecting the production process. Innovations may also be distinguished by their degree of novelty as primary and secondary innovations (see Davelaar and Nijkamp 1988). Secondary innovations are new only to the firm, whereas primary innovations are novel to the whole branch of industry (or sector) in the country.

A novel communication system may be a product innovation to its producer, but to the users of ACT the adoption of a new communication technology is likely to represent process innovation. The use of a new communication system may change the patterns of communication - i.e. the practices and the channels used for transferring information from and to the firm - as well as the ways information is handled and stored in a firm. The change may take place in the firm's *internal information flows* and/or in the firm's *external information flows* depending on the type of innovative technology adopted. Also, *the format of information* varies with the information systems. For instance, an advanced communication system may be used for the transfer of formal information of which form or language is task-specific and/or firm-specific (e.g., Electronic Data Interchange, (EDI)) or it may act as an informal discussion channel for the employees (e.g., internal e-mail)[1]. In line with these features, we classify the types of innovative communication technologies as follows:

1   Technological innovation altering a firm's internal communication patterns, e.g., internal e-mail.
2   Technological innovation altering a firm's external communication patterns, e.g., external e-mail.
3   Technological innovation which besides altering a firm's communication patterns also changes the firm's patterns of trade and/or patterns of production, e.g., EDI.

We assume that, as the different communication systems may serve divergent business and communication purposes, also the diffusion rate of the communication systems and the adopters of the different communication technologies are likely to differ from each other. Direct user interdependency, on the other hand, is typical to all communications systems and distinguishes the advanced communication technologies from most other innovative technologies. We will explore the determinants of the adoption of divergent forms of ACT below (see Section 17.4). Before going into the empirical results, we will briefly discuss the potential determinants of the firm's adoption of ACT in the light of previous theoretical and empirical studies.

Fischer (1995) distinguishes the following four major determinants influencing the firm's innovation behaviour: (i) the firm's innovation-relevant internal

---

[1]See Markus (1990) for discussion on critical mass issue and the firm's external and internal telecommunication needs.

characteristics (e.g., firm size, attitudes of management towards technological innovation), (ii) innovation-relevant locational influences (e.g., access to information and technological know-how), (iii) the techno-economic environment of a firm (e.g., industrial sector, communication and cooperation networks) and (iv) factors related to the political-institutional context in which the firm has to operate. We will make use of this framework in discussing the potential determinants of the firm's ACT adoption behaviour.

The firm's adoption of ACT and innovations in general is, thus, likely to be related to several spatial and firm-specific characteristics. We are inclined to think - on the grounds of the existing studies - that the following characteristics are likely to be (among) the decisive factors influencing the firm's ACT adoption decision[2]:

i firm-specific characteristics:
- firm size
- the number of business partners using ACT
- the expected number of the users of technology
- the decision maker's attitudes towards risk
- the ACT investment strategy of the firm's management

ii locational factors
- location of a firm, the firm's distance from the central (or metropolitan) area

iii techno-economic factors
- the branch of industry (or industrial sector) to which the firm belongs

The size of a firm is perhaps the most frequently mentioned firm-specific characteristic in connection with innovative behaviour. Larger firms generally appear to be likely to adopt a new technology sooner than smaller firms, due to their larger financial resources and for their better availability of skilled labour force and technical expertise (see, e.g., Alderman and Davies 1990; Mansfield 1968). It seems plausible that as the firm's size increases, the communication needs increase as well, thus larger firms are able to benefit more from an efficient communication technology than smaller ones. Also, as large firms may be composed of geographically dispersed business units, the importance of an efficient communication system is enhanced. Several empirical studies support the hypothesis of a positive relationship between the firm size and the innovativeness of a firm (see, e.g., Mansfield 1968; Karlsson 1995).

The presence of network externalities in the ACT market suggest that a critical factor in the adoption of a new communication technology is the number of business partners using the same or compatible technology (see, e.g., Capello 1994). Typically, an investing firm may achieve competitive advantage from the adoption of an innovative technology before its rivals. A potential user of network

---

[2]The literature presents an extensive list of potential determinants of the firm's innovative behaviour, but we will consider only the factors which are also tested in the empirical part of the paper below.

technology requires a minimum number of contemporary users, i.e. the private critical mass (PCM), before he is willing to adopt the new technology himself. This means that for exploiting the innovative communication technology in a strategic sense, a firm has to implement and use ACT *with* its business partners or economic network before its rivals. Thus, a firm has to take into account, besides the innovation behaviour of its rivals, also the ACT investment decisions of the network of its business partners and the networks of the rivals.

Due to uncertainty and the sunk costs related to the investment in ACT[3] expectations are likely to be an influential factor in the firm's ACT adoption decision. The interdependence of the ACT users indicates that expectations concerning the number of users of a technology (see Katz and Shapiro 1985) as well as expectations concerning the technological progress are important (see e.g., Dosi 1990; Rosenberg 1976). We assume that as the benefits accruing from the use of ACT depend essentially on the other users of the technology, the expectations of the network size are likely to have a particularly strong impact on the firm's decision making concerning the adoption of a new communication technology.

The management attitude towards technological innovation and managerial willingness to take risks, in general are also among the critical factors influencing the diffusion and adoption of the new technologies. The literature (see Ihamuotila 1994) points that the risk inherent in a firm's investment behaviour depends decisively on the following two factors: (i) who has control over the firm's investments (i.e. on the ownership structure) and (ii) how well the owners have diversified their investments. The firms in which ownership is concentrated (i.e. the largest shareholder(s) have control over the firm) and which are not the main investment object of the owners (i.e. the owners have diversified their investments) are less risk averse than firms which do not share these features.[4]

We suppose that the firm's adoption of a new technology - like the generation of innovations in a firm - depends essentially on the innovation or adoption strategy the firm's management opts for. This strategy may be aggressive (offensive) or defensive strategy (see Karlsson 1995; Kay 1988; Freeman 1982) or else a dependent strategy (see Kay 1988; Freeman 1982). The adopter of an aggressive ACT investment strategy aims to obtain competitive advantage or an increase in its market power by being among the early adopters. The defensive ACT investment strategy means that a firm monitors other firms' ACT investment decisions and then bases its own adoption decision on the other firms' behaviour. This strategy allows a firm to reduce the uncertainty related to the new technology and possibly avoid commitment to an inferior technology. The dependent ACT investment strategy implies that the firm's adoption depends more on the decision of some dominant firm (or firms) than on its own autonomous choice. For instance, the supplier of the intermediate products may follow the adoption

---

[3]The literature contains only a small number of empirical studies on the role of expectations in firm's investment behaviour (see e.g., Van der Ende and Nijkamp 1994 and Weiss 1994).

[4]See Ihamuotila (1994) for a detailed discussion on the subject.

behaviour of its largest customer(s).

The implementation and use of a new communication technology requires a costly learning process which means that the adoption costs of a new technology are far higher than the market price of the technology. Moreover, the costs of adoption and use of a communication technology depend on the technology choices of the other firms; the early adopters may reduce the firm's costs of learning to use the new communication system and hence late adopters have lower learning costs than early adapters (see Antonelli 1990). The exact learning costs are likely to be, nevertheless, unknown and thus a source of uncertainty. Other potential forms of uncertainty related to the new technology are, for example, the number of users and the level of profitability and use costs of a new technology. Whether a firm waits and gathers more information regarding the new technology or adopts ACT in spite of the uncertainty, depends on the management attitude towards risk and investment strategy of the firm

Studies on the spatial diffusion of innovations generally underline the important role of central areas in the generation of innovations and in early adoption of new technologies and innovations. The location of a firm or the regional seedbed conditions (see Kamann and Nijkamp 1990) may affect the firm's innovative behaviour via, for instance, the supply of skilled labour force and the quantity and quality of the information networks. Davelaar and Nijkamp (1988) suggest that the location of a firm may also affect the types of innovations generated, i.e. metropolitan or central areas facilitate the generation of primary innovations, whereas non-metropolitan areas are more suited for the adoption of secondary innovations. High quality innovations - like advanced communications systems - may be considered primary innovations and, thus, we can suppose that ACT is likely to spread outwards from central areas to the more remote ones.

The innovativeness of firms in different industrial sectors or branches of an industry may differ as well. Inter-sectoral disparities may arise, for example, from differences in the technological opportunities to innovate and in the market structures (Fischer 1995). Firms in high-tech industries with extensive R&D expenditures are likely to be more eager adopters of new communications technologies - due to their information intensiveness and need for the rapid exploitation of new information - than firms in the more traditional industries (see Macdonald 1992).

In this section, we have discussed certain firm-specific, locational and techno-economic factors which possibly influence the firm's adoption of a new communication technology. We shall now introduce the data which has been used for exploring the relationship between these potential determinants of adoption behaviour and the actual adoption decisions of firms in the Finnish metal industry.

17.3   Description of Data and Econometric Models

The data was collected by sending a mail questionnaire to 1020 firms in the metal industry.[5] We drew up two versions of the questionnaire[6] and the respondents were asked to choose the one they found more appropriate; one version was designed for the non-users of ACT and the other for the users of ACT. The firms were picked from a list of firms in the metal industry. The response rate amounted to about 22 percent. Over half of the responses (56.4%) - that is 124 questionnaires - came from users of ACT, the remaining 96 responses were from non-users. The firms returning the questionnaire were distributed within the metal industry as follows: 58.6% were from metal engineering, 18.6% from mechanical engineering and 18.6% from electrical engineering. In nine of the questionnaires this information was missing.

Table 17.1 lists the explanations of abbreviations and the definitions of the variables used in the estimations. Some of the variables require a closer look. Variable 'risk' is a rough measure of a firm's exposure to risk.[7] We assume that the firms of which ownership is concentrated and of which owners have (well)diversified their investments are less risk averse than the firms which do not share these features (see discussion above).

One of the questions concerned the relative importance of the sixteen factors affecting the firm's adoption decision of the ACT.[8] Use of all these variables would have led to a very large number of explanatory variables. In addition, the importance of certain factors appeared to be highly correlated, implying problems of collinearity. These two statististically undesirable properties led us to apply principal components method.[9] The aim of the principal component analysis is to extract from the matrix of explanatory variables those factors which can capture most (or all) of the variation in the matrix - i.e. the principal components which are linear combinations of the explanatory variables.

The statistical program used for the principal component analysis was SPSS 6.0. As a result of the factor extraction procedure we ended up with five principal components which were linear combinations of the observed variables. The principal components consisted, however, of all the variables through some weight, making the interpretation of these main factors a rather indistinct. To get rid of this undesirable property, we divided the variables into five groups according to their weights, i.e. the variables with the largest weights in a principal component were incorporated into the same group. After this step, we applied

---

[5]Including the three main branches of the industry: metal, mechanical and electrical.

[6]The questionnaires are available from the author.

[7]In forming this variable we have benefited from the work of Ihamuotila (1994).

[8]The respondents were asked to evaluate to importance of these factors by using the following scale: 0 = not important at all, 1 = not very important, 2 = quite important, 3 = very important.

[9]See for example, Koutsoyiannis (1973) for more detailed discussion on the principal components method.

principal component analysis to each group[10] separately, ending up with five main factors. These five main factors were the linear combinations of the (initial) variables with the largest weights in the principal components. By this procedure, we achieved a meaningful reduction of the parameters of the model, in the sense that all factors have a certain economic intepretation (see discussion below).

Table 17.1. Definitions of the variables appearing in the estimation results

use	=	1 if a firm is a user of ACT (ACT includes all technologies which combine information technology with communication technology)
	=	0 otherwise
in1	=	2 if a firm adopted internal e-mail before 1990
		1 if a firm adopted internal e-mail in 1990 or after that
		0 if a firm is a non-user of internal e-mail
out1	=	2 if a firm adopted external e-mail before 1990
		1 if a firm adopted external e-mail 1990 or after that
		0 if a firm is a non-user of external e-mail
in2	=	1 if a firm is a user of internal e-mail
		0 otherwise
out2	=	1 if a firm is a user of external e-mail
		0 otherwise
edi	=	1 if a firm is a user of EDI
		0 otherwise
pcm	=	private critical mass = (log) the number of ACT using business partners a firm requires before it is willing to adopt ACT
ex	=	1 if the number of ACT using partners is a critical factor in the firm's ACT adoption decision
		0 otherwise
risk	=	0 if the ownership of a firm is concentrated and the firm is not the main investment object of the owner(s) = 'less risk averse firm'
		1 otherwise = 'more risk averse firm'
llab	=	(log) number of employees
lkm	=	(log) firm's distance from the center of the province
loc	=	firm's location (province)
ind	=	firm's branch of industry
f2out	=	firm's expectations on the rate of external e-mail users within the next two years
f2edi	=	firm's expectations on the rate of EDI users within the next two years
tough1	=	linear combination of the variables describing the importance in the firm's ACT adoption decision of faster delivery of raw material, cheaper raw materials and extending the market
tough2	=	linear combination of the variables describing the importance in the firm's ACT adoption decision of staying in the business, increasing competivity and increasing the sales efficiency
soft	=	linear combination of the variables describing the importance in the firm's ACT adoption decision of rival's decision, co-operation with other firms and demand of some other firm(s)
ef	=	linear combination of the variables describing the importance in the firm's ACT adoption decision of faster transfer and delivery of information, number of services achieved by the system and other cost efficiency achieved by fast transfer of information
unc	=	linear combination of the variables describing the importance in the firm's ACT adoption decision of investment costs, decreased uncertainty on the use costs of ACT, decreased uncertainty related to the number of users and sufficient information regarding the benefits obtainable via the use of ACT

The variables 'unc', 'tough1', 'tough2', 'soft', and 'ef' represent the five main factors which are linear combinations of the relative importance of the factors

[10]The groups related to the original values of the variables, i.e. not the weights of the variables in the principal components.

affecting the ACT adoption decision of a firm. The economic interpretation of these main factors follows. We suppose that all the factors are closely associated with the firm's ACT investment strategy. The variables describing the importance of (i) decreased uncertainty regarding the use costs of ACT, (ii) decreased uncertainty related to the number of users and (iii) sufficient information in the firm's ACT adoption decision regarding the benefits obtainable, formed the first main factor, 'unc'. The factor 'unc' illustrates the importance of sunk costs of the ACT investment and and the importance of the reduced uncertainty related to the new communication technology in the firm's adoption decision of ACT. The significance of the properties embodied in the factor "unc" reflects the risk reluctancy of the firm's management and describes the firm's propensity for a defensive ACT investment strategy.

The second main factor, 'tough1', comprises the importance of the variables related to competitive advantage or market power a firm may achieve via the use of ACT, viz. the importance of (i) faster delivery of raw materials, (ii) cheaper raw materials and (iii) extending the market. The factor 'tough2' is also related to the ACT investment strategy of a firm. This factor is a linear combination of the variables describing the importance of (i) staying in business, (ii) increasing competivity and (iii) increasing sales efficiency in the firm's ACT adoption decision. The factors 'tough1' and 'tough2' both stand for an aggressive investment strategy, but the latter factor stresses even more directly the importance of the firm's competivity in the ACT adoption decision, whereas the former is more closely related to market power.

The factor 'coop' illustrates dependent investment behaviour; it is formed as a linear combination of the variables describing the importance of (i) rivals' decisions related to the use of advanced communications systems, (ii) production and/or technical cooperation with other firms and (iii) chain with some other firm(s) in the ACT adoption decision. It also captures the importance of the cooperative behaviour in relation to the ACT investment. In this sense, 'coop' groups the features of the ACT adoption which make a firm 'soft', i.e. the firm's investments in ACT are also likely to benefit some other firm(s). Whether these firms are rivals or not is, however, unknown.

The factor 'ef' reflects a type of ACT investment strategy which is related to the firm's aim to achieve cost reductions via the use of ACT. It is a linear combination of the importance of the following factors in the firm's ACT adoption decision: (i) faster transfer and delivery of information, (ii) the number of services achieved by the system and (iii) cost efficiency achieved by the faster transfer of information. This variable describes the importance of ACT as an efficient means of handling business activities and reducing the costs related to these activities.

Before we begin the exploration of the empirical results, we will briefly introduce the econometric models which we have used in the estimations. All models contain a discrete dependent variable. Since the firm size and the variables related to the firm's investments all varied considerably in the sample - i.e. the tails of the distribution of the variables are very long - the logistic distribution describes the distribution of the variables better than the normal distribution.[11] Thus, the use of the logit models seems to be an intuitively appealing choice, even though the estimation results of the probit and logit models do not usually differ much from each other.

We have estimated two types of the models in the following section:

i   Multinomial logit models of which the dependent variable
    = 2 if a firm is the early adopter of internal/external e-mail (e-mail adopted before 1990)
    = 1 if a firm is the late adopter of internal/external e-mail (e-mail adopted in 1990 or after that)
    = 0 if a firm is the non-user of internal/external e-mail

ii Logit models of which the dependent variable
    = 1 if a firm is the user of internal ACT/e-mail/external e-mail/EDI
    = 0 otherwise.

The estimated multinomial logit models can be written as follows:

$$\Pr(Y_i = j) = \frac{e^{\beta'_j x_i}}{1 + \sum_{k=0}^{2} e^{\beta'_k x_i}} \ , \text{ for } j = 0,1,2$$

For identifiability of the model, one of the categories of the multinomial logit model has to be dropped from the estimations. In our model, the non-users of ACT represent this reference category and thus - by using conventional normalization - we set $\beta_0 = 0$. This means that a positive and statistically significant coefficient in β_1 (β_2) indicates that the probability of late (early) adoption of ACT increases with the explanatory variable in question compared to the non-users of ACT. The logit model is just a special case of the multinomial logit with j=0,1 (see, for example, Greene 1993). Now, we will proceed to the examination of the results of the empirical estimations.

17.4   Determinants of the Adoption of ACT - Empirical Results

17.4.1   Differences between the Users and the Non-users of ACT

We will begin our analysis by examining how the users of ACT differ from non-users. We examine whether the following factors identified in the literature[12] are

---

[11]The logistic distribution is similar to the normal distribution except that it has clearly heavier tails than the normal (Greene, 1993). See also, for example, Agresti (1990) for the analysis of the logit models.

[12]See the previous section for a discussion of these variables.

able to explain differences between users and non-users of ACT: the firm size (llab), the presence of network externalities (ex), the firm's riskiness (risk), the firm's location (lkm and loc), the firm's branch of industry (ind), the size of the private critical mass (pcm) and the expected network size (f2out). The variables 'pcm' and 'f2out' both included a relatively large number of missing observations and thus - in order to obtain a sufficient statistical reliability in the models - we estimated the equations consisting 'pcm' (model 2) and 'f2out' (model3) separately.

Table 17.2 presents the results of maximum likelihood (ML)-estimation of the logit models. The statistical program used in all the estimations was LIMDEP 6.0. The dependent variable, 'use', was formed by dividing the sample into the users and into the non-users of ACT. The models seem to have a relatively high predictive power; the ratio of correct predictions to the number of observations exceeds 80% in all the models. The probability of a firm's adoption of ACT increases with firm size as expected. The firms in different branches of metal industry also seem to differ with respect to the use of ACT. The negative sign of the estimate of the variable 'risk' implies that the probability of adoption of ACT decreases as the firm's risk aversion increases. The estimate of the variable was not, however, statistically significant in the models (2) and (3).

All the models support the existence and importance of network externalities: the variable 'ex' captures the importance of the network size (i.e. the importance of the number of business partners using ACT) in the firm's ACT adoption decision.

Statistical significance and the negative signs of the estimates indicate that the probability of adopting ACT decreases as the importance of the user size increases. The private critical mass (pcm) turns out to have a statistically significant estimate in model 2. This means that the probability of ACT adoption decreases as the PCM increases (i.e. the minimum number of business partners a firm requires to use ACT before it adopts ACT). These results imply that the ACT market clearly exhibit network externalities as expected. Moreover, we may deduce that a low number of business partners using the advanced communication systems hinders the diffusion of ACT.[13]

The negative sign of the estimate of the firm's distance from the center of the province implies that probability of ACT adoption decreases as the distance from the province center increases. This result supports the theories of the spatial diffusion of innovations from central areas to the more remote ones. The firms in central areas benefit from conditions favouring the adoption of innovations, e.g., from the availability of skilled labour and an access to specialized information. The firm's location, on the other hand, was shown to be statistically insignificant variable: data didn't support any differences in ACT adoption behaviour by firms in different provinces. This may be to due the relatively homogeneous seedbed conditions the Finnish provinces provide to firms in terms of the quality of

---

[13]See Capello 1994 for a parellell result.

communication networks and the supply of highly skilled labour.[14]

Table 17.2. The estimates of the logit model of the firm's adoption of ACT

Explanatory variable	Model 1 lhs = use	Model 2 lhs = use	Model 3 lhs = use
constant	-0.88200 (-0.954)	-0.70165 (0.442)	**-2.2358 (-1.985)**
llab	**0.7116 (3.889)**	**0.75769 (2.625)**	**0.75710 (3.368)**
ex	**-2.0543 (-4.630)**	**-1.7156 (-2.719)**	**-2.2631 (-4.151)**
risk	**-1.2856 (-2.805)**	-0.96323 (-1.398)	-0.69648 (-1.255)
lkm	**-0.31180 (-2.338)**	**-0.41310 (-1.860)**	**-0.46183 (-2.722)**
loc	0.0032896 (0.047)	-0.19060 (-1.504)	-0.084625 (1.026)
ind	**1.1569 (3.394)**	**1.1578 (2.496)**	**1.2751 (3.287)**
pcm		**-0.55457 (-2.419)**	
f2out			**3.0118 (2.473)**
nobs	173	97	131
lnL	-70.69	-32.65	-49.19
$\chi 2$	96.35 (6 d.f.)	59.42 (7 d.f.)	81.51 (7 d.f.)
CPI	0.81	0.80	0.85

Note: t-values are in brackets.
CPI = correct prediction index = the ratio of correct predictions to the number of observations

We explored in model (3) whether a firm's expectations concerning the use of ACT by (potential) business partners had any role in its adoption of a new communication technology. Data supported the influence of such expectations: the estimate of the expected rate of external e-mail users within the next two years turned out to be positive and statistically significant.[15] This means that a more optimistic view of ACT diffusion among the firm's business partners or an increase in the expected network size increases the probability of the firm's ACT adoption. This result further supports the significance of network externalities in the diffusion of communication technologies and emphasizes the influential role of the expectations on the number of users in the firm's ACT adoption decision.

---

[14]In 1993, there were only minor differences in the education level of the population in the Finnish provinces. The only exception was the province of Uusimaa where the education level of the population was higher than the average in Finland. (Statistics Finland, Education, 1995:5).

[15]Further estimations showed that the estimate of the expected rate of external e-mail users within the next five years and the estimate of the expected rate of EDI users within the next two years were also positive and statistically significant.

17.4.2 Diffusion of Internal and External Communications Systems

We divided the sample into three types according to the date of adoption of internal and external e-mail: the early adopters (internal/external e-mail adopted before 1990), the late adopters (internal/external e-mail adopted in 1990 or after) and the non-adopters. The variable 'edi' separates the firms into non-users of EDI and into users of EDI.[16] We used the variables "in1", 'out1' and 'edi' (see Table 17.1) as dependent variables in the models I, II and III, respectively. Table 17.3 presents the estimation results as regards to the determinants of the diffusion of the different communication systems.

The adopters of advanced communications systems - both late adopters and early adopters of internal/external e-mail - tend to be large firms. It also seems that, in all but the last one, firms in different branches of industry differ with respect to the use of communication systems. The significance of the branch of industry also came up in the previous models (see Table 17.2). So we created a separate dummy variable for each branch, i.e. for metal, mechanical and engineering - to study the relationship between the branch of industry and the firm's adoption behaviour. The adoption of internal/external e-mail was positively related to firms in electrical engineering and negatively related to those in metal engineering. These intra-industry differences probably arise from the higher R&D intensity - or from the higher innovation intensiveness - of the firms in electrical engineering which represents a high-tech sector, whereas the metal engineering belongs to the medium-low technology group (see Virtaharju and Åkerblom 1993).

The adoption of internal e-mail seems to decrease with the firm's risk aversion. The estimate of a firm's riskiness is, however, statistically significant only in the case of the late adopters of internal e-mail. The early adoption of e-mail - both internal and external - is negatively related to the firm's distance from the provincial center. This result supports the view that the diffusion of innovations in general start from the central areas and spread gradually to the remote areas. The estimate of variable 'loc' is also statistically significant in model I; the late adopters of internal e-mail seem to differ from the non-adopters of internal e-mail with respect to the province where they are located.

The expectations on the number of users seem to be important in the firm's adoption of external e-mail. An increase in the expected number of potential partners adopting external e-mail increases the probability of late adoption (compared to non-adoption). The estimate of the variable describing the firm's expectations on the network size, however, is statistically insignificant in the case of the early adopters of external e-mail. This result seems reasonable since the early adopters adopted external e-mail in the 1980's and thus their expectations on the network size within the next two years do not necessarily reflect their future expectations of the network size at the time of adoption.

Somewhat unexpectedly, the users of EDI do not differ significantly from the non-users of EDI with respect to their expectations on the share of their (potential) business partners using EDI. At least two factors may explain this result. First,

---

[16]We did not divide the sample into the early and late adopters of EDI due to the relatively low number of EDI users.

the unexpectedly slow diffusion rate of EDI may mean that users now have different expectations from those they had at adoption time. Second, a firm may adopt EDI for certain specific business relations, for example, for subcontracting and benefit enough from the system even if other business relations are still handled by some other means of communication. Thus, more important than the number of business partners using EDI is *who* the EDI users are.

Table 17.3. The estimates of the (multinomial) logit models of the adoption of internal e-mail, external e-mail and EDI

Explanatory variable	model I lhs=in1		model II lhs=out1		model III lhs=edi
	late adoption	early adoption	late adoption	early adoption	
constant	**-3.7823** **(-3.644)**	**-8.1662** **(-4.980)**	**-4.8324** **(-3.837)**	**-4.9914** **(-2.680)**	**-6.8135** **(-3.797)**
llab	**0.86823** **(4.186)**	**1.6017** **(5.427)**	**0.74418** **(3.406)**	**0.81606** **(2.632)**	**1.0437** **(3.706)**
risk	**-0.94704** **(-1.943)**	-1.2565 (-1.434)	-0.11559 (-0.205)	-1.2072 (-1.038)	0.24679 (0.320)
lkm	-0.19372 (-1.394)	**-0.44286** **(-2.274)**	-0.28153 (-1.761)	**-0.46820** **(-1.970)**	-0.13445 (-0.659)
loc	**-0.15395** **(-2.011)**	-0.070087 (-0.705)	-0.014451 (-0.183)	-0.22748 (-1.504)	-0.012128 (0.122)
ind	**1.3378** **(4.468)**	**1.1397** **(2.784)**	**1.2212** **(3.757)**	**1.0574** **(2.256)**	0.28102 (0.700)
ex			-0.61321 (-1.163)	-0.78998 (-0.987)	-1.2473 (-1.733)
f2out			**3.1697** **(2.959)**	1.2380 (0.722)	
f2edi					
				2.4900 (1.212)	
nobs	174		130		106
lnL	-107.81		-77.59		-34.04
χ2	105.91 (10 d.f.)		69.44 (14 d.f.)		31.61 (7 d.f.)
CPI	0.77		0.75		0.88

Note: t-values are in brackets.
CPI = correct prediction index = the ratio of correct predictions to the number of observations

The EDI adoption rate is much lower than that of either internal or external e-mail, but only the firm size is able to explain - in a statistically satisfactory manner - the differences between the users of EDI and the non-users of EDI. The network size seems to have some relevance in the firm's adoption of EDI, even though the estimate of variable 'ef' does not quite reach the level of clear statistical significance. This gives further support to the view that EDI use by certain of partners counts more than the total number of business partners using EDI. The early/late adopters of external e-mail do not differ from the non-users of external e-mail with respect to the importance of the network size in the firm's ACT adoption decision.

17.4.3 Investment Strategies Related to the Adoption of the Advanced Communication Systems

The previous sections indicated that apart from the differences between users and the non-users of ACT, there also exist some deviations among the adopters of different communication systems. We suppose that these disparities are closely related to the ACT investment strategy the management of a firm is willing to commit itself to. In this section, we will explore whether there exists a relationship between the different investment strategies and the form of ACT a firm adopts. Table 17.4 introduces the estimation results which show that the predictive power of the models I-IV do not differ greatly from those presented above. This means that we can separate the users of ACT from the non-users by the types of investment strategy as well as by the conventional background characteristics (e.g., firm size, location etc.).

We consider first the variable describing management interest in the aggressive ACT investment strategy 'tough1'. The estimate of this factor is statistically significant and negative in all models except the third one. It seems that the firms which gave importance to cheaper raw materials, faster delivery of them and extending the market via the use of ACT are less likely to be users of internal and external e-mail than firms which do not value these features as much. Internal and external e-mail are probably not expected to serve the aims of the ACT investment strategy related to variable 'tough1'. A possible reason for this is that both internal and external e-mail are generally used more for informal communication than for market transactions (e.g., for ordering or selling the products).

The next result concerning the adoption of EDI is even more interesting: the importance of factor 'tough2' increases the probability of EDI adoption. The estimate of the variable 'tough2' is statistically insignificant in all other models. The last model implies that the probability of adoption of EDI increases with the importance of the strategic aspects included in 'tough2'. The aggressive ACT investment strategy aiming to achieve competitive advantage is positively related to the use of EDI. This implies that the implementation and use of EDI for market transactions, such as subcontracting, before rivals, is expected to improve the firm's position in the market.

Table 17.4. Investment strategies related to the adoption of ACT

Explana-tory variable	model I lhs = in2		model II lhs = out2		model III lhs = edi		model IV lhs = use	
constant	-0.26573	(-1.577)	**-0.48479**	**(-2.990)**	**-2.1043**	**(-7.890)**	**0.45201**	**(2.236)**
tough1	**-1.0346**	**(-4.180)**	**-0.70170**	**(-3.160)**	-0.22408	(-0.810)	**-1.5782**	**(-5.218)**
tough2	-0.070775	(-0.337)	-0.032523	(-0.162)	**0.54695**	**(1.926)**	-0.21419	(-0.873)
coop	**0.43729**	**(2.113)**	0.18807	(0.990)	0.087962	(0.359)	0.22103	(0.964)
unc	-0.25305	(-1.394)	-0.18789	(-1.098)	**-0.89218**	**(-3.585)**	**-0.69240**	**(-3.130)**
ef	**0.97088**	**(4.735)**	**0.71323**	**(3.854)**	**0.54693**	**(2.201)**	**1.4832**	**(5.644)**
nobs	192		192		192		193	
lnL	-104.86		-112.80		-68.20		-83.21	
χ^2	53.44 (5 d.f.)		31.30 (5 d.f.)		23.11 (5 d.f.)		99.97 (5 d.f.)	
CPI	0.72		0.68		0.84		0.80	

Note: t-values are in brackets.
CPI = correct prediction index = the ratio of correct predictions to the number of observations

The only dependent variable which is clearly related to the factor 'coop' is the use of internal e-mail. The probability of adoption of internal e-mail increases with the importance of the characteristics linked with the dependent ACT investment strategy. The pressure to adopt a new internal communication system may come from the other business units and/or arise from the firm's own interests, i.e. for co-operation with the other business units. Also, rivals' decisions related to ACT may also affect the firms' adoption of internal e-mail.

The probability of adopting EDI - and ACT in general - decreases as the importance of the reduced uncertainty regarding ACT increases. This result supports the view that the adoption of ACT, and particularly the diffusion of EDI, is hindered by the sunk costs of ACT investment and by uncertainty related to the new communication technology. The last factor, 'ef', is a statistically significant explanatory variable in all models. Clearly, the cost reduction object acts as a stimulus facilitating the firms' adoption of ACT.

The estimation results presented above indicated that some divergent and also some similar ACT investment strategies are related to the adoption of different ACT systems. These findings, with the estimation results presented in Sections 17.4.1 and 17.4.2 gave us some new information regarding the firm's ACT adoption behaviour. In the next section, we will summarize our empirical findings and briefly discuss their implications.

17.5   Discussion

In this chapter we have investigated the potential determinants of the firms' adoption of advanced communication systems. We used data from the Finnish metal industry for exploring the firms' adoption of ACT in general and also for examining the diffusion of certain specific forms of ACT - i.e. internal e-mail, external e-mail and EDI - serving the divergent business purposes or communication needs. The estimation results stress the importance of two features affecting both the adoption of ACT in general as well as the diffusion of the divergent forms of ACT. First, the probability of firm's adoption of ACT increases with the firm size. Second, the firms in the different branch of industries differ with respect to the use of ACT. The innovative high-technology industries seem to offer a favourable ground for the diffusion of ACT.

The diffusion of ACT in the Finnish metal industry seems to follow, at least to a some extent, the diffusion patterns conventionally related to the spatial diffusion of innovative technologies. The probability of the early adoption of internal/external e-mail was negatively related to the firm's distance from the provincial center. The intra-industry diffusion of these communication technologies started with the firms located in central areas and spread gradually to the more remote areas. The location of a firm was not, however, able to explain the differences between the users and the non-users of EDI.

The data supported our view that the direct user interdependence influences the firm's adoption of a new communication technology; the ACT market clearly exhibit network externalities. Two essential results emerged arising from the user interdependence of ACT: (i) a low number of business partners using ACT seemed to obstruct the diffusion of ACT and (ii) the expectations on network size are of critical importance in the firm's ACT adoption decision. We found that, in general, the more optimistic the view of the diffusion of ACT among the firm's business partners the greater the probability of adoption. However, data did not support any statistical differences between the users of EDI and the non-users of EDI with respect to their expectations on the share of their potential business partners using EDI.

We used a rough measure of the firm's risk propensity to investigate the relationship between the firm's adoption of ACT and its attitude towards risk. Our data indicates that the probability of the firm's adoption of ACT decreases with the firm's risk aversion. The variable describing the firm's attitude towards risk seems to be a factor which distinguishes - to some degree - the users of ACT from the non-users of ACT, but does not succeed as well in explaining the differences regarding the adoption of the different communication systems. We may deduce that the firm's reluctance to undertake risky investments to some extent also hinders the firm's adoption of ACT.

The estimation results hint that the firm's ACT investment decision involves strategic behaviour. The data indicates that some divergent and also some similar ACT investment strategies are related to the adoption of different ACT systems. The most obvious objectives facilitating the firm's adoption of ACT seem to be the gains in efficiency and the cost reductions obtainable with the utilization of

ACT. This result applies to all the communication systems referred to above. An aggressive ACT investment strategy was negatively related to the adoption of e-mail. Internal and external e-mail probably do not provide sufficient advantages to the firms which seek a competitive advantage or an increase in market power by the use of ACT. The adoption of internal e-mail was also related to a dependent ACT investment strategy: the adoption of internal e-mail seems to arise from the co-operation and communication between autonomous business units and also from the pressure or demand of the other (dominant) business units.

The most interesting findings regarding the firms' strategies concerning the adoption of EDI. Besides the firm size, only the ACT investment strategies managed to explain the differences between the users and the non-users of EDI. Two ACT strategies facilitate the adoption of EDI: (i) the strategy related to cost reductions or efficiency gains by the use of ACT and (ii) the strategy related to the achievement of competitive advantage or market power by the use of ACT. The third ACT investment strategy, concerning the importance of the reduction of uncertainty related to the new technology, was negatively related to the adoption of EDI. This result indicates that the diffusion of EDI is hindered by the sunk costs of the investment and by the uncertainty related to the new communication technology. It seems that - at least in the Finnish metal industry - the larger firms exploit the benefits of EDI whereas the smaller firms regard EDI as too risky an investment.

This chapter has provided some new information as regards firms' adoption of new communication technologies and their investment strategies related to different communication systems. The results suggest that the firm's strategic behaviour related to ACT adoption decision deserves further examination.

Acknowledgements

I am grateful to Peter Nijkamp for his useful comments and encouragement. I would also like to thank Rauli Svento and the Department of Economics, University of Oulu, for their support. Finally, I wish to acknowledge the financial support of the Finnish Postgraduate Programme in Economics.

References

Agresti A (1990) Categorical Data Analysis. John Wiley & Sons

Alderman N and S Davies (1990) Modelling regional patterns of innovation diffusion in the UK metalworking industries. Regional Studies 24:513-528

Antonelli C (1990) Induced adoption and externalities in the regional diffusion of information technology. Regional Studies 24 31-40

Bertuglia C S, Fischer M M and G Preto (Eds) (1995) Technological Change, Economic Development and Space. Springer-Verlag

Capello R (1994) Spatial Economic Analysis of Telecommunications Network Externalities. Avebury Aldershot

Davelaar J E and P Nijkamp (1988) The incubator hypothesis: Re-vitalization of metropolitan areas? The Annals of Regional Science XXII:48-65

Dosi G (1990) The research on innovation diffusion: An assessment. In: Nakicenovic N and A Grübler (Eds) Diffusion of Technologies and Social Behavior. Springer-Verlag

Van der Ende M A and P Nijkamp (1995) Industrial dynamics and rational expectations in a spatial setting. In: Bertuglia C S, Fischer M M and G Preto (Eds) Technological Change, Economic Development and Space. Springer-Verlag

Fischer M M (1995) Technological change and innovation behaviour. In: Bertuglia C S, Fischer M M and G Preto (Eds) Technological Change, Economic Development and Space. Springer-Verlag

Freeman C (1982) The Economics of Industrial Innovation. MIT Press

Greene W H (1993) Econometric Analysis. Macmillan Publishing Company

Ihamuotila M (1994) Corporate ownership, capital structure and investment. Helsinki School of Economics and Business Administration A:97

Kamann D-J F and P Nijkamp (1990) Technogenesis: Origins and diffusion in a turbulent environment. In: Nakicenovic N and A Grübler (Eds) Diffusion of Technologies and Social Behavior. Springer-Verlag

Karlsson C (1995) Innovation adoption, innovation networks and agglomeration economies. In: Bertuglia C S, Fischer M M and G Preto (Eds) Technological Change, Economic Development and Space. Springer-Verlag

Katz M and Shapiro C (1985) Network externalities, competition and compatibility. American Economic Review 75:424-440

Kay N (1988) The R&D function: corporate strategy and structure. In: Dosi G, Freeman C, Nelson R, Silverberg G and L Soete (Eds) Technical Change and Economic Theory:282-294. Pinter Publishers

Koutsoyiannis A (1977) Theory of Econometrics. The Macmillan Press

Macdonald S (1992) Information networks and the exchange of information. In Antonelli C (Ed) The Economics of Information Network. Elsevier Science Publishers:51-69

Mansfield E (1968) Industrial Research and Technological Innovation. An Econometric Analysis. W W Norton & Company

Markus L (1990) Critical mass contingencies for telecommunications consumers. In: Antonelli C (Ed) The Economics of Information Network. Elsevier Science Publishers:431-450

Nijkamp P and A Reggiani (1995) Space-time synergetics in innovation diffusion: A nested network simulation approach. Research Paper. Free University Amsterdam

Rosenberg N (1976) On technological expectations. Economic Journal 86:523-535

Stoneman P (1983) The Economic Analysis of Technological Change. Oxford University Press

Virtaharju M and M Åkerblom (1993) Technology intensity of Finnish manufacturing industries. Statistics Finland. Science and Technology 1993:3

Weiss A M (1994) The effects of expectations on technology adoption: Some empirical evidence. The Journal of Industrial Economics XLII:341-360

18 Transport Behaviour and Diffusion of Telematics: A Conceptual Framework and Empirical Applications

Peter Nijkamp, Gerard Pepping, George Argyrakos, David Banister and Maria Giaoutzi

$| \text{Europe} \rangle$

$R41$

18.1   Introduction

Anyone who regularly uses Europe's roads will recognize that traffic congestion is often the normal state of affairs rather than an occasional inconvenience. The most obvious effects of congestion are: increased journey times for both private and commercial motorists; escalating traffic accident levels and environmental damages resulting from higher pollution levels. In order to deal with these problems, (super)national and local authorities have implemented various traffic management schemes. The development of new technologies in the transport sector may offer additional solutions. In this context the interest in the blend of telecommunication and informatics, referred to as telematics, is noteworthy. Telematics technologies, which are widely available, are gaining increasing importance world-wide. Witness the stimuli provided by major research programmes like DRIVE (Europe), IVHS (United States) and VICS (Japan). The range of transport telematics options comprises inter alia:

- route planning (e.g. in the case of individual travellers by in-home/office travel and traffic information, and in the case of logistic operations by fleet management systems)
- substitution of physical transport (e.g. teleworking, teleshopping, tele-education)
- automatic debiting (e.g. for toll roads or road charging)
- traffic guidance (e.g. by motorway control and signalling systems and car navigation systems).

The *potential impacts* of these transport telematics systems can be found in four main areas. First, they will have a significant influence on *network efficiency*. The efficiency can be improved by diverting traffic from bottlenecks in the network. To illustrate this, a framework plan in the Netherlands for the implementation of transport telematics shows some optimism on infrastructure management: an increase of traffic throughput of 10% in 1995 and 15-25% in 2010 seems to be a realistic target (Rijkswaterstaat 1992). Second, telematics may have impacts on the environment by a *substitution of physical transport* (e.g. Quaid et al. 1992 and

Vanderschuren et al. 1993) and/or an *improvement of flow efficiency* of traffic (e.g. by motorway signalling), leading to less pollution. Third, telematics may improve *safety*. Accidents are one of the most severe implications of large-scale transport systems. Information on weather conditions or on traffic jams are obvious examples of tools reducing social costs of transport. In the longer run, board computers, speed/distance keepers and vehicle guidance systems may lead to significant reductions in fatalities (Malaterre et al. 1993). Finally, the use of telematics will have a favourable impact on *energy use*. The energy use of traffic is formidable. The use of telematics may lead to more energy-efficient transport systems, saving both the environment and the earth's natural resources. Besides a better use of cars (e.g. driving style) made possible by advanced computer-controlled vehicle technology, also the choice for more energy-saving transport modes may be favoured by telematics. Avoidance of traffic jams is another beneficial factor, as the energy consumption of a slow-moving car is relatively high.

Considering the above potential impacts, the promise of transport telematics is considerable. However, as is the case with any new technological innovation, at the basis of success of transport telematics lies the acceptance by potential users. User acceptance (and diffusion potential) manifests itself through attitude, usage, change in travel behaviour and willingness to invest (intermediate users) or buy (private users) with respect to certain types of telematics. It should be stated a priori that the interrelationships between these indicators are often complex and rather unpredictable. For instance, the diffusion of individual route guidance equipment could be dependent not only on the actual need for the information provided to make travel decisions, but could also turn into a status product. Fashion might in the end become the main market driving force.

Given the need to explore the interaction between human behaviour and transport telematics, the aim of this paper is to investigate behavioural factors and their interrelationships in the use and diffusion of new telematics technologies. We will focus our attention on those systems which provide information to various categories of users, viz. public transport users and private car users. The next section will start with a general description of the dynamic interaction between travel behaviour and the diffusion of transport telematics technologies. Subsequently, in Section 18.3, a conceptual framework will be presented that covers all variables of interest for the investigation of user responses to these technologies, followed in Section 18.4 by a discussion of a theoretical concept of information use by travellers. The concepts used will be illustrated by two case studies recently carried out in Europe. One case study concerns a real time passenger information system for buses in Southampton in the UK, the second one focuses on a motorway driver information system which has been implemented in the Netherlands. Finally, some important lessons drawn from the previous experiences are presented in the last section.

18.2   Some Dynamics of Travel Behaviour and Diffusion of Transport Telematics

The character of the diffusion process of transport telematics can be outlined in the conceptual model presented in Figure 18.1. First of all, the introduction of transport telematics will have an impact on individual travel behaviour in a *direct* sense and an *indirect* sense. Direct impacts are envisaged by systems aimed to influence travel behaviour (e.g. route guidance), while, indirectly, impacts on travel choices can also be expected from systems improving travel conditions (e.g. driver assistance facilities and public transport information). If a direct impact of new technologies on travel choices occurs, then together with the influence of user segmentation factors on travel choices, there will be an impact on the awareness level of the users. Moreover, if a significant indirect impact of new technologies also occurs, this will affect via improved travel conditions the awareness level of the users (Argyrakos et al. 1994).

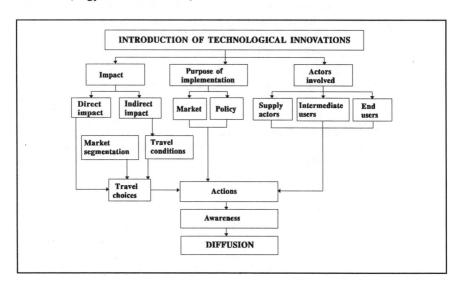

Fig. 18.1. The dynamics between travel behaviour and diffusion of technological innovations in transport.

Table 18.1. Technological innovations and the impact on travel choice.

Group of telematics	Impact on travel choice
Demand management Travel & traffic information Urban traffic management Inter-urban traffic management Public transport management	Direct
Driver assistance Freight and fleet management	Indirect

A functional classification of the wide range of transport telematics applications currently being developed is shown in Table 18.1, together with the type of impact expected. Furthermore, it is in the interest of the system's manager that the purpose of the implementation will achieve maximum public awareness, especially where it concerns system-wide public applications (e.g. environmental area licensing in cities). The purpose will depend on the whether the applications are market driven or policy driven, as outlined in Table 18.2.

Finally, there are various actors involved with the introduction of new technologies. The public awareness levels are limited by the type of actors involved with specific groups of applications (Table 18.3). There are supply actors or developers (producers of hardware and software, the automobile industry and also road managers and public transport operators), intermediate or collective users (national and local authorities, road managers, public transport operators) and end- or individual users (car drivers, public transport users and freight forwarders).

Table 18.2. Technological innovations and type of driving force.

Group of telematics	Type of main driving force
Demand management Public transport management	Policy driven
Urban traffic management Inter-urban traffic management	Policy and market driven
Travel & traffic information Driver assistance Freight and fleet management	Market driven

Growing awareness levels will in turn favour the diffusion rate of transport telematics, as far as individual systems for the end-users and public systems for intermediate users are concerned.

Table 18.3. Technological innovations and actors involved.

Group of telematics	Actors involved
Demand management Public transport management	Intermediate users End users
Travel & traffic information Driver assistance Freight and fleet management	Supply actors End users
Urban traffic management Inter-urban traffic management	Supply actors Intermediate users End users

18.3   A Nested Approach

In the complex and dynamic process which will ultimately lead to the diffusion of transport telematics as outlined above, a critical success factor will be the acceptance by the end-users. Relatively little conceptual analysis has been undertaken on assessing the social impact, including changes in travel behaviour, which result from the introduction of transport telematics. Such investigation requires coverage of a wide range of relevant (dynamic) behavioural issues. A comprehensive (nested) approach to deal with all relevant aspects should include the impacts of a full range of areas of operational interest in telematics (BATT 1992). Some of the expected benefits of applying such an approach are that it makes it possible to:

- provide behavioural parameters to those involved in the development of telematics systems;
- inform authorities of the best ways to implement the new technologies and to increase the success of ATT technology;
- inform the industry of better ways to promote and market ATT technologies to meet user requirements.

The nested approach integrates the elements described above, allowing the impacts of telematics to be measured at three separate *levels of reference*: first, at a strategic level, second, at the level of market potential and third, at the level of market responses.

At the *strategic level*, we are concerned with the system wide impacts, given certain types and certain levels of introduction of ATT. Within this framework, various changes can be assessed in terms of user and producer benefits, the direct and indirect environmental impacts, reductions in accidents, energy savings and more efficient use of the infrastructure. The assessment may cover the total performance of the system, the distribution and equity implications as well as the technological achievements. At the *market potential level*, we are interested in the means by which the potential for ATT can be maximized in terms of acceptability and penetration of the various parts of the market. One of the aims of this marketing is to access market awareness of the product, while the other is to identify which segments of the market are likely to represent the greatest potential for telematics. It is realized that some people will be more positive about the use of telematics than others and that not all people will use it in the same way. The identification of different markets is a very important part of applied research and will provide a link between research and the telematics industries. At the *market response level*, we are concerned with the costs of the technology, changes in individual behaviour and the scale of implementation. The focus is on the cost effectiveness and the direct benefits to the individual users of ATT, the range, scale and timing of introduction of ATT and the rate of behavioural change which might follow. Much hypothetical research has already been carried out on the impact of telematics, often in terms of the most optimistic scenario, assuming saturation of the technology is achieved over a very short period of time.

Table 18.4 brings together these three levels into a composite table. The three levels of reference are shown against three categories of evaluation criteria: namely technical, socio-economic and political/dynamic. The cells of this table contain the main areas of investigation.

Table 18.4. Head elements of a nested approach.

Evaluation criteria	Levels of reference		
	Strategic	**Market Potential**	**Market Response**
Technical	Performance	Marketing	Cost of technology
Socio-economic	Distribution & Equity	Segmentation	Behaviour
Political & Dynamic	Technological Perspective	Awareness	Diffusion

It should be noted that not all telematics applications are likely to feature in all elements. For instance, traffic information and public transport information will be evident in all cells, but other telematics applications, e.g. demand management, will only relate to 'market response', since it is system wide and affects all users.

Given the emphasis on the end-user side, some elements of this table need further refinement: segmentation, behaviour, awareness and diffusion. Here the focus is on the behavioural response in combination with segmentation factors that are necessary
to establish the market potential. These four issues are elaborated below.

Behaviour
The argument advanced here is that user behaviour will change as a result of the introduction of ATT, but that changes may vary according to the individual, the situation and the type of ATT being tested. We have identified a range of behavioural responses which might be anticipated for a particular journey at one point in time. These include: mode shift, departure time, change in route and destination, trip generation/suppression, trip scheduling, parking choice and adherence to advice (see Table 18.5). They are listed below:

Mode shift: The impact of ATT may be to cause users to shift mode in order to gain time or to save costs or meet their constraints.
Departure time: A shift may occur in departure time, given the ATT information on the current level of congestion or the generalized cost of the prospective trip if individual utility is to be maximized or to meet specified preference constraints.
Route: Provided that the technology is available, route choice may be modified. Route diversion or adherence to advice supplied by ATT may be influenced, not only by items cited in the segmentation (such as familiarity), but also by reliability of the information provided.
Destination: Decisions may be made to select alternative destinations if the route previously selected is congested or if the ATT system can give information on alternative opportunities. Destination choice is clearly relevant to some types of discretionary trips.

Trip generation/suppression: Technology may influence the decision whether to make a trip or not, as advice on congestion may result in trip deferral or cancellation within the decision period considered. The purpose and need for the journey will be decisive in establishing possible changes with respect the trip.

Trip scheduling: This arrangement within a determined user dependent time period may be considered if satisfactory or non-acceptable alternatives are suggested by ATT (e.g. route, trip timing, parking).

Parking choice: parking decisions may be influenced by access to ATT information regarding the location of car parks, the availability of space, and the route followed.

Adherence to advice: Adherence to the information provided may be influenced by many factors including the quality of the information, reliance on such information, familiarity with the network, previous experience and user characteristics.

Segmentation

Here we are concerned with the main socio-economic characteristics which might influence both the decision to acquire a particular form of ATT and the actual use of that ATT at one point in time. The argument is that not all people require access to the same technology and that even if they have that technology, use patterns will vary. Meaningful segmentation factors would include:

Car availability: the availability of a car identifies not only social groups, but also provides information such as existence of alternatives or dependency on public transport. It may be important to select segments of both car owners and non car owners. This segmentation has proved in many previous studies to be useful.

Age/sex: It can be expected that the penetration of ATT will be differentiated by age and sex of potential users. Younger people may be more likely to respond to innovation than older people and men may be more responsive than women.

Social group: Socio-economic group, type of employment and some measure of class may all affect patterns of use of ATT, both in terms of actual takeup and in terms of marketing.

Income levels: Income is generally closely related to the social group and is likely to be the main factor in the decision to acquire the ATT technology or to obtain access to it.

Experience: positive or negative experience may modify the usage of ATT. Past experience has been found to be traded off against ATT information supplied on the current situation. A high level of reliability of current information must be maintained and improved. This factor is related closely to user familiarity and awareness of the alternatives available.

Familiarity: The issue of familiarity has been identified in previous research as important in determining whether pre-trip information is required in the home (unfamiliar trips) or during the trip (familiar trips).

Purpose: Trip purpose may also help to identify which types of activities have the greatest potential for ATT. Discretionary trips (e.g. social, leisure and shopping) may present greater opportunities than regular trips (e.g. work and education) where there is a much greater degree of familiarity.

Table 18.5. Important market potential and market response parameters.

	Market Potential	Market Response
Socio-economic	**SEGMENTATION** 1. car availability 2. age 3. social group 4. income group 5. experience 6. familiarity 7. purpose	**BEHAVIOUR** 1. mode shift 2. departure time 3. route 4. destination 5. trip generation /suppression 6. trip scheduling 7. parking choice 8. adherence to advice
Political & Dynamic	**AWARENESS** 1. exposure to ATT 2. acceptability 3. publicity	**DIFFUSION** 1. pre-conditions 2. take-off 3. saturation levels 4. adaptation

Awareness
It takes time for people to become aware of innovations. Awareness depends on exposure or experience, but can also be influenced by publicity. Equally important is the public acceptability of innovation and the perceived necessity and benefits. We look at these factors in turn.

Exposure to ATT: Previous knowledge and exposure to the technology may be decisive in the usage of any future application. This exposure relates to knowledge, experience and acceptability as well as to user characteristics.
Acceptability: Apart form awareness of technology, there is a considerable problem concerning the public acceptability of technology (e.g. the debate on road pricing and privacy). Innovation takes time to become accepted and the market response may be seriously affected if social attitudes are not favourable.
Publicity: Awareness and acceptability can be raised by publicity and marketing which will both promote ATT technology and help to allay any concerns that people may have.

Diffusion
Innovation diffusion also takes time as the market does not respond instantaneously. Even when all conditions are favourable, responses have to be monitored and evaluated over a significant period of time, as standardization becomes possible and substantial economies of scale prevail. Critical diffusion parameters are:

Pre-conditions: These are the necessary political and technical conditions which have to be in place prior to any large scale application of ATT and relate to a willingness to address environmental and traffic problems.
Take off: as diffusion takes place, initial interest begins to snowball and market penetration expands at a faster rate after reaching a critical acceptance threshold which depends largely on the conditions.

Saturation levels: With maturity, a saturation level is reached, but suppliers then identify new markets to ensure the total market for all ATT continues to expand. *Adaptation*: The closely linked dynamic process outlined in the move from pre-conditions through take off to saturation is not a unidirectional process. There are also important feedback effects, as individuals and companies modify their behaviour patterns and change habits.

18.4   Travel   Information   and   Travel   Behaviour: Some Conceptual Issues

The framework of relevant variables discussed above does not yet provide insight into the *individual decision chain* which determines the individual need and use of more advanced types of travel and traffic information. Modelling such individual travel behaviour is a difficult task and requires some critical assumptions on the basis of human behaviour. Such assumptions conform to the classic economic model of utility maximization where the decision to make a trip is followed by a trip planning stage which results in a ranking of the key trip characteristics according to individual utility before the choice is made (Banister et al. 1994). This choice could include all of the trip planning variables or a subset thereof. The assumptions made here include:

- rationality in choice: if the same choice set is presented again the same decision will be made;
- complete knowledge: the individual decision-maker has access to information concerning the accepted alternative together with knowledge of any rejected alternative.

Individuals act so as to maximize some benefit or utility, and each individual exercises his or her choice over the full range of available options, limited only by constraints of time and money. In transport, this choice is normally represented as discrete alternatives between mutually exclusive modes, but it can be used to analyze other parts of the travel decision. The addition of ATT information systems (e.g. on public transport, VMS or route guidance) reinforces the knowledge assumption in the utility maximization model, as decisions would be based on the best available information prior to the trip being made and during the actual trip.

An alternative model would argue that decision makers are not utility maximizers but satisficers (Banister et al. 1994). The individual makes choices in a situation of partial knowledge. When certain thresholds are reached (e.g. significant foreseeable delays), action will take place. Such a procedure explicitly involves feedback with the result that each new trip has been modified by previous experience, which could be positive or negative.

Such a model gives rise to a more complex decision process and would suggest a more selective use of any pre-trip or in-trip ATT information, as all information is modified by previous experience. In turn, it would suggest that ATT information has to be selective and targeted to individual users, as information of a more general nature may not be relevant.

Information given to ATT users also has to be accurate, as any failure in the system would result in the strengthening of the individual's own experience as opposed to the experience from the information system. Any such reduction in the quality of ATT information would reduce reliance on it, the market for it, and the price that would be paid for it. Selective use of ATT information is a key research area about which little is known.

Information can either be provided at no direct cost to the user (Type I), in which case utility will be increased, or at a cost to the user (Type II) in which case the increased quality and value of that information will have to be balanced against its cost. Similarly, on the supply side, information can either be provided to users in general (e.g. VMS and public transport information systems) or to the user on an individual basis (e.g. route guidance). Examples of pre-trip and in-trip information product types are shown in Table 18.6.

Because of the different user and supplier constraints, there will be different levels of adherence to advice. Using the utility maximizing framework, when there is no direct cost to the user, utility should increase and a greater use of both pre-trip and in-trip information results. If there is a direct cost, then either no use of ATT will be made (no change in utility) or there will be some use of pre-trip and in-trip information, with full adherence or partial adherence to advice amnd hence some increase in utility. It is here that the changes in utility aznd the conditions under which advice is accepted or rejected may be difficult to assess.

Table 18.6. Examples of pre-trip and in-trip information supply.

Pre-trip		
	Type I - No cost	Type II - Cost
General	CEEFAX ORACLE	In home terminals PROMISE MINITEL
Particular	Timetables	Route planning Telephone inquiry

In-trip		
	Type I - No cost	Type II - Cost
General	Variable message sign Passenger transport information system	AUTOGUIDE TRAFFIC MASTER
Particular	Radio data system	Route guidance

Using the satisficing behaviour framework, some further differences may be observed. If the trip is a new one, then behaviour is similar to utility maximizing, but if the trip is one which has been made before, then satisfaction and feedback become important. Selective use will be made of ATT if there was dissatisfaction with the previous trip, but if that trip was perceived as satisfactory, then there is no need to have any new information. Again, adherence to advice can be either complete or partial, dependent on the type of information required, previous experience and levels of satisfaction. This is where concepts of experience and familiarity become important - behaviour may become routinized.

In both models, the determinant factor in the success of different forms of ATT is the quality of the information and when and how it is presented. In-trip information has to be more general when it is limited by the size of a display available. To expect major changes in behaviour resulting from such information may be optimistic, when the level of advice cannot be detailed. Alternative routes are not given on VMS or public transport information systems. Unless the traveller has good knowledge of the alternatives (the assumption in utility maximization), the perception might be that there is no choice and so the action would be to remain in the queue at the bus stop or on the road. The benefits of such systems are that they are provided free at the point of use and they are likely to raise levels of satisfaction with the service being provided: a reassurance utility.

There is a much greater potential with respect to pre-trip information where options and choices can be made on a much wider variety of behavioural variables. It is here that decisions on trip timing (including trip scheduling or trip suppression), alternative modes, different destinations and the best route can all be made. The range of information and the personalization of the relevant parts of that information can be made available to the user, so that the assumptions of rationality and knowledge can be met. However, this requires people to spend time prior to the trip extracting the relevant information. This again involves a cost in the time spent, and may only be appropriate when a new or exceptional trip is being made or when problems are expected. Similarly, many people may argue that their degrees of freedom in terms of mode, destination, start time and even route are limited. Route guidance systems attempt to address some of these problems, but again the information given relates to route choice, not to the other factors which make up the trip. For example, it does not suggest to the driver to park the car and take the train to the destination, or to give a range of alternative destinations (e.g. shops).

It seems that there are a series of major conceptual issues which need to be thought through on ATT systems. To some extent they depend on the theoretical model being used, but more generally they relate to the type of information being given to the traveller, when and where the information is given, and the relatively narrow range of options available to change behaviour. To expect major behavioural changes resulting from the types of ATT currently being developed seems optimistic. Despite these limitations (e.g. on measurement and change), it is still meaningful and possible to set up experiments and collect data for analysis and evaluation.

18.5   Some European Case Studies

Measurement of user responses to transport telematics presents certain conceptual difficulties. These include the large range of responses (see the above nested approach), the small response scale and the long time reaction time to change. These three problems represent a classic dilemma for social research: the range, scale and timing/diffusion of innovation make it very difficult to measure. Even if it could be measured, the difficulties of sample capture and identification make it hard to obtain sufficient data to place the results on a solid statistical basis.

The use of the nested framework requires the possibility of monitoring behavioural changes over a sufficient period of time. The measurement of these changes is limited by the character of empirical field trials, which are small scale applications in a limited time of major technological innovations. In order to obtain a longer-term insight under these conditions should not only current travel behaviour resulting from the introduction of the applications be measured, but also the changes of perceptions and preferences (stated or revealed) with regard to the applications. This holds in particular for the assessment of the dynamic aspects of awareness and diffusion.

Table 18.7. Parameters of the nested approach covered by two European case studies.

	SOUTHAMPTON	NORTH WING RANDSTAD
ATT AREA	Public Transport Management	Integrated Inter-urban Traffic Management
Nested Approach Parameters	*Behaviour:* Trip generation; Mode; Adherence	*Behaviour:* Route; Departure time; Adherence
	Segmentation: Car availability; Age/sex; Social group; Income; Familiarity; Purpose/Distance	*Segmentation:* Age/Sex; Profession; Purpose/Distance
	Awareness: Exposure to ATT; Acceptability; Publicity	*Awareness:* Exposure to ATT; Acceptability
	Diffusion:* None	*Diffusion*:* None

* among end-users

Recently, some interesting case studies have been carried out in Europe. The aim was to assess behavioural responses of end-users to transport telematics applications (Argyrakos et al. 1994). Efforts have been made to systematically assess the impacts of a range of information telematics, using the nested methodology. Two case studies using this approach have been carried out in conjunction with field trials.

1 In *Southampton* (UK), a public transport traveller information system has been tested for buses. The main objective of this trial was to create conditions in which people would switch to public transport for their journeys. This trial consists of real time bus information displays at bus stops along a single corridor.

2 In the *North Wing of the Randstad* (Netherlands), an inter-urban motorway driver information system using Variable Message Signs was tested. The aim of this application was to relieve parts of the motorways that are highly congested during traffic peak-hours, by re-routing traffic to underutilized parts.

User surveys have been carried out at each of these sites. The surveys followed different strategies, since the types of behavioural change aimed for and expected varies between both types of applications. Where appropriate, cross-sectional and before/after surveys were carried out. Table 18.7 illustrates how these two case studies are linked to each other by the common conceptual approach.

We now present the finding of these two respective case studies, giving the main results in the case of Southampton and, greater detail for the North Wing of the Randstad in the Netherlands.

18.6 Passenger Transport Information (STOPWATCH) in the UK

One of the key problems for passengers using urban bus services is the unreliability of the service, caused mainly by congestion within the system. Real time information systems allow passengers to make decisions based on the actual arrival time of the service and it may also provide a greater satisfaction with the service being provided.

The trial corridor in Southampton is seven kilometres long and covers twelve different bus routes being operated by two bus companies. The bus location is transmitted from in-bus units to roadside beacons and then transmitted to a central computer. This data, together with historic journey times, is used to calculate the expected arrival time of the bus at particular stops. The system has been installed on 114 buses and displays are now in operation at 44 bus stops on the main bus corridor into the centre of the city from the north.

Behavioural surveys have been carried out on bus users before and after the installation of the real time information system (STOPWATCH). About 11% of the before sample (1538 respondents in total) were aware of the plan to introduce the system, with those who were frequent users having greater knowledge. About 16% stated that the STOPWATCH system would increase their use of the bus, but thought that this would be most significant among occasional and first time bus users. The most positive responses came from those who used the bus for social and shopping purposes, particularly among the young. However, in the survey carried out after the installation of the passenger information system, very different results emerged. Of 1702 respondents, only 4% had actually increased their trip making as a result of the STOPWATCH system. There seems to have been a considerable overestimation of the impact on the trip making patterns of bus users. There is a clear difference between behavioural intention and actual behaviour.

Table 18.8. Summary of the STOPWATCH effect in Southampton.

Travel Characteristics	New Bus Users	New Trips from Existing Bus Users	STOPWATCH Switchers	All Sample
. Trip Purpose				
- Work	20.7%	25.9%	21.5%	23.5%
- Education	14.6%	15.7%	13.6%	8.4%
- Shopping	34.1%	45.1%	40.2%	39.6%
- Other	30.6%	13.4%	24.7%	28.5%
. Car availability	47.5%	19.7%	13.2%	17.8%
. Non car owing households	52.5%	68.2%	57.9%	55.5%
. Never used the car	15%	24.7%	18.5%	14.8%
Socio-economic Characteristics	**New Bus Users**	**New Trips from Existing Bus Users**	**STOPWATCH Switchers**	**All Sample**
. Women	56%	60.7%	63.6%	63.1%
. Under 34 years	69.7%	67.7%	62.2%	50.3%
. Full-time employment	28.8%	40.7%	34.1%	38.9%
. Full-time student	37.2%	25.1%	40.1%	23.3%
Sample size	**406**	**55**	**214**	**1702**

Nevertheless, there were three main ways in which STOPWATCH seemed to have had an effect on travel behaviour. Firstly, there has been a substantial number of new users of public transport since the introduction of STOPWATCH. These new users are young, usually travelling for education or leisure purposes, and they often have high levels of car availability. They have a very positive view towards STOPWATCH, but make a less than average amount of use of the bus service. The new users seem to form the basis of a revitalized interest in new types of public transport systems. The provision of high quality real time information is important to this group of users, therefore the effective marketing of STOPWATCH to this group can contribute to the success of the information systems and hence in generating greater use of buses (Table 18.8).

The second STOPWATCH effect has been to encourage existing bus users to make more trips on the improved service. This 4% mentioned earlier have a high level of knowledge of STOPWATCH (73% as against the sample average of 59%), and they actually make use of the electronic display information (35% as against the sample average of 22%). All of this group's respondents used the route before the electronic displays were installed, and over 80% looked at the displays

374

on several occasions (sample average 53%). This group of users came predominantly from non car owing households and they were never likely to use the car. They were young people on shopping or education purpose trips. It seems to be possible to increase the use of buses, even among captive users and heavy users of the existing services through marketing a better quality services with real time information systems. These increases in bus use are small, but as the overall quality improves, further trips should be generated. The third effect is the STOPWATCH switchers. These are bus users who have left the bus stop during the last fortnight because the electronic display showed that the bus would not arrive for some time. Some 214 (12.5%) responded positively in this way and they either switched route to another bus or stop, or went to a local facility and then returned to same stop or another stop (Figure 18.2). The total mode shift effect resulting from these figures is 45%.

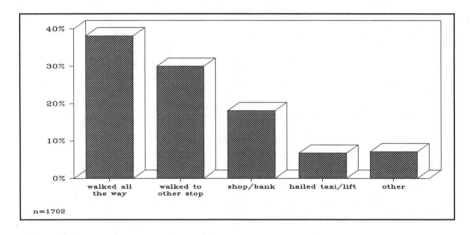

Fig. 18.2. What users did when they left the stop.

So actions can involve modal changes, route changes, timing changes, activity changes or destination changes. The introduction of the STOPWATCH system has allowed a new flexibility in trip making that seems to have been well received, particularly by young school-goers. However, there is also concern over the accuracy of the information: the most important issue for all bus users is a reliable and frequent service.

The use of technology in bus services provides a better quality service, particularly in a deregulated bus market such as that in the UK. One of the major losses caused by deregulation has been the poor quality of information about the bus services available. The numbers of services running in the Southampton area and their frequency has increased, but this is information not available in the bus. Good quality real time information does not seem to improve the perceived quality and reliability of the bus service. Nevertheless, even after only 10 months of the STOPWATCH system in Southampton, users are making more journeys, although on a modest scale.

18.7   Route Choice Information (RIA) in the Netherlands

The Northern Wing of the Randstad in the Netherlands (the Greater Amsterdam area) suffers from severe traffic problems on its inter-urban roads. The major roads connecting Amsterdam with surrounding towns are heavily congested, especially during peak-hours. One of the main traffic problems is the crossing of the river IJ which splits Amsterdam in two parts just north of the inner city. Every day, a large flow of commuters travels from the residential areas north of the river IJ to employment in the southern part of the agglomeration. This development caused the need for the orbital motorway, which is a very important link for all regional and through motorway traffic (Buijn et al. 1994). Major parts of the western and southern side were already completed in the 1970s and 1980s. In September 1990, the last part of the Amsterdam orbital motorway was completed, opening up the northern and eastern segments.

The completion of this orbital motorway provided new routing alternatives for a considerable number of users of the regional inter-urban road network, owing to the new capacity to lead traffic along the eastern and northern side of Amsterdam. A dynamic traffic management application consisting of Variable Message Signs has been implemented by the Dutch road manager (Rijkswaterstaat) to support users of the ringroad in selecting their route. The system is called Route Information Amsterdam (RIA). RIA provides users approaching the ringroad with information on traffic queues (including the length of the queues) and with information about closure of tunnels or driving lanes. The type of information provided is specifically meant for those road users who are familiar with the network, knowing their route possibilities when passing the VMS signs. Furthermore, the longer the distance driven on the ringroad, the less the difference between going clockwise or anticlockwise around the ringroad. The information is thus expected to be useful for through traffic and a certain part of traffic with Amsterdam as destination.

In November 1991 the first variable message sign was put into use at the most strategic location, namely on the motorway from the north before the junction with the ringroad. In April 1994 another three identical signs were installed on the three access motorways from the south, each just before the respective junction with the orbital motorway. It is interesting to find out the degree of acceptance of this particular technology.

A behavioural survey was carried out among users of the motorway network approximately three months after the implementation of all four variable message signs. The survey target consisted of car drivers who made use of the ringroad and arrived in Amsterdam via one of the four main motorway access roads where VMS signs have been installed. The survey had a sample size of 826 observations. I was possible to made an in-depth investigation of the market potential and as the survey used the nested approach with segmentation of the survey sample and the measurement of attitudes and actual behavioural changes to this operational telematics system.

It was hypothesized that important segmentational variables should be sought in the the age, gender and income/social group of the respondents as well as their travel characteristics, such as their experience with dynamic driver information, frequency of travelling (and inherently the familiarity with alternative routes) and trip purposes. Considering the possible effects on travel behaviour of the kind of information provided, emphasis was given to choice options related to route followed and trip departure time. The relatively long-distance character of the car trips implied that it was unlikely that there would be an impact on other travel behaviour parameters (like changes in destination choice or trip rescheduling).

Several considerations affect the possible impact of the information on route choices. For instance, the change from a planned route under influence of dynamic information would depend in the first place on the existence of any possible alternatives. Secondly, the demand for alternative routes would be determined by the expected duration and cause of the queueing ahead on the motorway. Personal preferences of motorway drivers with respect to traffic delays and rerouting and travel features like distance and time restrictions play an important role here. If, for example, the expected delay ahead on the followed route is equal to the detour time of any other possible route, a certain share of drivers would change route. The uncertainty which stems from using an alternative route may also deter a certain share of drivers, particularly those who are less familiar with the area.

Table 18.9. Survey characteristics.

	per cent *(1)*		per cent *(1)*
sex		**age**	
male	83.8	<24	4.4
female	16.2	25-34	44.1
		35-44	26.9
		45-59	22.8
		>60	2.8
trip purpose		**frequency of using ringroad**	
commuting or work-work	56.2	≥ 5 days a week	47.8
business	35.2	3-4 days a week	17.4
otherwise	8.6	1-2 days a week	18.0
		< once a week	16.8
flexibility of arrival time		**average trip distance**	
impossible to arrive late	43.9	< 10 km	1.0
possible to arrive late	56.1	10-25 km	10.5
		25-50 km	35.0
		> 50 km	53.5
alternative routes available		**passing frequency of RIA sign**	
yes	57.1	> once a week	77.9
no	42.9	once a week	8.4
		< once a week	13.6

(1) Missing values were omitted.

In Table 18.9 some important characteristics of the survey are shown. There were significantly more males than females in the sample; 84% of drivers were male. It was remarkable that the females were slightly younger than the males. The majority of the respondents, 80%, had full time employment. In short, the population surveyed can be characterized as mainly male, full time workers, whose age was between 25 and 60, and who had much driving experience. An important feature is the usual purpose of the trip. Business appointments were mentioned by nearly 35% of the respondents as one of the dominant purposes of trips. They were relatively stronger represented in the male group. Around 56% were commuters or freight or delivery drivers.

It appeared that 55% of the respondents had more than one route alternative which would not cause a significant delay to their arrival time. Of these drivers approximately 31% regularly took an another route. This high route choice flexibility than one route was quite encouraging for the possible usefulness of the driver information provided.

The attitudes towards the system were in general very positive. In total 90% of the respondents said that they were pleased to have the information provided. This percentage was the same for both those drivers who had route alternatives as well as those who did not have route alternatives. This indicates that besides travel time, drivers may obtain other forms of benefits from the information, like for example a reduction of uncertainty with respect to the traffic situation.

The impact on route choices of drivers seemed to be considerable. Some key figures were the following: about 72% declared to have been affected in their route choice by the VMS information. About 23% changed route regularly and 9% often (Figure 18.3). It may be concluded that a very significant number of drivers has been affected by the information in their route choice since the VMS system was installed.

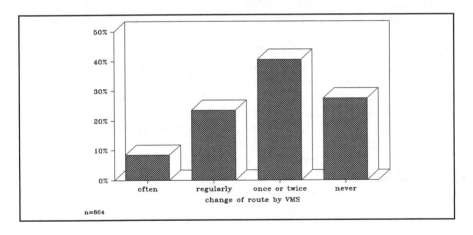

Fig. 18.3. The impact of VMS queue information on driver's route choice (% of sample).

The survey also investigated what impact different VMS stimuli had had impacts on route choice. Of those drivers affected by the information, 30% declared to have decided to change route once or more when seeing a message warning of a queue length between 0 and 2 km. About 50% took this decision only when seeing a message of a queue of at least 2 km. To 16% a minimum queue length of more than 4 km was reason to change route and 4% change route only when queues of more than 6 km were indicated. The satisfaction rates with alternative routes were encouraging. Nearly 38% of the route switchers declared to be better off with the other route, 13% felt not to be better off. Almost half of the respondents (49%) did not know.

The alternative route for just over half of the route switchers was usually motorway. The other half declared that they took alternative routes which were partly off the motorway. The high number of route switchers using secondary city roads, which is not the intention of the VMS application, suggests that the information might also potentially generate some negative side-effects that might conflict with targets to keep motorway traffic as much as possible outside urban areas.

Route changes and driven detour distances appeared to depend on trip purpose. A significant relationship was found between the frequency of route changes and the trip purpose. The respective shares of business drivers, commuters and discretionary drivers who often changed route were 10%, 8% and 6%, respectively. In the case of those who never changed route, these shares were 20%, 31% and 36%. Thus business drivers seemed to be slightly more sensitive to route change than others, while commuters also changed more frequently than people with discretionary journeys. This result is not unexpected since business drivers are likely to place higher value on travel time and have a lower perception of travel costs.

It seemed that men were in general more sensitive to route change than women. They changed route more often, and also reacted more readily to messages of smaller queue lengths. One possible explanation can probably be found in psychological differences between the two sexes, but there is also some correlation between gender and trip purpose (it was found that men travelled more for business purposes than women). This difference between males and females confirmed the results of other behavioural studies (e.g. Mannering et al. 1993).

The impacts on route choices also appeared to be influenced to a certain extent by the frequency of driving (and, inherently, the familiarity with the road network). Of those who drove very frequently (five days a week) 32% appeared to reroute their trips often or regularly, while only 19% of those who drove less than once a week changed route frequently.

Furthermore, there appeared to be a positive relationship between the average distance of the trip and the length of detours made. Of those travelling distances of over 50 km, 71% make detour distances of more than 2 km, while this percentage was substantially lower (53%) for those who travel shorter distances (below 25 km). Thus it seems that the reluctance to additional kilometers decreases as the journey length increases. But some correlation also existed here with the trip purpose: those travelling longer distances consisted more often of business drivers. Of those travelling more than 50 km, 40% had business motives, while of those travelling less than 25 km, this percentage was only 26%.

Besides the impact on route choice, another aspect of the information provision which may not be discarded out of hand is the possible effect on departure time of drivers. It may be hypothesized that in some cases habitual behaviour related to departure time choice may change because drivers may get used to a reduction of uncertainty or, more directly, because individual travel times may be repeatedly reduced. Figures show however that from the whole sample, the departure times of most respondents (82%) were never affected by the information system.

The VMS system concerned was a public system, so there was no cost to the end-user. In order to obtain an indication of the potential market for individual systems providing the same type of traffic information, the willingness to pay was investigated for in-car systems continuously providing this kind of information. It appeared that the willingness to pay for the information was in general quite low (Figure 18.4). Almost half of the respondents would not be prepared to pay anything. Only about 7% would have paid more than 40 ECU a year.

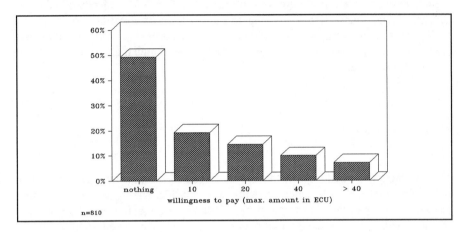

Fig. 18.4. The willingness to pay (% of sample).

As expected, a relationship was found between the willingness to pay and the usual trip purpose. Respondents with regular business appointments were more prepared to pay higher amounts. Approximately 13% would pay more than 40 ECU a year, against 4% and 1% for commuting and discretionary purposes respectively. The last group was very unwilling to pay anything: 62% did not want to pay anything, against 44% and 51% for business and commuting purposes, respectively.

The relationship between the willingness to pay and the availability of route alternatives was similarly unsurprising. Respondents without an alternative route available were less prepared to pay for road information than respondents with an alternative. However, a considerable number, 43%, of the respondents without route alternatives was still willing to pay something. This result re-affirmed that individual benefits were also obtained from the VMS information by those who had no route alternatives or did not actively react to the information by rerouting their trip.

Another important aspect which emerged from the survey was that the information provided by the VMS was mainly used by those drivers who had experience with other (more conventional) means of traffic information, for example traffic information broadcast by radio. Of those who regularly or often changed route on the basis of the VMS information, 90% once or often also rerouted their trip as a result of traffic information provided by radio, while for those never changing route this percentage was 48%. Furthermore, the willingness to pay for the continuous availability of the VMS information was higher among those who were frequent listeners to radio traffic information. Of those never using radio traffic information, 37% were prepared to pay a certain amount for the VMS information compared with (56%) for those often using radio traffic information.

This led to the conclusion that for the user group with a high propensity to listen to conventional traffic information, the VMS information was a substantial complement to these other sources of information. If the VMS information did not add any value, then drivers listening to radio traffic information would not be willing to pay anything for the additional VMS information. Consequently, the level of acceptance of a new invention may also be complementary to the use of a related invention.

18.8 Lessons

The potential beneficial impacts that may be expected from the introduction of new advanced transport telematics systems are considerable. However, positive technical and operational test results are often seen as a guarantee of the success of the developed systems. In this paper the focus has been on another critical success factor, namely the interaction of the technologies with the potential *users* of these systems, since acceptance and diffusion are important issues.

The process of innovation and diffusion of transport telematics is dynamic and complex, like any other new technology. The behavioural responses of transport users in terms of changes in travel choices are very important factors in this process, but also the range and types of actors involved and the purpose of the project affect the extent and rate of the potential diffusion of the respective technologies as these influence the levels of awareness. The fundamental role played by the potential users suggests that more emphasis should be placed on a coherent and systematic investigation of the user side. To this end a framework which covers the full range of dynamic issues related to the strategic aspects, the market potential and the market response of transport telematics applications has been presented. This framework has the advantage that it is applicable to a wide range of telematics functions.

It should be noted that the effects of technological innovations will vary between various groups of transport users. This emphasizes the importance of a thorough user segmentation. For instance, it can be hypothesized that certain user groups repeatedly making the same trips (e.g. commuters) may adopt a satisficing travel behaviour, in which case use will only be made of telematic travel information when there is dissatisfaction with previous trips made.

The usefulness of the framework presented was illustrated by a set of two case studies in conjunction with pilot tests of new telematics applications in Europe. Our empirical work focused on a real time passenger information system for buses in Southampton, England and a motorway driver information systems on

Amsterdam's orbital ringroad. User surveys provided various interesting conclusions, both on the potential impact of such systems on travel choices and the potential market for disseminating travel and traffic information by means of private equipment.

The awareness of the STOPWATCH passenger information system in Southampton was moderate. Before the system was installed about 11% were aware of the plan to introduce the system, with those who were frequent users having greater knowledge. There were three main ways in which the system seemed to have had an effect on travel behaviour. In the first place, the introduction of STOPWATCH generated a substantial number of new users of public transport. These new users were young, usually travelling for education or leisure purposes, and they often had high levels of car availability.

Secondly, existing bus users were encouraged to make more trips on the improved service. This group of bus users had a high level of knowledge of STOPWATCH (73% as against the sample average of 59%), and they actually made use of the electronic display information (35% as against the sample average of 22%). It therefore seems to be possible to increase the use of buses, even among captive users and heavy users of existing services through marketing a better quality service with real time information systems. The actual increases in bus use have so far been small, but as the overall quality improves, further trips should be generated.

Thirdly, the system resulted in 'switchers'. These were bus users who had left the bus stop because the electronic display showed that the bus would not arrive for some time. About 12% responded positively in this way and either switched route to another bus or stop, or went to a local facility and then returned later to same stop or another stop. The total mode shift effect was 45%.

The introduction of passenger information systems like STOPWATCH seems to allow a new flexibility in trip making, and is particularly well received by young school-goers. However, good quality real time information did not seem to improve the perceived quality and reliability of the bus service, which is the most important factor for bus users.

The awareness and attitudes towards the driver information system adopted in the Netherlands were in general very positive. In total, 90% of the sample said that they were pleased to have the information provided. This percentage was the same for those drivers who had route alternatives and for those who did not. This indicates that besides travel time, drivers might obtain other forms of benefit from the information, like for example a reduction of uncertainty with respect to the traffic situation.

The impact on route choices was considerable. The propensity to change route was strongly influenced by the purpose of the trip made. The group of drivers regularly travelling for business purpose more frequently followed alternative routes; it also seems that men were in general more sensitive to route change than women.

About half of the respondents declared that they would be willing to pay for having the dynamic traffic information provided by the VMS continuously available in the car. This willingness was higher for business drivers. A high percentage of drivers who did not have reasonable route alternatives and who had never changed route, still declared their wilingness to pay for the information, re-affirming that other kinds of benefits apart from saved travel time, may be obtained.

The survey also indicated that the information provided by VMS is mainly used by those drivers who already use other kinds of traffic information. This confirms the conclusion that for this group of drivers the VMS information may be a substantial complement to other sources of information.

A final lesson from our empirical work, in the light of the generally low willingness to pay, may be that the scope for commercialization of traffic information among the broad public might be more limited than initially expected. It also emerged however that clear user segments can be distinguished for which the market penetration is likely to vary significantly. Such socio-psychological and economic factors have important consequences for the future diffusion rates of new telematics technology.

References

Argyrakos G, Giaoutzi G and K Petrakis (1994) Travel Behaviour and the Diffusion of New Technologies. Proceedings of the 7th International Conference on Travel Behaviour. Valle Nevado Santiago Chile 13-16 June. 771-785

Banister D, Nijkamp P, Camara P and G Pepping (1994) The Analysis of User Response to Advanced Transport Telematics - Measurement, Methodological and Conceptual Issues. BATT.DRIVE II Programme. EC DG XIII Brussels

BATT (1992) Identification of requirements (deliverable 2). DRIVE II Programme. EC DG XIII. Brussels

Buijn H and W Schouten (1994) Route Choice Information Amsterdam. Proceedings of the 27th ISATA Conference. Aachen Germany. 31 October - 4 November. 143-150

Catling I and H Keller (1993) The LLAMD Euro-project: Expected Impacts of Dynamic Route Guidance systems in London, Amsterdam and Munich. Advanced Transport Telematics. Proceedings of the Technical Days. EC Brussels. 70-74

Malaterre G and H Fontaine (1993) The Potential Safety Impacts of Driving Aids. Recherche Transports Sécurité (English issue) 9:15-25.

Mannering F, Kim S, Barfield W and L Ng (1993) Statistical Analysis of Commuters' Route, Mode and Departure Time Flexibility and the Influence of Traffic Information. Research report. University of Washington Seattle

Quaid M, Heifetz L and M Farley (1992) The Puget Sound Telecommuting Demonstration: Program Description and Preliminary Results

Rijkswaterstaat (1992) Dynamic Traffic Management in the Netherlands. Ministry of Transport, Public Works and Water Management. Transportation and Research Division. Rotterdam The Netherlands

Vanderschuren M, Vlist M van der and E Rooijmans (1993) Verkeersmanagement, Energie en Milieu: De Effecten van Telematicatoepassingen. INRO-VVG Report. Delft Netherlands

19 Towards a New Theory on Technological Innovations and Network Management: The Introduction of Environmental Technology in the Transport Sector

Harry Geerlings, Peter Nijkamp and Piet Rietveld

|U,5| Ō3l   Q2⁰

19.1   Introduction   L9l

Environmental problems have become quite urgent in many countries. Changes in consumption patterns appear to be insufficient to bring about the necessary reductions of emissions of various types. Therefore, a large contribution to the solution of environmental problems is expected from technological change. This raises the issue of how technological change takes place and, more particularly how to promote technological change that is beneficial to long term sustainability objectives. Since market incentives are a major driving force of firms involved in the development and adoption of new technologies, it is not certain that the necessary developments will take place at sufficient speed in those cases where markets fail, i.e. in the case of environmental problems. Therefore the issue of technological change in the case of environmental problems deserves special attention.

In this chapter we will discuss this topic in the context of the transport sector. This is a relevant sector, because it contributes significantly to various types of pollution at different spatial levels. From technology large contributions are expected to reduce the pollution levels.

We will approach the issue of environmentally friendly technological change from the perspective of evolutionary theories of technology dynamics (Section 19.2). We will argue, however, that additional theoretical insights are needed, especially for the study of technological change in the context of environmental problems. Therefore we discuss in Section 19.3 various theories on strategic cooperation of firms in networks aiming at technology development. Special attention is paid to the role of the government in such networks. The role of governments is important, not only because governments can formulate technological standards and give financial incentives to stimulate technology development; governments can also play an important role in the networks of actors that are formed to develop the technological innovations. By way of example, the Clean Air Act of the state of California is discussed in Section 19.4.

The Californian case, aiming to achieve technological innovation associated with environmentally friendly transport (with the government playing a well defined role in networks and the use of a new type of instrument involving interaction processes between actors), indicates that the approach is promising.

19.2 Evolutionary Theory of Technology Dynamics

In recent years growing attention has been paid to the role of technology in society. From an optimistic perspective, technology is seen as the key to solve a number of problems (especially related to the natural environment) and technological innovation is considered as a motor for economic welfare. This optimistic view is prevalent in the transport sector (OECD 1988; IIASA 1991; EC 1992; US 1993, UNEP 1993)). It is indeed indisputable that technological development has led to a more efficient use of energy, materials and capital which, in turn, has led to higher productivity. Moreover, technology in particular has permitted a more efficient use of the time factor (Grübler 1991). Transport technology is also considered an important tool to bridge the gap between accessibility and the quality of life (SVV II 1989).

Yet we have to look carefully to see if this optimistic view is fully justified in practice. For many of the proposed technological measures, especially those that aim at reducing pollution (like strict emission standards, CO_2-tax, traffic reduction, etc.) there is great difficulty both in the E.U. and the U.S. in obtaining the necessary parliamentary majority.[1] In addition to political feasibility, technical feasibility may also be a bottleneck. Certain large infrastructure projects, like underground railways or new energy systems such as nuclear fusion, are near the border of what is technologically feasible. Even when the technologies are available, large problems of implementation may occur. Moreover, the complexity of decision-making and the many parties involved often hinder the energetic development of new possibilities. These examples illustrate that there are several types of bottlenecks that should be removed before technology can be put in place. Technological developments cannot be taken for granted. A clear conclusion is that a better understanding of technology dynamics is needed.

Many books and articles have attempt to predict the future. Some of the predictions have been wide of the mark. In 1920 the Scientific American printed an editorial which made some forecasts based on the developments of the last 75 years. Although, in retrospect, the article seems to have made strikingly accurate predictions, two omissions can be distinguished. None of the authors had foreseen that only one month after publication of the article radio

---

[1] During the debate in the U.S. Congress on the budget of the Ministry of Economic Affairs (DOE), the notion of a stricter CAFE-norm (Corporate Average Fuel Economy) has never come up for discussion. During the decision making process an advertising campaign was shown on radio and television to promote the 'Coalition for Vehicle Choice', funded by the car industry and the insurance business, which claimed that lighter and smaller (and consequently more fuel efficient) cars were less safe.

telegraphy would be invented, neither did they predict that in the same year sound would be added to films. Both developments meant an important breakthrough in those days.

Another example is the report published in 1937 by the National Resource Committee in the United States entitled Technological Trends and their Social Implications (NRC 1937). This report, still worth reading, sketches several technological trends which were predicted in a number of sectors, including transportation. The committee failed however to predict developments in aviation. A large barrier for the further development of aviation in those days was fog. More than 25 research methods for solving this problem are mentioned, some of them imaginative, such as the use of bombs to dispel the fog. The study 'predicted' that the visibility problem would be overcome, but did not foresee that only a year later radar would be invented, making further efforts redundant.

In the last few years, the process of innovation and diffusion of technological developments has received much attention in economics as well as the social sciences. A major impulse can be found in the work of Nelson and Winter (1977) and Dosi (1982, 1988). Since then the process of innovation and diffusion of technological developments has received much attention inside as well as outside the economic sciences (economic and stochastic modelling, dynamic growth theory) and social sciences (sociology, environmental science, public administration, etc.). Some authors elaborate on the theories of Kondratieff (1926) and Schumpeter (1939). Kondratieff focuses attention on the interaction between economic and technological development, seen in the long term perspective[2] Schumpeter developed the proposition that important innovations occur at the beginning of an economic recovery. He maintained that, particularly in times of economic crisis, companies are prepared to take risks and open new channels for trade. He identifies three important conditions that determine a technological change: the concept of innovation, the concept of interpretation and the concept of technological dynamics.

Nelson, Winter and Dosi, elaborating on the Schumpeterian addition to neo-classical economics, have focused attention on the role that social-cultural and institutional factors play in processes of innovation and diffusion of technology. This approach is known as the evolutionary theory of technology dynamics. The recently developed theoretical models in technological change by Callon (1986) and Latour (1987) go even further. They stress that the so-called 'social constructionism' or contextual approach to technological change is crucial.

Nelson, Winter and Dosi interpret technological changes as a constant succession of variation and selection processes aimed at solving technologically defined problems. These processes are not random, but clearly structured. There is a certain rigidity and inertia in the degree of change in technology, as

---

[2]Kondratieff wrote that, with respect to wholesale prices, the world economy followed cycles lasting from 45 to 55 years. The bottoms of the cycles generally coincided with severe worldwide depressions. Kondratieff did not develop an explanation for the cause of those waver; rather, he speculated that they resulted from periodic overexpansion in large industrial capital projects and infrastructure.

a result of which the variation is not unrestricted. This leads to a certain regularity in the development of technology, which can be indicated as 'routines' or 'trajectories'. Routines differ from trajectories because they are related to the process of decision making and learning. Routines are in many cases persistent and hard to change, partly due to the fact that not all relevant variables can be included in the decision making process. Social systems also consist of routines. Trajectories, on the other hand, concern the process of technological development. A technological trajectory contains the technological changes which take place in the framework of a technological regime or paradigm. It represents, in other words, the 'direction of progress' within a technological regime[3].

The relationship between technological development and the selection environment is a complex one. The study of the forces driving technological innovation is mainly done ex-post. This means that besides deepening our insight into the factors that influence the technological development (ex-ante), it is important to understand that technology development has an evolutionary character. Most of those who follow the evolutionary view, however, interpret the development of technology as a firm-related activity. In the next section, we will show that this perspective, though valuable, is too narrow.

The group that has formed around Nelson, Winter and Dosi, who were particularly concerned with developing the evolutionary theory of technology dynamics, appears to focus in fact on intra and inter firm policies. This vision assumes the existence of trajectories (the development of technologies along fixed patterns which exclude a technological jump) routines and an R&D policy based on known technologies and experience. An important aspect of this research programme is the notion that competition is the driving force for innovation and inter and intra firm relationships.

Arthur (1988) argues that past technological developments leave a lasting mark on future developments. Existing technological knowledge, specific infrastructure, vested interests and long-term research programs make it hard to deviate from the chosen path. This is highly relevant with respect to the transport sector and raises doubts about the real effects of technological innovation.

In the last two decades the forces that influence the innovation process have been thoroughly investigated. In the current view, the innovation process is seen as a complicated and lengthy process by a system of 'drivers of innovation' (Janszen 1993).

---

[3]A technological paradigm or regime is interpreted as the dominant cultural matrix of technology-developers and includes a limited number of scientific principles, understanding and heuristics (search directions) and a limited number of material technologies. The central idea of the technological regime is the basic technology by reason of which further adaptation and development takes place. Departing from the basic technology the search direction of the variation processes in the technological trajectory is shaped by heuristics. The development of technology within a certain problem area takes place inside the borders of a similar technological regime for quite some time and is as such pre-structured (Mol 1991).

19.3   Network Management and Cooperation

19.3.1   The Concept of Network Management

After the high hopes raised by of new management tools and the government policies developed in the sixties and seventies, the results actually achieved were quite disappointing (In 't Veld 1989; Koppejan 1993). Policies were less effective than expected, their benefits were often addressed to the wrong groups, procedures and legislation took more time than was expected. The failures were reflected in new types of problems, such as the economic recession, unemployment and environmental problems.

There are various explanations for the failure of government policies. Firstly, the problems society is confronted with are of an increasingly complex nature. Secondly, there is an increasing number of actors involved in decision process. Thirdly, there is an increasing interdependence between (related) policy areas.[4]

These governmental failures stimulated the establishment of the 'new right movement' (Reaganism) in the U.S.A. and Thatcherism ('less government, more market') in Great Britain. This initiated new thoughts on the (evolutionary) role of government policy (independent from the new existing evolutionary theory that was later applied in economics and technology dynamics). The network approach, as practised in business administration and public policy sciences, has therefore become more important during the last ten years.

With the erosion of faith in government, more attention was given to the role of private firms as drivers for employment, economic growth, etc. This trend became even stronger after political changes in the Eastern European Countries and Russia and the economic success of Japanese firms. These observations are our starting point for a closer look at the function of networks for more efficient government action and cooperation with private firms.

The so-called 'network approach' provides important support for formulating collective responsibilities and converting these into operational policies. The concept of the network approach originates from logistics (production networks). In the political sciences the concept of policy network is used for the cooperation between businesses and governments. The core of the definition is formed by the reference to the mutual dependence between actors, which results from the insecurities in the policy field. Information, aims and means

---

[4]An example is the development within the transport sector: the legitimacy of the involvement of many actors in the steering processes can be found in the competence and expertise of the participants (technological rationality) and the interests that are involved indicate a political rationality. There is a discussion on the role of the central rationale and central directing role of government in society. For this reason the government reduces her ambitions and stresses the importance of the market (EU 1993). In many cases the implementation fails because of a wrong assessment or contrast of objectives and instruments, a lack of information on the pursued policy, opposition within society, and lack of control of the institutes that carry out the policy.

are exchanged to help reduce this uncertainty (Koppejan et al 1993).[5]

Network management can also be characterized as an activity aimed at raising the effectiveness of steering instruments or as a method of problem solving (cf. De Bruijn and Ten Heuvelhof 1991). The problem of weighing the factors of success or failure of policy can to a large extent be influenced by managing and stimulating so-called *'fields of common interest'* or *'win-win situations'* (cf. Geerlings and Hommes 1994). Network processes and network management are judged successful when those involved consider the new situation an improvement on the existing one[6]. By way of selective network management new forms of cooperation can be initiated[7].

A correct formulation of these targets can provide this cooperation with a strategic character. The selective activation of a network is also important to stimulate strategic cooperation in the field of the technology dynamics.

In our opinion network management can be considered as *the process of looking for areas of common interest and creating win-win situations when there are many actors with diverse and conflicting interests. The aim is to solve a mutual problem.* An important question in this context is to what extent actors succeed in tuning in their principles and resources to be able to take common action and bring about a situation which is an improvement for everyone involved. A few comments must be made here:

- creating win-win situations must not be interpreted as the realization of consensus on a common objective. Win-win implies the central thought that actors have agreed or will agree on objectives and resources of policy. The result is a set of measures everyone can agree with. When generating win-win situations the point is to realize an outcome which holds an attractive element for every actor. This does not necessarily mean that every party should agree with the total outcome: 'agree to disagree' will play an important part in win-win situations. This is, however, very important as there are risks attached to striving for consensus, such as excluding criticism, neglecting non-represented interests or ignoring dangers;
- there is the problem of delimiting the network and the process involved. By

---

[5]Network management can have several meanings. A traffic expert will probably initially think of a physical network of infrastructural buildings or an information network. A sociologist may think of a personal network. The concept of production network originates from logistics. In the political sciences the term policy network is used for the strategic cooperation of businesses and governments, given the insecurities of the policy field.

[6]De Bruijn (1994) stresses the difficulty or even impossibility of giving an exhaustive list of the instruments the actors call forth. In a horizontal context these instruments are reduced to four core concepts, namely networks, incentive-steering, covenanting and communication.

[7]Important criteria to develop a satisfactory policy for the actors involved are balancing one's own interests, openness, carefulness, reliability, transparency, unambiguity, consistency, long term perspective and democratic legitimacy (Geerlings 1994).

390

means of selective activation, potential non-winners can be excluded. The external effects of interaction processes within networks should be included in the judgement. The term win-win situation does not justify costs of the interaction being passed on to individuals, groups or organizations outside the network.

19.3.2 Network Management and Private Firms

Private firms operate in a dynamic environment, where demand, legislation, technology, and so on, are continuously changing. In order to cope with these changes firms try to anticipate such developments. Firms prefer a position in the dynamic environment (playing field) which makes them less vulnerable to these changes and which offer opportunities to benefit from these changes. As an integrated part of their policy, firms will anticipate environmental legislation. This is not only due to noble motives or a sense of responsibility, but can have to do with public image, market shares, competition, etc.[8] Most firms will benefit from clear, transparent and consistent governmental policies, with a long term perspective. Consequently there appears to emerge a tension between the trend towards deregulation and subsidiarity of governmental policies (despite the fact that environmental policies are dealing with collective interests) and the trend towards globalization in the world economy as can be observed for example in the car manufacturing industries.

Hence, it is in the interest of private firms to communicate with governments so that each can express their long term views and take the other's views into consideration. Watanabe (1994) states in this context that there is a great need within Japanese private firms to reduce risks and uncertainties with respect to the dynamic (outer) environment.[9] Governments have to realize that expressing long term objectives and policies to private firms will contribute to the image of a trustworthy partner and hence stimulate cooperation.

Environmental standards and consumer demand, although increasingly uniform, are mainly inspired by national interests. The trend we can observe within the car manufacturing industry is, in contrast with environmental policies, oriented towards globalization. Mainly inspired by the economic success of Japan in the eighties, we see at the moment a tendency towards

---

[8]There are firms which prefer to operate (quite often from the perspective of profit maximization) on the border of the 'playing field', and some firms even benefit from operating just over this border. What the consequences are for the continuity of the firm is not always clear.

[9]Interview with professor Watanabe, deputy secretary-general of MITI, 19th August 1994. He states that this can be explained by the conservative attitude of managers and that this attitude is a great concern to the government because it forms a barrier for innovative actions.

strategic cooperation[10] (Contractor and Lorange 1988), technological cooperation (Reich 1991), globalization (Porter 1986, 1990; Ohmae 1993), and changing specializations and performances of the industries (Nelson and Winter 1982).

Cooperation between firms and governments also seems to be becoming an increasingly common phenomenon. There are some significant (synergetic) effects. The apparently growing number of collaborative agreements between firms and an increasing rate of transnational alliancing imply a need for more intensive and pro-active attitudes of governments.

The cooperation between firms is given an impetus by the apparently greater complexity and costliness of technological development. Commandeur (1994) identifies seven factors that are the driving forces for private firms to cooperate: 1. increasing capital intensity of technological processes and products; 2. increasing speed of R&D developments; 3. increasing costs of R&D activities; 4. shortening of product-life cycles; 5. increasing global competition and neo-protectionism; 6. increasing diversity of products due to market-pull and technology push and finally 7. increasing uncertainty about competitive relationships caused by strategic alliances between competitors.

Interactive relationships connect individual companies into structures that can be analyzed by means of network concepts. Littler et al (1994) underline that the terminology of forms of inter-firm cooperation is confusing with little agreement on terms.[11]

According to Hakansson (1987), a network aiming to encourage innovation processes contains three basic element: actors, activities and resources. This

---

[10]Not everyone is convinced about the success of cooperation. In an article in The Economist entitled "Don't collaborate, compete." of June 1990, the author stresses that the weaknesses of the EU-policy is the fact that governments are encouraging alliances while there is a lack of dynamism to compete on the world market.

[11]Harrigan (1988), for example, differentiates between joint ventures and 'cooperative agreements', which embraces all other means of alliancing. Lorange and Roos (1992) consider alliances along a continuous scale, ranging from market based transactions (market) to total internalization (hierarchy). Business International (1990) distinguishes between 'trading' alliances in which companies typically remain at arms length and simply exchange assets and 'functional alliances', involving a pooling of assets and more intensive integration. Jorder and Teece (1989) regard strategic alliances as relationships involving the commitment of two or more partner firms to achieve a common goal, although the goal may evolve over time, as may the scope and member- ship of the alliance. Faulkner (1992) similarly distinguishes between focused alliances, set up to achieve specific objectives normally defined at the outset, and complex alliances, which occur between two partners who recognize that they are stronger together than apart, and envisage a wide range of possible cooperative activity with each other. Littler and Wilson (1991) suggest a division based on 'hard' collaboration being founded on an explicit, relatively constrained, common purpose agreed by parties, often backed by some contracts and 'soft' collaborations which are more informal, tacit and 'casual'.

approach implies two important theoretical extensions. First, the parties involved are no longer restricted to the buying and selling firm. Although the relationship between the manufacturer and the user can still be of major importance in developing an innovation, many other parties can be involved as well. The government can stimulate innovation via specific incentives, like subsidies. Universities and other research institutes can carry out basic research that leads to new technologies, governmental agencies can bring the relevant parties together and competitors can share the risks and costs of large development projects, etc. Through the network concept, the individual buyer-seller relationship is put into the context of national and international relationships.

The second conceptual extension concerns relationship. Every firm has a certain position in a network that can be defined by (1) the function performed by the firm, (2) the relative importance of the firm in the network, (3) the strength of the relationship with other firms, and (4) the identity of the firm with which the firm has direct relationships. The present network positioning can be regarded as the firm's 'strategic situation'.

Because of these tendencies in industrial markets, manufacturers are increasingly developing innovations as a result of cooperation with other organizations. These potential partners are not limited to the potential users of the innovation. Many other organizations, such as competitors and research institutes, can contribute to the product development process (for example an industry standard may be developed by cooperation between a number of competitors and major customers; an emission standard on air quality may be developed through cooperation between the government and manufacturers, etc.). By means of these interactive relationships manufacturers can shorten the duration of the total development process, share the costs and risks, obtain the necessary technical, market and/or application knowledge, gain access to the international market, and create industrial standards.

Despite these developments, researchers of industrial innovation processes have paid only limited attention to the network concept. This could partly be due to the abstract and theoretical nature of most publications on networks. In a broader perspective, however, the increasing number of articles and books concerning networks provides evidence of the recognition of the relevance of the network concept (Groenewegen en Bije 1993 and Commandeur 1994).

It is obvious that all these factors can lead to more emphasis on engaging alliances, partnerships or networks in order to gain access to specialist technology, to spread the costs of development and marketing, and to gain rapid and effective access to knowledge and international markets. But it is also important to state that cooperation can never be aim by itself. Porter (1990) states "Alliances are not a panacea; most alliances are unstable, difficult to manage (and anyway risk creating a rival). Only alliances that are highly selective will support true competitive advantage".

Despite such sobering considerations, collaboration is accepted as an integral component of strategic recipes in many sectors. It is argued that firms need to engage in strategic alliancing in order to ensure their continued competitiveness (Devlin et al 1988 and Jarillo 1988).

For governments strategic alliancing with private firms can contribute to influence the development of technologies.

19.3.3 Network Management and the Government

At present there is a debate on the role governments should play in creating conditions for effective policies. In 't Veld (1994) concludes that many products of policy and governance are valued diversely, which may indicate that the government is not able to keep track of the fast changes in society. In 't Veld discusses the so-called 'incapacity to innovate' and concludes that the government should start to fulfil another role. We claim that the government should propose itself in the role as **conductor**, and that results, performance and quality should be its main concern.

In other words the government should play an important part in the formulation of aims which can be characterized as valuable for society. This would create an opening for a new way of defining environmental policies. In a society characterized by the contradiction between individualization and internationaliszation, only governments are capable of defining collective responsibilities such as sustainable development. As conductor the government would use its position to determine the values and standards connected with sustainable development and help them to be internalized by the target groups, by indicating that this objective goes beyond the interests of an individualized collective and propagate this concept. It could also guarantee and integrate the interest of sustainable development[12]. In that sense, the government may self-consciously consider itself conductor and **co-owner** of the theatre, which seems to be easily forgotten, given the trend towards deregulation and privatization. This does not remove the fact that the policy has to come about in an interaction between citizens, business, science etc. For environmental policy to be credible, the government will have to find approval by the target groups; when collective interest is not recognized the policy seems doomed to fail beforehand.

The changing social situation (increasing individualization) is forcing the government to claim a second role i.e. as **co-maker**. By introducing new concepts for regulation and steering, the government can to a certain extent anticipate this new individualized culture. Preferably by using positive stimuli, joining in with individual preferences and the market.

What does this mean for the position of the government? Policy networks are sometimes perceived negatively in connection with the position, role and the task of governments due to the supposed opacity and unverifiability of processes that take place inside networks. It is useful to make a distinction

---

[12]In its report entitled *The first global revolution* (1991) the Club of Rome claims that national governments fail to reduce the environmental pollution and the development problem. They argue in favour of composing teams of 'independent thinkers and free decision-makers' who could assist the governments with binding advice to guarantee collective interest at an international level.

between the tasks that have been assigned to the government, the problem of the resources the government disposes of and how they are managed.

With respect to the tasks, it can be observed that the government should be seen as keeper of the common good (for instance where ecological or social aspects are concerned). In addition to this, the government can play an active part in the construction and participation in the network. As far as the availability of resources is concerned, it should be noted that the government has extensive budgets available, large numbers of workers, plus specific, far-reaching and unique authority, media access and democratic legitimacy, which give it a special position in the networks.

The position of governments in policy networks, and the way in which they operate within them requires some attention. There are a large number of risks attached to their role. An overly pragmatic way of working could threaten of the government's legitimacy, which is its power source. Another danger is the undermining of representative organs, such as parliament, as a result of negotiation processes between executive government bodies and private and semi-private partners.

Finally, it can be noted that a judgement on the position of governments should indicate whether it concerns a normative or empirical statement. A more explicit confrontation of actual observations with ingrained postulates on the position of the government would be desirable.

19.3.4 Network Management and Environmental Policies

In this section we will discuss the role of environmental aspects in network management. We focus on the transport sector in which the environmental problems are relatively new and very challenging. In spite of the considerable variety in the environmental effects of transport, there is the joint characteristic that the 'collective' effects are caused by the cumulative effect of a large number of individual emissions. Sustainable development involves dealing with this collective responsibility; individuals are part of that larger collective. In many environmental policies, the collective responsibility of a group is acknowledged by the formulation of its target group policy.

In environmental policies, processes such as deregulation, privatization and decentralization lead to a break down of collective responsibility into the individual interests of various target groups. Environmental policies are in this respect characterized by a 'top-down' approach. But target groups do have individual objectives which are in many cases differ from the collective responsibility[13]. For example, in the transport sector, there is a gap between

---

[13]Target group policy is aimed at the activities of different collectives causing environmental pollution. It is used to try and create a basis for the realization of policy measures. A possible classification of target groups is: transport, consumers, industry, energy supply, building industry, waste-disposal industry, environmental production industry, institutes for research and education, social groups and environmentalists (Van Ast et al. 1993)

the perceived freedom of mobility at the individual level and environmental objectives; in the agricultural sector the individual farmer prefers to maintain the level of his personal income, although there is already overproduction in the agricultural sector, etc. This process pushes individuals forward and the effect is that greater interdependence arises between actors. Breaking down collective responsibility into individual interests makes the implementation of efficient policies a difficult process. The involvement of a multitude of actors in policy fields is the main reason for the increasing number of actors participating in steering processes.

A target group is not a homogenous whole, but made up of a large number of actors. Different levels of aggregation can be distinguished within a target group (for instance in the transport sector we can distinguish the individual car-owner, a cluster of car drivers, the sub-sector passenger transport and the transportation sector as a whole). Literature in political science suggests that great importance is currently attached to the concept of social complexity (Bruijn et al. 1991; Teisman 1992; Termeer 1993). In this view, modern society is made up of networks rather than hierarchical relations (In 't Veld 1994). The network is a set of actors with a public, semi-public or private character. There can be of a number of individuals or collectives, who operate on the basis of their own values, aims and interests.

As a consequence, although the 'target group approach' is particularly suitable, transport policy is a policy area where measures are difficult to establish. Policies aiming at the reduction of environmental pollution frequently seem to lack a coherent basis.

Intervention is commonly experienced at the individual level as a violation of personal freedom. At the higher aggregation levels (or the national economy) freedom is qualified as a more important good. This is due to the contrasting aims and interests: the individual freedom of the actors seems to be more appreciated than the collective responsibility for society. Stressing the importance of target groups in environmental policies can be counter productive in reaching objectives related to collective responsibility. It seems that appro-aching the car manufacturers prevents individual dealings with drivers.

In attempting to implement environmental policies, it is important to realize that firms are operating in a dynamic institutional set-up. The shift of R&D priorities and investments is a global (mega) trend. In car manufacturing, important new processes, like Just-In-Time (JIT, Kan-Ban), Quality Circle, Total Quality Control (TQC), and Computer Integrated Manufacturing (CIM), are being implemented in many industrial countries.

To return to the position of the government in relation to technology policy, one important task is its key role in developing a basis for a policy which operationalizes the concept of sustainability, and also tries to gain support among those involved[14]. Technology development can, if steered correctly,

---

[14]This is in line with the ideas of the Brundtland-Committee which states: *"The fulfilment of all these tasks will require the reorientation of technology - the key link between humans and nature. First, the capacity for technological innovation needs to be greatly enhanced so they can respond more effectively to the challenges of*

contribute to a more sustainable development.

As indicated in section 19.3.3, the government plays two different roles at the same time. In the first role, as 'conductor', the government acts, as a 96 lender of the collective interest' and presents the structure of environmental policies. In many cases these policies are based on deregulation and subsidiarity. This includes the danger of over estimation of individual responsibility, whereas what is needed is collective action. In the role of conductor the government also sets limiting conditions to influence the dynamic environment in which the company functions. When the formulated conditions are met, business will internalize these new demands.

The government's second role as co-maker is to create a stimulating and dynamic environment in which technology developments will flourish.

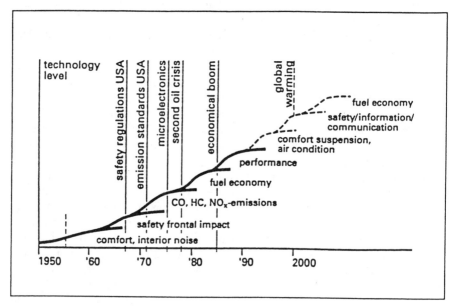

Fig. 19.1. Trends in manufacturing

(Source: TNO 1994, based on Seiffert and Walzer (1991))

Traditionally a distinction is made between direct regulation, indirect regulation and self regulation (Van Ast et al 1993). These are 'vertical instruments': a top-down approach from government(s) to influence the behaviour of target groups. Steering a network which reaches the objectives mentioned above requires, on the other hand, an integrated approach and open cooperation, deliberation and negotiation between all participants. In other words, the emphasis has to switch from vertical instruments to more

sustainable development. Second, the orientation of technology developments must be changed to pay greater attention to environmental factors". (WCED 1987)

'horizontal' instruments (Figure 19.2). This requires a reorientation of the instruments currently used in public policies. It is necessary to identify new types of instruments with a horizontal structure (cooperation in a network on an equal basis). A characteristic of these instruments is the interaction between actors, the mutual dependency and the interaction processes, in which information, goals and resources are exchanged.[15]

Collective responsibility and the recognition of the network approach are not yet fully recognized in technology policy. This is partly due to a lack of understanding of the meaning of sustainable development.

As the government can formulate societal and/or social desired aims, it could benefit from 'drivers of innovation' for the purpose of steering technology policy.

The government could adopt a favourable policy and act regulatory. In this respect it adopts once again the conductor's role. An important point of departure is that the technology policy should come about in a dialogue between business and government. This can be translated as the need to form coalitions with the character of strategic alliances. These alliances could take the form of public-private cooperation. This implies that the government should conceive its role as co-manufacturer, participating in a creative process as conductor/actor. The government can also set conditions, by using market mechanisms, to let technological development take place in the desired direction. These are good reasons to commission the government to develop a basis for a right operationalization of the collective interest when a target group is unable to do this. Government can encourage and regulate.

When the trends of governmental and business cooperation are put together, a dialogue and cooperation should become important objectives. In private companies in the automobile industry, a reorientation is already taking place. It is typified by strategic collaboration between companies, in some cases initiated by the government, and by processes of globalization.

To stimulate this new development, governments have to reconsider their task. A government acting as co-maker, with insight into the way in which technologies tend to develop, can make a constructive contribution. Fixing prior conditions and active participation can influence a dynamic environment in which a social sector functions. This means that the government will have to develop new types of instruments based on stimulating strategic cooperation. These instruments distinguish themselves from the classical steering paradigm by their vertical structure. Creating so-called 'win-win' situations becomes the decisive factor.

From this angle there is no objection to a government that self-assuredly and

---

[15]De Bruijn and Ten Heuvelhof name the specific instruments. They distinguish: a- indirect regulation, b- fine-tuning (differentiation of steering capacities dependent on the group characteristics), c- serendipitism (anticipation of the complexity and dynamics of processes), d- interactive steering (steering as a communication process), and e- network management and constituting (influencing and changing the dependency relationships).

actively sets norms and standards. An important condition for this role, however, is that policy intentions are characterized by openness, carefulness, reliability, transparency, unambiguity, consistency, long term perspective and democratic legitimacy.

Steering paradigms of the government		
Dimensions	*Classical Steering*	*-New Steering Paradigm*
Level of analysis	policy maker/firm relationship	network of actors
Perspective	central steering agency	interaction between actors
Characterization of relations	hierarchical	interdependent
Characterization of interaction	neutral implementation of pre-stated objectives	interaction and exchange of information
Indicators of success	achievement of formal objectives	common attempt to solve a problem
Indicators of failure	- vague objectives - too many actors - lack of information and control	- blockades - lack of incentives
Recommendation for improvement	- coordination - centralization	- network management
Example	- reactive environmental policies - taxation	- pro-active environmental policies - strategic alliances

Fig. 19.2. Steering paradigms of the government

Based on Koppejan et al. (1993)

If we consider the position of the government in relation to technology policy, a conclusion must be that the government can anticipate this niche in the market. By developing clear, consistent and long term policies, the government could provide support and direct technological innovations. Setting limiting conditions to the dynamic environment in which the company functions is not a priori a negative development, as long as all participants are affected. It can be stated that a strict standard is set in a major market (like the Clean Air Act in California), it is in the interest of all firms that this should become a world-wide standard. Because of the advantages of mass production the costs for the private firms decrease rapidly and dramatically.[16] When the conditions of the required policies are met, business will internalize these new demands. This is an ongoing process that requires constant fine-tuning of the limiting conditions, through consultation between private firms and the government. This is an important conclusion with respect to the development of a robust environmental policy.

19.4 Environmental Technology in the Transport Sector

At the global level, there have been significant differences in experience with collaboration between government and business concerning the steering of technology. In Japan, government and business are traditionally interwoven. The development and implementation of technology policy is undertaken in close cooperation between the government and private firms.[17] MITI (1992) for instance has presented a technology plan to improve the state of the environment up to the year 2100. The Japanese government proposes in *New Earth 21* a trajectory for the environmental technologies to be developed. The competitive position of Japanese firms is partly due to their close cooperation (on basis of the market principle, including competition) within networks, largely dominated by the government.

The image of the European technology policy is far less coherent, in fact a clear EU-technology policy is lacking. There are a number of independent programs, such as JESSE, JOULE and DRIVE, but only limited coordination between the different programs. Little coordination exists between technological policies of the individual European member states. The system is characterized by fragmented production systems and standards, with a strong dependence on national policy programs. It seems that Europe has no clearly defined common interest.

---

[16]In this context Watanabe states that it is the objective of Japanese firms and the government to stimulate technological innovations and the application of these innovations, so that firms will reach with quality products and there is an incentive (namely reduction of the costs) to continue with innovative strategies (Watanabe ibid).

[17]An interesting observation is made by Linstone (1991). He advocates an analysis of innovation and diffusion from multiple perspectives. The different perspectives consist of the technical perspective (T), the organizational or societal perspective (O) and the personal or individual perspectives (P). He states that the great difference in Japanese and American approaches to strategic planning can be traced to cultural traits. The Japanese have tended to submerge the personal in favour of the societal view, the Americans the societal in favour of the personal view. In Japan O and T have become tightly bonded, while P is minimized. In America P plays a much stronger role. The relative strength of P in the U.S. is reflected in its individual creativity output of basic research: 135 U.S. Nobel Prizes in science to 4 for Japan.

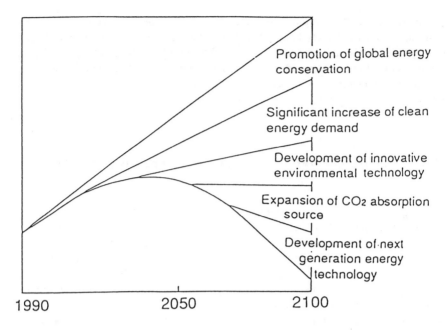

Fig. 19.3. Japanese R&D-efforts till 2100

(Source: MITI 1992)

A striking change, however, has recently taken place in the U.S. Partly due to the strong competition of Japan and other counties in South-East Asia, new forms of cooperation have been established between the government, business, society and scientific institutions. In a number of areas, a technology policy was effected in the form of newly created 'occasional coalitions'. Following successes in technology policy in the fields of micro-electronics (SEMATECH), new forms of cooperation were initiated in which environmental issues play an important role. Recent examples are the State of California Clean Air Act and the federal Fuel-cell program.

19.4.1 An Illustration: The Clean Air Act

A new approach to the encouragement of strategic cooperation between government, business and consumers' organizations can be found in the amended Clean Air Act (CAA) which became effective on 15 November 1990.[18]

---

[18]This legislation consists of seven subtitles (titles I-VII). The CAA is generally seen as one of the most significant examples of environmental legislation in the American history. Title II of the CAA contains stricter emission and fuel norms and announces programs to promote the development of cleaner cars.

For many years the quality of the air in urban areas in the United States has been a matter of concern. The CAA was first passed in 1964 in order to improve air quality in the State of California. It included a large number of measures, including standard setting. In this paper we will concentrate on the amended version of the CAA.

Firstly, the government has developed an introduction schedule for environmental performance standards of new cars up to the year 2010. These standards will reduce the gaseous emissions.

The government has also drawn up targets for the introduction of alternative energy or zero-emission vehicles. From 1996 onwards car manufacturers should sell annually 150,000 vehicles using alternative energy sources. In the State of California in 1998, 2% of new cars must comply with the so-called 'zero-emission' requirement; increasing to 10% in 2003 (Technieuws 1993). It is interesting to see that these two demands can stimulate parallel developments from the start, in which the environmental effects of the different options can be followed accurately.

Secondly, 40 government and private organizations are working together to guide the process of development, implementation and introduction of electric cars. Via stringent legislation, the Federal government is stimulating programs for the development of vehicles that have been especially designed for using alternative fuels. Moreover, this legislation enables the government to influence the composition of motor-fuel and determine the most suitable type with a view to protecting the environment.

Joint public-private research into the design and introduction process is investigating, for instance, the possibility of increasing the power of batteries to give cars a larger radius of action and ways of improving the environmental performance of batteries to make them more customer-friendly, since at present charging takes quite some time. Another area being explored is the noise aspect. Finally, investigations are being made as to how the introduction into the market should be achieved. The whole undertaking is comparable to the building of a new town ward. At the initiative of vice-president Gore (Clinton 1993), Congress has been presented with a plan for the further investigation of the application of the fuel cell in the transport business. The costs of this program are estimated at roughly 2 billion dollar.

Following the success of technology policy in the field of micro-electronics and the car industry, some optimism about these projects seems justified.

So, technological developments can be influenced by setting limiting conditions on the dynamic environment in which the private firm operates. To protect themselves against changes in their environment, private firms anticipate and consequently profit from a government with a clear and consistent policy which proves to be a trustworthy partner. In that respect, car-manufacturers are now already anticipating developments in the area of alternative fuels that could take place after 2025.

The introduction of the CAA is a good example of a new type of government policy linked to strategic cooperation. This is evident from the role played by the government and the choice of instruments. The instruments are characterized by the steering paradigm described in Figure 19.2: the amended

CAA is to a large extent based on the construction of horizontal networks. The government initiated a process in which those involved (private firms, R&D organizations and government-related institutes) were able to reach mutual agreement by negotiation. This process, known as Regulatory Negotiation (Reg-Neg), is the new playground for those who make the rules for environmental agreements. All parties involved (thirty or more in the case of the legislation for oxygenous fuels) join in the negotiations and can put forward their interests, but have to promise beforehand that they will accept the final compromise.

The Reg-Neg process helps the legislator draw up the measures to be taken. The government ascertains its part as the party that defines the playground. It does this by setting clear targets (typified by transparency, unambiguity and long term character) and hence creating the conditions for cooperation with the players.

The government also plays an active part in achieving the goals by using a set of (horizontal) instruments aimed at (a) stimulating strategic cooperation by means of network management and active participation, (b) offering a consistent policy that contributes to a certain stability in the dynamic environment by finding areas of common interest, (c) stimulating the market, (d) making use of direct regulation.

We will discuss these points in more detail:

1 Stimulating strategic cooperation/network management.
The car itself and to a lesser degree the car manufacturing industry, are the targets of the above mentioned measures. As stated above, although the oil companies are responsible for the composition and supply of sufficient (alternative) fuels, the responsibility for complying with the norms on exhaust emissions mentioned in the amended CAA lies mainly with the car manufacturing industry. The costs, however, will be mainly borne by the consumers. This solution fits in with the principle of reg-neg on an equal basis so that a direct confrontation with 'individualized' target groups is prevented. Consulting with the car producers also avoids having to make separate transactions with individual drivers.

On the government side technological cooperation is encouraged and supported by the Office of Transportation Technologies (OTT) part of the Federal government's Department of Energy (DOE). The OTT stimulates the development of those advanced technologies which are not expected to take off without government intervention e.g. technologies aimed at increasing the efficiency of energy consumption, the possibility of a flexible use of different types of fuels and reducing the environmental pollution.

Other government-related agencies playing a part in this policy include the Department of Transportation (DOT) and the General Services Administration (GSA). There are also several committees attempting to tackle the problem of fuel consumption in the future. For instance in 1988 the Federal government set up the Alternative Motor Fuels Act in accordance with which the Interagency Commission on Alternative Motor Fuels was established. The task of this committee is to make long term proposals for the fuel consumption of

American cars. Members of the DOE, DOT, the Department of Labour (DOL), the Department of Defence (DOD), EPA, GSA and the US Postal Service all have seats on this committee. The committee is advised by the US Alternative Fuel Council in which industry, universities, State bodies and interest groups have seats. The council advises on all kinds of alternative fuels, including electricity, and looks at the feasibility of possible measures. The council has, among other things, tackled the requirements of the amended CAA.

The Environmental Protection Agency (EPA), an environmental agency affiliated to the Federal Government, is responsible for the drawing up and observance of the legislation. The EPA will start a technology project in cooperation with the car industry aimed at introducing vehicles suitable for alternative fuels (Light Emission Vehicles, LEV's). In addition to this, oil companies should provide the LEV's with the alternative fuels the law desires. The EPA have determined how many cars the different manufacturers will have to produce (a total of 150,000 in 1996, going up to 300,000 in 2000).

2 Creating 'win-win' situations by stimulating the market

An impediment to R&D expenditure in the transport sector is the financial risk involved for car manufacturers. One important aspect of policy therefore lies in reducing this risk. For California and 39 so-called 'non-attainment areas,'[19] the CAA includes guidelines for owners of large fleets of cars including, for example, the U.S. Postal Service. It is established that fleets over ten cars have to switch over to alternative fuels and that the emissions of these cars have to be 70% lower that those allowed in the basic norms for normal cars[20]. By fulfilling these conditions, the government provides the manufacturers a guaranteed quota for the production of LEV's.

3 Steering of the external network (consistent policy)

The CAA needs to be seen in the context of the National Energy Strategy, which was laid down after a dialogue between the Department of Energy (DOE) and President Bush in 1991. The plan is based on five goals, partly concerning energy policies and partly with a more active government policy, namely:

- to reduce the dependency on oil supply from foreign countries,
- to guarantee a sufficient supply of energy against reasonable costs,
- to maintain a strong economy,

---

[19]A non-attainment area is a region in the USA where the air quality does not comply with the fixed health norms, due to the intensive use of cars. On 1 January 1992 39 N.A.A.'s were distinguished, including: Los Angeles, New York, Greater Connecticut, Baltimore, Philadelphia, Chicago, Milwaukee, Houston and San Diego (total population 73 million people).

[20]As such car fleets are common, the expectation is that as a result of this legislation the use of alternative fuels, such as reformulated gasolines (RFG), oxygenous fuels, alcohols and such as LNG and LPG will increase considerably.

- to protect the environment,[21]
- to ensure minimal use of government regulations and maximum use of market-oriented instruments.

A number of ambitious objectives[22] have derived from these goals, including the obligation by the Federal government: (1) to buy a given quota of cars that run on alternative fuel. This should stimulate the demand for such fuels (for individual States similar conditions have been included); (2) to finance demonstration projects for electrical vehicles and research into the necessary infrastructure for their use and maintenance.

Other measures concern energy saving in public buildings, subsidies for investigating energy-intensive branches and norms for insulation, window systems and lighting systems. With the CAA the government employs strict regulations. In the non-attainment areas (25% of the total American fuel sales) car-fuel dealers are obliged to sell fuels that contain a minimum of 2,7% oxygenous components. This assumes that by mixing gasoline with a small percentage of oxygenous components, carbon monoxide (CO) air pollution will be considerably reduced.

Moreover, nine areas that do not comply with the fixed limits for concentrations of benzene will have to start using other aromatics. The oil companies will be forced to offer gasoline of the desired composition.

As far as the protection of the environment is concerned, the strategies listed above e.g. using market leverage and stimulating cooperation between the parties involved, are strongly supported by the energy plans of the amended CAA.

In other words, the strategies used by the USA federal government in the Clean Air Act are to a large extent in line with the strategies mentioned in paragraph 19.2.2. They are also consistent with the thoughts presented in the paper concerning strategic cooperation and the desirability of expanding the model of Nelson and Winter on technology dynamics towards the concept of sustainable development. This does not imply that these concepts can be applied everywhere without modifications. Further research be undertaken on the results of the approach followed in the U.S. and the bottlenecks encountered.

---

[21]The reduction of SO-emissions and stabilization of the NOx-policy in 2010 conform to the CAA. In comparison with the present situation CO_2-emissions will be about 25% higher in 2010 as a result of the intended policy. This is better than the 40% growth that would have occurred without the energy-plan, but is far from the stabilization and reduction objectives that have been accepted in Europe and the other OECD countries.

[22]The formulated objectives contain among other things a decline in oil usage of 40%, to a total of 39% in 1995 and 33% in 2010. Energy efficiency also has to increase by 10% in 1995 and 40% by 2010. These are measures that directly affect the development of the transport sector.

19.5   Strategic Conclusions

There is an urgent need for the application of technological innovation to help solve the environmental problems in the transport sector. The well known evolutionary theory of technology dynamics proves a useful starting point for the analysis of this issue, but we find that some additional inputs are needed to analyze innovative behaviour in the context of environmental technology. These inputs can be found in theories on network management and on the way governments define their role in such networks. Governments have an impact by means of traditional instruments in the form of regulation and taxation. However, in order that networks receive the necessary momentum, governments need to reconsider their role and begin to participate in a more 'horizontal' way. The experience with the Clean Air Act in California, in which government and industry are cooperating to reach a mutual goal, seems to indicate that such an approach can indeed be successful.

Further research on the role of governments in networks of actors aiming to stimulate technological innovation (and in environmentally friendly transport) is needed. In order to improve our understanding of the critical success factors, it will be important however to study not only successful cases, but also cases where technological innovation failed to materialize.

References

Arthur A (1988) Competing Technologies: an Overview. In: Dosi et al (Eds) Technical Change and Economic Theory. Pinter Publishers London

Ast J A van and H Geerlings (1995) Milieukunde en Milieubeleid; een Introductie. Samsom H D Tjeenk Willink Alphen aan de Rijn

Bije P, Groenewegen J and O Nuys (1993) Networking in Dutch Industries. Garant/Siswo Leuven/Apeldoorn

Bruijn J A de and E F ten Heuvelhof (1991) Sturingsinstrumenten voor de Overheid; Over Complexe Netwerken en een Tweede Generatie Sturingsinstrumenten. Stenfert Kroese Uitgevers Leiden/Antwerp

Bruijn J A de (1994) Sturing en Goederenvervoer. In: Goederenvervoer over korte afstand. Stichting Toekomst der Techniek. 's Gravenhage

Clinton W J and A Gore (1993) Technology for America's Economic Growth; a New Direction to Build Economic Strength. Washington

Callon M, Law J and A Rip (1986) Mapping the Dynamics of Science and Technology. MacMillan London

Commandeur H R (1994) Strategische Samenwerking in Netwerkperspectief; een Theoretisch Raamwerk voor Industriele Ondernemingen. Proefschrift Alblasserdam

Contractor F J and P Lorange (1988) Cooperative Strategies in International Business. Massachusetts/Toronto

Devlin G and M Bleakly (1988) Strategic Alliances-Guidelines for Success. Long Range Planning 21:18-23

Dosi G (1982) Technological Paradigms and Technological Trajectories: a Suggested Interpretation of Determinants and Directions of Technological Change. Research Policy 11:147-162

Dosi G (1988) Sources, Procedures and Microeconomic effects of Innovation. The Journal of Economic Literature 26:1120-1171

European Commission (1992) Towards Sustainability, a European Community Program of Policy and Action in Relation to the Environment and Sustainable Development. COM(92) 23 Brussels

Geerlings H and R W Hommes (1995) Technologische Ontwikkeling en Nieuwe Vormen van Samenwerking. Syllabus Milieu & Verkeer en Vervoer. Erasmus Universiteit Rotterdam

Grübler A (1991) Diffussion: Long Term Patterns and Discontinuities. In: Nakicenovic N and A Grübler. Diffussion of Technologies and Social Behaviour. Springer Verlag/IIASA Berlin

Gwilliam K M en H Geerlings (1992) Research and Technology Strategy to Help Overcome the Environmental Problems in Relation to Transport. Overall Strategic Review. E.C. Brussels/Rotterdam

Hakansson H (1987) Industrial Technological Development: a Network Approach. New Hampshire/Stockholm

Janszen F H A (1993) Onderzoeksprogramma Technologie en Innovatie. Vakgroep T&O. Erasmus Universiteit Rotterdam

King A and B Schneider (1991) De Eerste Wereldwijde Revolutie; een Rapport van de Raad van de Club van Rome (Eng: The First Global Revolution). Samsom H D Tjeenk Willink Alphen aan de Rijn

Kondriatieff N D (1926) Die Langen Wellen der Konjunktur. Archiv für Sozialwissenschaft und Sozialpolitik. Band 56

Koppejan J F M, Bruijn J A de and W J M Kickert (1993) Netwerkmanagement in het Openbaar Bestuur. Vuga 's Gravenhage

Latour B (1987) Science in Action. Harvard University Press

Linstone H A (1991) Multiple Perspectives on Technological Diffusion: Insights and Lessons. In: Nakicenovic N and A Grübler. Diffussion of Technologies and Social Behaviour. Springer Verlag/IIASA Berlin

Littler D, Leverick F and D Wilson (1993) Collaboration in New-technology-based Product markets. Technology Analysis & Strategic Management 5 3:211-233

Lorange P and J Roos (1992) Strategic Alliances. Oxford

MITI (Industrial Technology Council) (1992) A Comprehensive Aproach to the New Sunshine Program: New Earth 21 Tokyo

Mol Tuur (1991) Technologie-ontwikkeling en Milieubeheer. In: Mol A P J and G Spaargaren (Eds) Technologie en Milieubeheer; Tussen Sanering en Ecologische Modernisering. Sdu Uitgeverij 's Gravenhage

Nakicenovic N (1991) Diffusion of Pervasive Systems: A case of Transport Infrastructures. In: Nakicenovic N and A Grübler. Diffussion of Technologies and Social Behaviour. Springer Verlag/IIASA Berlin

National Resource Committee (1937) Technological Trends and their Social Implications. Washington

Nelson R R and S G Winter (1982) An Evolutionary Theory of Economic Change

Nelson R R and S G Winter (1977) In Search for a Useful Theory of Innovation. Research Policy 6:36-76

Ohmea K (1989) The Global Logic of Strategic Alliances. Harvard Business Review 67:143-154

OECD (1988) Transport and the Environment. Paris

Porter M E (1985) Competitive Advantage. The Free Press New York

Porter M E (1990) The Competitive Advantage of Nations. McMillan New York

Reich B (1991) The Work of Nations. Knopf New York

Schumpeter J A (1939) Business Cycles. A Theoretical, Historical and Statistical Analysis of the Capitalist Process. MacGraw-Hill New York London

Technieuws Washington (1993) W.93-04. Published by the Ministry of Economics (in Dutch)

Teisman G R (1992) Complexe Besluitvorming; een Pluricentrisch Perspektief op Besluitvorming over Ruimtelijke Investeringen. VUGA Uitgeverij 's Gravenhage

Termeer C J A M (1993) Dynamiek en Inertie rondom Mestbeleid; een Studie naar Veranderingsprocessen in het Varkenshouderijnetwerk. Vuga, 's Gravenhage

United Nations Environment Programme (UNEP) (1993) Industry and Environment 16 1-2

Veld R J In 't (1994) Door Spiegels van Utopie en Waarheid; Een beschouwing over de Toekomstige Rijksoverheid vanuit het Lokaal Bestuur. Wiardi Beckman Stichting

Watanabe C (1994) Sustainable Development by Substituting Technology for Energy and Environmental Constraints: Japan's View. Paper presented at the Maastricht workshop on the transfer of environmentally sound technology. 14-15 April

Watanabe C and T Clark (1991) Inducing Technological Innovation in Japan - The Mechanisms of Japan's Industrial Science and Technology Policies. Journal of Science & Industrial Research 50:771-785

World Commission on Environment and Development (WCED) (1987) Our Common Future. Oxford University Press Oxford

In the paper references are made to the Scientific American (1920), The Economist (1990), and American Economic Review (1993).

20 Innovative Behaviour, R&D Development Activities and Technology Policies in Countries in Transition: The Case of Central Europe

Rolf H. Funck and Jan S. Kowalski

20.1 Introduction

In the early 80s Professor Zbigniew Brzezinski, former President Jimmy Carter's National Security adviser prophesized that the decade would either result in a war between the superpowers, or lead to the demise of the communist system and of the Soviet dominance in Eastern Europe. His main argument was that the Soviet socio-economic system was unable to meet the challenge of the new wave of innovations in communications and computers, based on research and development in microelectronics and related fields. This wave of innovations had already started to reshape the way business was done and the way society and its communication channels were controlled. It rewrote the military-defence framework and accelerated the pace of change in societal arrangements. It also made the information-control mechanisms used by the totalitarian communist regimes increasingly inefficient and irrelevant, so that they would become untenable. Brzezinski's fear of a major confrontation between the superpowers stemmed from his belief that the Soviet generals had been aware of the innovation weaknesses and deficits inherent in their system, and could have been tempted to launch a surprise preemptive strike, before the free-market technological superiority established itself.

If one recalls the heated discussions and tense superpowers relations in the wake of the decision by President Reagan to deploy the Pershing rockets and launch the R&D-programme known as the 'Star-Wars project', the spectre of the last effort on the part of the waning 'evil empire' seemed not so far fetched. Luckily, as we now know, the more pessimistic variant of Professor Brzezinski's forecast did not happen, it was his other alternative - the demise of the Soviet system which in fact occurred. This was certainly not only because of the advent of the new technologies, but to a large extent also due to them.

Indeed during the last two decades the innovation and productivity crisis has become perhaps the most characteristic feature of the so-called centrally planned economies (CPEs), or more aptly named Soviet-type economies (STEs) (on the meaning of these terms see Kowalski 1983). The ability to overcome it, to accelerate the pace of technological and societal change will to a large extent determine the success prospects of the countries in transition from planned to market economies. To some degree the very fact of transformation from the old to the new system. Through evolution from the state-owned bureaucratic enterprises to the private firms competing in the market must naturally increase the pace of innovation in these countries. But, on the other hand, negative or positive attitudes of economic actors towards R&D, towards innovation and change, coupled with the institutional and legal arrangements, either fostering or hampering innovation, exert accelerating or dampening effects on the transformation processes.

In this contribution we shall first illustrate briefly the extent of the technology gap between the developed market economies and the STEs, and looking at the situation in the ex-USSR and Soviet influenced countries, we indicate the reasons for the inherent weaknesses of the STEs in innovation. We then proceed to explain how the transformation of Central European economies into market oriented systems is connected with research and development activities, innovation generation and spread, and how the transformation processes could be affected by policy measures aiming at diminishing the technology gap. In the last part of this paper we present some evidence on innovative behaviour research and development activities, their sectoral and regional aspects in the 'core' transformation countries.

20.2   Some Evidence on the Technology and Innovation Gap

The literature on the subject is very broad and will be presented only very selectively here. It can be roughly divided into two types. Firstly there are contributions aiming at a more general comparison of the productivity differences between the STEs and the market economies, and reasons for these differences (the reasons being not only of a technological but more general systemic character). Secondly, there is research focusing more specifically on measurement of R&D and innovation activities and the resulting technology gaps in various economic sectors. Both areas in the literature have intensified since the early 80s, due to the visible retardation of growth in the STEs since the middle of the 70s.

The first strand in the literature already has a relatively long tradition, being most prominently represented by Abram Bergson (Bergson, 1964, 1978 and 1987). His principal conclusions are that output per worker under socialism was found to systematically fall short of that in Western mixed economies (Bergson 1987 p 355), the shortfall ranging from 25% to 34% in the 70s and early 80s. Bergson's results are similar to those presented in another very influential study on the subject by Kravis et al. (1982). Numerous other investigations are available pointing to an even larger productivity gap (for example Blum, Kowalski 1986, Gomulka 1986, 1990, Desai 1986, Poznanski 1987, Terrell 1992, among many others) whose estimates suggested a two to three fold difference in productivity between the STEs and the market economies.

Sectoral technology time-gap studies (for example, Hanson 1981, Kux 1979 or

Matusiak 1992) estimate a range of 12 to 20 years as the time lag between the state of the art technologies in the mature Western market economies and the STEs. The exact results depend of course on the assumptions, the quality of the data, and the sectors considered (See Table 20.1 and 20.2 for some results).

Table 20.1. Technological gap between the Polish firms investigated and the market leaders

Technology specification	total	Time lag in years			number of answers
		modern	sectors average	traditional	
1 technical production level	14.8	10.2	17.1	15.8	90
2 applied technology	12.2	19.5	15.1	10.8	86
3 quality and modernity of production	11.0	9.3	14.4	8.0	87
4 organization & management	15.6	13.4	17.0	15.7	75
5 computing techniques	14.2	12.5	15.8	13.7	84
6 R&D information system	13.2	11.3	15.2	12.4	79
7 communication with domestic firms	14.3	14.4	16.5	12.0	63
8 communication with foreign firms	15.2	15.6	16.4	13.5	75
9 market knowledge	14.9	11.7	17.0	14.9	72

Paradoxically, despite the dismal innovation and R&D performance of STEs, the status of scientific activities and basic research was in general quite elevated in the communist countries. The lag in research and instruction at the major universities in the Soviet block was certainly much less than in the economic applications of the scientific results. Academic careers, although not better rewarded financially than other disciplines, promised in many cases more political and intellectual independence than other professions. Soft financing with respect to domestic currency applied also to research institutions (although all institutions suffered from a permanent shortage of 'hard money' to buy equipment from the West).

Table 20.2. Time lag and rate of diffusion of oxygen process and continuous casting steel technologies in selected countries

Country	Time Lag	(Diffusion Speed) Oxygen Process 20% Share	40% Share	From 10 to 50% Share
Austria	0	1	4	6
Canada	2	5	18	17
United States	2	12	15	7
France	4	13	16	9
Sweden	4	9	19	17
Japan	5	5	7	5
West Germany	5	9	12	6
Netherlands	6	1	1	4
Soviet Union	8	14	-	-
Great Britain	8	5	12	11
Belgium	10	4	7	5
Luxemburg	10	5	10	9
Spain	11	5	8	8
Italy	12	1	9	-
Poland	14	6	12	-
Czechoslov.	14	7	-	-
Bulgaria	15	1	1	-
Rumania	16	2	10	-
East Germany	21	-	-	-

Country		Continuous Casting (Ingot Output) 5% Share	10% Share	20% Share
Austria	0	18	21	24
West Germany	2	15	18	21
France	2	19	21	22
Soviet Union	3	18	26	-
Canada	3	13	15	25
Great Britain	6	17	19	-
Italy	6	12	14	17
Japan	8	11	12	14
Spain	9	9	11	16
Hungary	9	-	14	16
United States	10	8	15	18
Czechosl.	10	-	-	-
Poland	12	-	-	-
Sweden	12	4	5	11
Belgium	13	13	16	15
East Germany	16	5	12	-

Source: Poznanski (1987:163)

On the whole the academic level of the major institutions of learning and research was good, especially in natural sciences and basic research. Even in social sciences Poland and Hungary, and to a lesser extent Czechoslovakia, lagged much less behind the West than did their enterprises with respect to technology and innovation levels.

In most former communist countries another characteristic of basic research was its strong concentration in a few large urban centres (with the exception of the secret military establishments located somewhere in the wilderness). Usually the capital city held a predominant position (the so called 'capital city effect', see Kowalski 1986).

20.3 Factors Explaining Low Productivity and Poor Innovation and Technology Performance of the STEs

Many reasons for the lackluster performance of the STEs are provided in the literature (for an extensive review see Poznanski 1987, and for a concise one see Desai 1986). Most authors agree however that in the last instance it was the way STEs functioned in general which was responsible for the productivity, innovation and technology gap. Because the STEs attempted to steer a very large number of economic actors from a central 'command', it must have resulted in an adverse attitude towards change. Only in a very static system, in which relatively few changes are necessary, is it possible for central planning, either of the purely command or of the 'reformed' kind to work (for an extensive discussion of these issues see Lavoi 1985, Kowalski 1992).

Innovation of any kind (be it a product, process or organizational innovation) destroys the status-quo, making centrally formulated plans for the whole economy obsolete, and making plans for individual enterprises impossible to be implemented in their original form. If we consider that the incentive mechanisms of STEs reward quantitative production-plan implementation, it is not surprising that innovations of any kind were seen as a danger for the wellbeing of the firms, their employees and managements. Coupled with the phenomenon of 'soft-financing' or 'soft-budget constraints' for firms (see Kornai 1980, 1992, Kowalski 1983, 1990), which resulted in wide-spread shortages in inter-industry relations and a very pronounced 'seller's' market, the incentive to modernize, to adopt new technology and to introduce innovations was greatly reduced. This general feature of STEs which explains on the 'meta-level' the adversity towards innovation and technical change, was reinforced by a number of specific factors. Some of the most important were (Poznanski 1987):

- the limited access of STEs to Western technology supplies
- difficulties in the practical application of new technologies
- avoidance of innovation as a device for minimizing uncertainty of supply of inputs (Berliner 1976:72-73)

The inherent weaknesses and deficits outlined above set the framework for the determination of the main policies concerning research, technology and innovation and their institutional, sectoral and spatial spread in the transforming economies. In a recent analysis of the main challenges in this field, they are summarized as follows:

i   There is a pressing need for product innovation in Central and Eastern Europe. The seller's market. i.e. low pressure from the consumers and the monopolistic position of many suppliers has been the cause of product obsolescence. In order to be able to compete in free markets and with Western competitors product upgrading is necessary.

ii   Process innovation is also needed. For the reasons described above, the production lines are obsolete and not up to Western standards.

iii   Under communist regimes, innovation and technology transfer were largely concentrated in branch-level institutes and design bureaus. R&D activities were relatively weak at the level of the individual firm (Freeman 1994). Science and technology research tended to be carried out separately, both in physical and organizational terms (the former as a rule in the so called 'academy' institutes, i.e. research establishments run by the national academies of science, the latter in 'applied' institutes, i.e. units belonging to the ministries). Because research was specific to an enterprise or branch of industry, there was no route for technology transfer across industrial or company boundaries.

iv   There was no perceived need for an innovation or technology transfer infrastructure in Central and Eastern Europe under the former regimes. The compartmentalization of industrial and research activity, combined with political repression created insular cultures, not conducive to collaboration or free flow of ideas. Consequently, there was a severe lack of international and internal interaction.

v   Central and Eastern European countries therefore have inappropriate institutional structures for generating a high rate of innovation, and there are strong cultural barriers to innovation (on these issues see also Funck, Kowalski 1993a, 1993b). Innovation support has low priority in public policy.

vi   There is a strong danger that the old R&D infrastructure, much of which could still provide a basis on which to build, is being weakened by the funding cuts which took place after the transformation to a market system began.

20.4   Transformation of the Economic Systems in Central and Eastern Europe and the Innovation Processes

It is already about five years since the countries in Central and Eastern Europe started to plan and seriously implement the transformation of their economic, social and political regimes towards a democratic, market-oriented system. It is therefore possible to venture some cautious evaluations of the transformation processes and their impacts and relevance for the innovation activities. In our contribution we limit ourselves mainly to the 'core-reform countries', by which we understand Poland, Hungary, the Czech and Slovak Republics, Estonia and Latvia. Our considerations apply however also to the other former communist countries, even though their commitment to transformation and prospects of success are less certain.

It becomes more and more apparent that the transformation of the Eastern European economies is a multidimensional process involving many spheres of action at the same time (for a more extensive discussion of these issues see Böttcher, Funck, Kowalski 1993). It is impossible to restrict reform efforts to only some areas and leave other mechanisms of the old system intact. The precondition for success of economic reforms is that they must be comprehensive, deep and irreversible. As the reforms are necessarily connected with the loss of power by some groups which commanded the economy in the past, it is not surprising that reluctance and resistance to the transformation is common. But it should be recognized that also in case of the other social groups in the societies in Eastern Europe, the unavoidable stress and hardships connected with the change of the traditional system like open unemployment (instead of the hidden unemployment in the form of sub-minimal activity of employees within the soft-budget enterprises) and elimination of many forms of subsidized collective consumption, must result in tensions and political problems, in the implementation of the reform programmes.

What are the necessary elements of the reform packages aimed at the transformation from Marx to the market?

a Privatization. An efficient market economy with hard budget constraints calls for an elimination of the predominance of state ownership of fixed assets, for the formulation of bankruptcy laws and the establishment of stock exchange activities. This element of transformation is probably the single most important one in all reform programmes in the post-communist Europe and one which promises to be burdened with numerous obstacles and theoretically thorny issues of the resulting control and incentive patterns (Kowalski 1992, Raiser, Nunnenkamp 1983, Raiser 1994).

b Elimination of the planning apparatus and institutions administering the traditional economic system. Experience from the early reform countries of the region (Poland, Hungary, the CSFR) shows that despite initial massive resistance from the employees of these institutions, it was not too difficult to implement this element of reforms due to the extensive participation of the employees of these institutions in privatization processes.

c Introduction of convertibility of currencies connected with the reform of the price system, based on market mechanisms.

d Establishment of a private banking system. Recent experiences in the countries in transition show that the lack of an efficient system for the execution of payments can be a major bottleneck on the way to a market economy.

e Introduction of a tax system concomitant with the market economy.

f Reforms of the laws governing activities of foreign capital and of the regulations with respect to foreign trade, coupled with the general economic opening of the transition countries to much more intensive international interactions. In particular, the wish to become full members (after an association phase) of the European Union, brings enormous challenges to the countries in question, with respect to their international competitiveness.

g Macroeconomic stabilization, which must precede the other elements of transition. The hardening of financing-mechanisms, elimination of most of the subsidies to large state enterprises, price liberalization, elimination of the excessive central budget deficits, all these steps must be implemented in order to provide a basis for the institutional, legal transformation measures.

A glance at the above list of preconditions of success in economic reforms shows that the elements are parts of an interrelated system and must be carried out either simultaneously or according to certain logical sequence (for more on this issue, see Kowalski 1992). Experiences gained during the first five years of implementation of the reform programmes point to the deep barriers that exist on the way to a market economic system, stemming mainly from the ways of behaviour implanted in economic actors during the long decades of the prevalence of a Soviet type economy, and from the nonexistence of the institutional and legal frameworks that form the organic fabric of market economies in the West (Funck, Böttcher, Kowalski 1993).

It should be stressed that these psychological and mental barriers as well as the lack of experience with market-economy institutions and rules of behaviour are of utmost importance even in such relatively well informed Eastern European countries as Poland and Hungary and the Czecho-Slovak Republics. Obviously, the situation in the ex-Soviet Union, Rumania and Bulgaria, i.e. countries which remained for a long time far more insulated and ideologically indoctrinated, is even more complicated.

Which aspects of the transformation processes in Central and Eastern Europe are directly relevant to the generation, existence and spread of innovation-related activities? It would seem useful to distinguish between the direct short term impact of the transformation on basic research activities and the more general and persistent effects on the behaviour of economic agents.

In the first group, the most obvious impact connected with the macroeconomic stabilization undertaken since 1990 in the core reform countries, is the curtailing of public funding for all kinds of R&D activities and related institutions. Indeed, the funding for research centres, universities, institutes was reduced drastically and in many instances massive redundancies followed (see for example the Polish case, which is typical for all transforming countries, in Jasinski 1994 and Table 20.3). There were two main motivations behind this reduction. A more general, 'philosophical' one concerned the tendency to regard all activities which were not directly profitable as less relevant for the main purpose of the transformation, i.e. the creation of a market economy and increase in the standard of living of the population. We do not need to argue at length in this chapter that this is a very simplistic argument and mistaken in its premise. The second reason, which is less readily dismissable, was the need to bring the state's finances into some semblance of equilibrium. This inevitably involved the reduction of money available for learning and research. This financial squeeze has been very hard for the institutions in question, as in the emerging, weak market economies, no well-functioning mechanisms have yet been established for the private funding of research and higher learning.

Table 20.3. R&D potential, 1989-1992

	1989	1990	1991	1992
% share of GERD[a] in GDP (estimates)[b]	1.3	1.2	0.9	0.92
% share of government R&D in GDP (estimates)	n.a.	n.a.	0.77	0.66
% share of government R&D in GERD (estimates)	n.a.	n.a.	85	72
Total number of R&D workers (in thousands)	66.2	65.1	65.2	63.2
in higher education institutions (HEIs) (all academic staff)	50.5	50.0	51.3	50.7
in Polish Academy of Sciences (PAS) institutes	4.6	4.4	4.4	4.0
in industrial R&D units	10.9	10.5	9.3	8.4
Number of HEIs	92	96	117	124
Number PAS institutes	81	77	75	81
Number of industrial R&D units	297	260	296	252
Gross value of research appliances (current prices, bn zls)	n.a.	3591	3576	2988
Number of domestic patents	2854	3242	3418	3443
Polish inventions patented abroad	190	126	150	103

[a]Gross expenditure on research and development
[b]Unreliability of government statistics mean that these are subject to a wide margin of error

(Source: KBM (1992); KBN (1993); GUS (1992); GUS (1993))
(Source: Jasinski (1994:120))

The other, systemic impacts of transformation on innovation and R&D activities are connected with the privatization programmes and creation of workable competition in the Central and Eastern European markets. Here, as expected, the transformation is exerting a positive influence by creating institutional arrangements conducive to R&D and innovation (see Figure 20.1). Privatization of state enterprises and the establishment of an ever larger number of small, medium and large new private units is gradually transforming the innovative behaviour of economic actors in the former communist countries. These private enterprises are well financed and compete against each other. Thus, they are strongly motivated to introduce R&D, apply the results, create new products, produce more cheaply, sell more efficiently. Enterprises which fail in this respect disappear from the market.

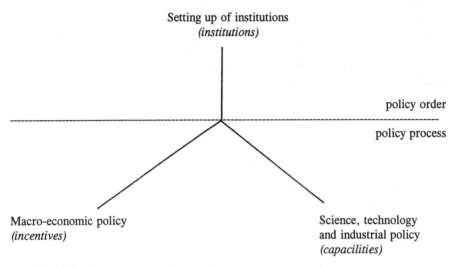

Setting up of institutions
(institutions)

policy order

policy process

Macro-economic policy
(incentives)

Science, technology
and industrial policy
(capacilities)

Fig. 20.1. The elements of transition policy

(Source: Radosevic (1994:90))

The opening to foreign competition and to inflows of foreign capital also influence the innovation mechanisms and innovative behaviour. Foreign direct investment often involves diffusion of new technologies, but even more importantly, new ways of behaviour, new business routines, new mentalities, i.e. phenomena which can either be interpreted as innovations per se or as a basis for innovative behaviour (see Welfens 1994). On the other hand, there are cases of negative innovation effects where the Western 'parent' companies have closed R&D centres in the firms they took over in Eastern Europe, or have bought them with the expressed aim of disactivating certain patents or technologies.

20.5 Innovative Behaviour, R&D Activities in the Transition Economies: Some Evidence of the Emerging Spatial Patterns

The five years since the beginning of the transformation process in Eastern and Central Europe is actually a very short span of time in which to examine the issue of innovation. Some changes in innovative behaviour, (e.g. those caused by the evolution in the competitive environment of private companies) can occur very rapidly, but many need longer to generate observable changes (e.g. those dependent on the existence of scientific hardware). As is well known from experience in Western countries, it is extremely difficult to conceptualize, operationalize and measure innovative activities (see Meyer-Krahmer 1984, Funck,

Kowalski 1987). Available statistics usually provide information about input into R&D activities (such as the number of research staff, figures on research spending, number of innovative centres, patent applications, etc.) but are unable to measure the real innovative content or results of research activities.

These difficulties are even more pronounced in the transforming countries of Central and Eastern Europe, since the traditional focus of data-gathering was directed towards production and employment and has tended to pay even less regard to qualitative factors than in the West. Even R&D input data of the aforementioned (quantitative) kind is difficult to obtain in a regional and sectoral disaggregation. For this reason it is practically impossible to provide a coherent picture of the innovative activities in the countries in transition. We are however of the opinion that there are some proxies which are useful for explaining innovative behaviour and that these permit useful insights into the emerging regional patterns. They concern:

- the extent of the implementation of privatization
- the scope and scale of the foundation of new enterprises
- scope and scale of inflow of various forms of foreign investment
- international competitiveness
- spatial pattern of R&D institutions

In other words, they are indicators of the readiness to accept change in general, to create new institutions and to increase the pace of transition.

Table 20.4   Some relevant indicators

	GDP 1990-1993	Share of private sector in GDP 1994 in %
Bulgaria	-30%	20
The Czech and Slovak Republics	-21%	60-80
Poland	-14%	60
Rumania	-36%	25
Russia	-41%	80
Hungary	-21%	50

Tables 20.4, 20.5, 20.6 and 20.7 show some of the relevant indicators at the national level. They point to the rather large differences between the transforming countries, with Poland, Hungary and the Czech Republic clearly 'leading the pack'. The surprisingly high level of privatization in Russia should not be misinterpreted. Most of the Russian enterprises, while privatized on paper, remain in reality 'quasi-state' property. It is clear that the competitive position of the transforming countries (Table 20.6) is in general still relatively weak; they still continue to depend on labour intensive, non-technical goods for export. In Poland and Hungary however the rapid increase in labour productivity permits some hope that this will change in the future (Table 20.7). The apparently weak position of the Czech Republic is probably due in part to the relatively high productivity there at the beginning of the transformation process, and in part to the fact that at the time of measurement (the end of September 1993) the privatization programme was not yet fully implemented.

Table 20.5. Foreign direct investment into transition economies

	Net inflows in US$ million				Annual per capita net inflow (average 1992/93)	Number of registered joint-ventures (3)			
	1990	1991	1992	1993 (1)		1990	1991	1992	1993 (2)
Central and Eastern Europe (A)									
Albania			21	22	6			1200	1300
Bulgaria	4	56	42	44	5	140	900	1200	1200
Former CSFR	199	594	1054	661		1600	4000	5995	7684
Czech Republic			983	561	75			3120	3700
Slovac Republic			71	100	16			2875	3984
Hungary	337	1459	1471	1200	130	5693	9117	13218	15311
Poland	88	117	284	580	11	2799	4796	5740	6300
Roumania	-18	37	73	50	3	1501	8022	20684	26249
Former Yugoslavia	67	119	93	176					
Croatia			16	36	5				
Slovenia	-2	41	113	140	70		1000	2815	3050
Total above (A)	**677**	**2382**	**3038**	**2733**	**23**	**2905**	**3920**	**15290**	**20290**
Former USSR (B)	**100**	**186**	**1580**	**1897**	**6**				
Estonia			58	86	46		1100	2662	4052
Latvia			43	50	18		295	2621	2700
Lithuania			10	40	7		220	2000	2638
Kazakhstan			200	300	15			540	
Russia		100	700	666	5		2022	3252	5249
Ukraine			200	225	4		400	2000	2400
Uzbekistan			100	100	5			570	
Total transition c (A+B)	**777**	**2568**	**4618**	**4630**	**11**				

(1) For many countries; (2) Figures as of July 1993; (3) The high figures reported by Rumania probably reflect a methodological problems in the statistical definitions used by this country.
(Source: European Economy. Supplement A, no 3, March 1994 p.5)

Table 20.6. Share in exports to the EU and revealed comparative advantage of Visegrad countries in trade with the European Union, 1989 and 1993*

		CR		SR	PL	HU
immobile Schumpeter	1989		12.1		8	8.2
industries	1993	19.3		12.7	13.6	13.2
	1989		-115.9		-138.2	-158.1
	1993	-92.7		-115.3	-105.7	-108.2
mobile Schumpeter	1989		12.4		9.4	13.6
industries	1993	15.4		10.7	10.4	17.8
	1989		-45.7		-64	-67.8
	1993	-49.7		-60.0	-97.5	-53.3
resource intensive	1989		34.3		48.6	38.5
industries	1993	16.9		21.4	28.2	20.6
	1989		150.5		49.2	75.4
	1993	94.9		129.6	71.3	66.0
capital intensive	1989		17.0		10.4	9.7
industries	1993	13.5		12.3	11	9.2
	1989		73.6		-24.3	-29.5
	1993	47.0		-20.3	-62.4	-60.0
labour intensive	1989		17.1		16.7	23.2
industries	1993	28.4		41.2	33.4	36.1
	1989		35.6		31.2	21.3
	1993	39.9		89.8	49.4	30.8

*exports shares are for May 1993

(Source: Welgens 1994)

Table 20.7. Changes in labour productivity (gA) in manufacturing and economic growth (gY) of selected countries in % per year

	ΔA				ΔY		
	1990	1991	1992	1993*	1992	1993	1994
Czech Repub.	-0.3	-14.4	-2.3	-1.3	-7	-1	3
Hungary	-3.9	-10.2	-1.4	13.8	-5	-2	1
Poland	-20.3	-5.2	13.7	11.0	1	4	4
Slovac Repub.	-1.1	-14.6	-10.2	-8.6	-6	-6	-2
Bulgaria	-12.0	-5.5	-7.0	-	-6	-5	-2
Rumania	-4.4	-15.0	-13.4	6.8	-15	0	0
Russia	3.4	-6.4	-16.5	-	-19	-12	-10
Ukraine	2.6	-2.3	-3.4	-	-14	-9	-10

*Jan-Sept

(Source: Welgens 1994)

Turning now to the spatial aspect, the regional pattern of modernization in Central and Eastern Europe, the emerging 'modernization surfaces' permit us to identify two phenomena (Maps 20.1 to 20.4). First, a strengthening of the existing polarization between the strong and weak regions can be observed. By 'strong' regions we mean those which used to dominate industrially and were political administrative centres under the Soviet regime (Kowalski 1986). The polarization process may be the inevitable price for speeding up the overall development but it would appear to pose some dangers. It should be carefully monitored and discussed. In particular the dominant position of the capital cities could lead in the future to spatial problems. Second, a 'western border neighbourhood effect' is clearly discernible. Again, this phenomenon may be unavoidable, but policy makers should at least be aware of it. Thirdly, a few of the old industrial regions have become acute 'problem' regions due to the monostructural character of their economies and their inability to remain competitive. The Hungarian old industrial core, the Silesian industrial and mining areas and the old textile region of Lodz in Poland provide prominent examples of declining regions (Dziembowska-Kowalska 1995).

20.6   Technology Policy Recommendation for the Countries in Transition

With the development of market economies, many beneficial changes are occurring. New institutions, similar to the former applied research centres are emerging and generating market-oriented innovations. New mechanisms for funding R&D are being developed and seem to be functioning fairly efficiently, although it is too early to venture a final judgement (see Jasinski 1994 and Figure 20.2). With the help of international bodies, such as the EU, World Bank, EBRD and EIB, institutions for facilitating and accelerating R&D transfer and fostering innovation and new establishments are beginning to be set up.

Most importantly, after the initial phase of lack of concern about investment in R&D and innovation related activities, awareness of the overwhelming importance of long-term policies for science and technology seems to be growing (Freeman 1994). With the current financial difficulties, however, these policies are subject to very tight budget constraints. But even with limited financial resources, formulation and implementation of policy is possible and necessary.

The elements of such policy could encompass:

- promotion of the development of small technology-oriented companies
- assistance in the restructuring of applied research institutes
- promotion of interaction between SMEs and technology organizations
- provision of training in activities related to the innovation process
- creation of national and regional transfer channels and policy

422

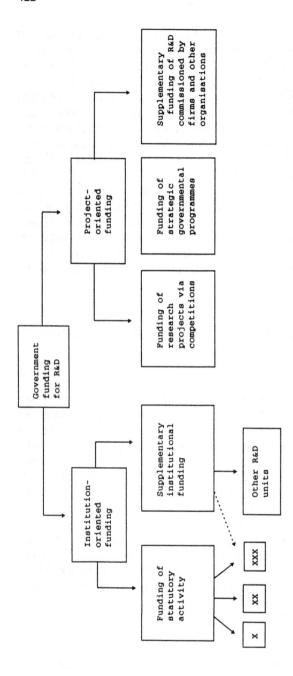

X = research units of the Polish Academy of Sciences; XX = Chartered R&D institutions; XXX = university departments (Source: Jasinki 1994 p. 127)

Fig. 20.2. The new system of R&D financing in Poland

The results of research conducted by the authors of this contribution for the European Commission (see Funck, Kowalski 1993a, 1993b) indicate that regional policy aimed at encouraging the setting up of small and medium enterprises and strengthening their research, innovation and market potential is of particular importance for transition countries. For various reasons, STEs have traditionally had few SMEs, and those which existed were weak in R&D activities and had low innovation levels. The weak regions were in particular disadvantaged in this respect.

In the recommendations for a specific regional policy (see Funck, Kowalski 1993b) aimed at supporting SMEs, it was proposed that the main thrust should be to build up an innovation supporting infrastructure to help individuals (entrepreneurs) and firms in their innovation activities, because success in innovation depends essentially on the existence of an 'innovation-friendly' socio-economic environment. It was suggested that this infrastructure should include high-level institutions for education, good communication structures, inter-firm and firm-university-government networks (within the region but also with partners from outside), research institutions, institutions providing support for information and technology transfer, and for transfer of information on the management of enterprises, their marketing and finance. However, experience in Western Europe shows that there are no easy general prescriptions for creating this kind of infrastructure, but that the use of endogenous resources and existing regional institutions is a necessity.

It is also of utmost importance that the measures introduced to improve the frame conditions for enterprises and overall regional development should constitute a coherent 'package' including economic, legal, infrastructural, cultural and socio-political elements. The aim of the package must be the definition of a 'regional profile', stressing and taking advantage of specific features of each local area

Particular attention should be devoted to the development of 'local industrial networks' of SMEs or 'industrial districts', a concept developed in regional policy in northern and central Italy, and in rural areas of other west European countries. The main idea is for SMEs belonging to a similar branch of industry and competing with each other in terms of their end product, to develop formal and informal institutional and logistic networks in order to promote common research, marketing, acquisition, transportation, publicity and advertising. This approach, which requires willingness to cooperate, rather than investment in hardware installation would seem to be a promising one for the Central and Eastern European economies also.

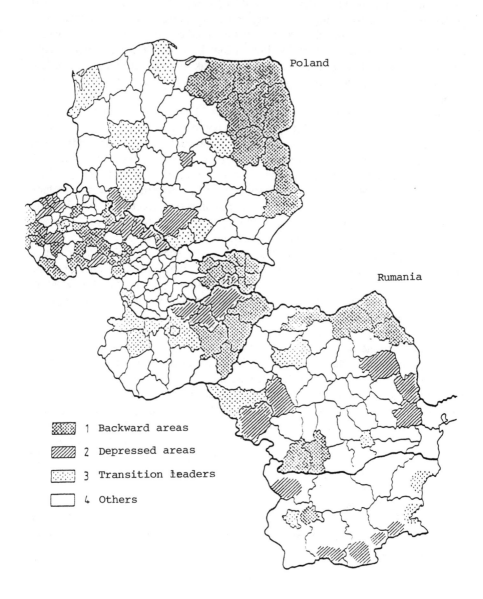

Map 20.1. Typology of regions in East-Central Europe

(Source: Horvath 1994)

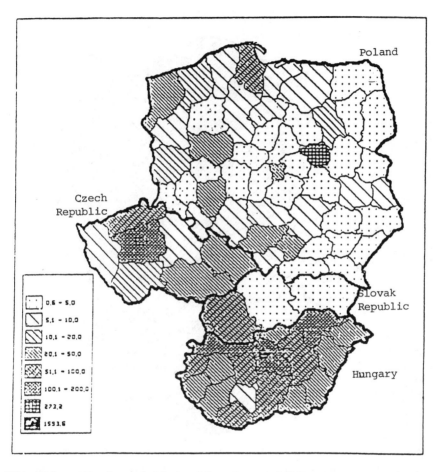

Map 20.2.   Foreign capital invested in regions, end 1992

(Source: Gorzelak 1994)

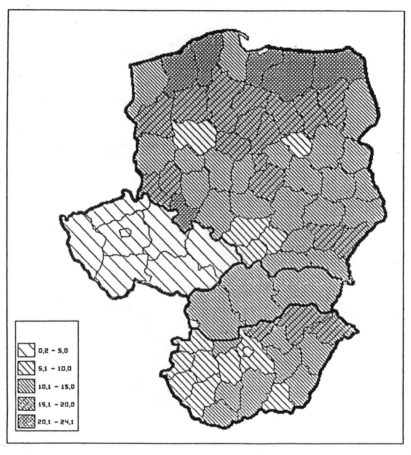

Map 20.3. Unemployment in regions, 1992 (Poland, Czech and Slovak Republics and Hungary)

(Source: Gorzelak 1994)

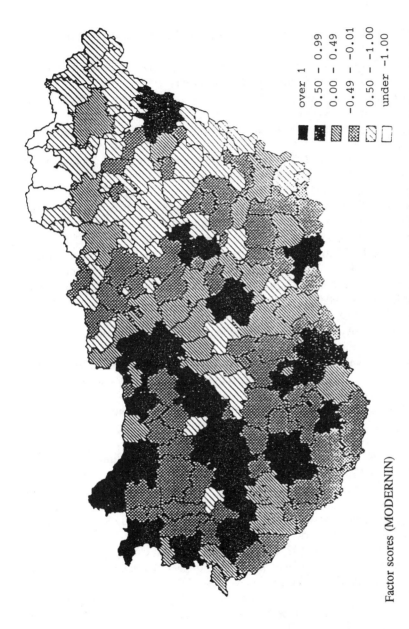

Factor scores (MODERNIN)

Map 20.4.   Hungary, regional patterns of economic modernization (new companies, joint ventures)

(Source: Nagy Ruttkay 1994)

References

Bergson A (1964) Economics of Soviet Planning. Yale University Press New Haven

Bergson A (1978) Productivity and the Social System - The USSR and The West. Harvard University Press Cambridge Massachusetts

Bergson A (1987) Comparative Productivity: The USSR, Eastern Europe and The West. The American Economic Review. 77 3:342-357

Berliner J (1976) Innovation Decisions in Soviet Industry. MIT Press Cambridge Massachusetts

Blum U and J Kowalski (1986) On the Efficiency of Regional Production in Poland, 1976-1982. Annals of Regional Science. 1:12-32

Böttcher H, Funck R H and J Kowalski (1993) The New Europe: Political, Social and Economic Changes in Eastern European Countries and their Impacts on the Spatial Division of Labour. In: Hirotado Kohno and P Nijkamp (Eds). Potential and Bottlenecks in Spatial Development. Springer Verlag Berlin. 89-107

Cappellin R and R H Funck (1993) Kleine und mittlere Unternehmen als Träger der Entwicklung im Ländlichen Raumn - Infrastrukturelle Erfordernisse und wirtschaftliche Chancen. In: Grosskopf K and K Herdzina (Eds.) Der ländliche Raum im Europa der 90er Jahre. Stuttgart (Europäischer Forschungsschwerpunkt Ländlicher Raum) 48-66

Desai P (1986) Soviet Growth Retardation. American Economic Association Papers and Proceedings. 76 2:175-179

Dziembowska-Kowalska J (1995) Theo, wir fahren nach Lodz: Chancen der Restrukturierung einer alten Industrieregion in Polen. Jahrbuch für Regionalwissenschaft. Forthcoming

European Commission Brussels (1994) Survey of the Innovation Infrastructure in Central and Eastern Europe - Executive Summary. European Innovation Monitoring System. November

Freeman C (1994) 'Postscript' to 'Problems of Science and Technology Policy in Countries in Transition' in Economic Systems. 18 2:215-218

Funck R, Böttcher H and J Kowalski (1993) Gegenwartseignung und Zukunftsfähigkeit der Sozialen Marktwirtschaft. In: Diewert W E, Spremann K and F Stehling (Eds). Mathematical Modelling in Economics. Springer-Verlag Berlin/Heidelberg/New York

Funck R H and J Kowalski (1987) Innovation and Urban Change. In: Brotchie J F, Hall P and P W Newton (Eds.) The Spatial Impact of Technological Change. Croom-Helm London/New York/Syndey. 229-239

Funck R H and J Kowalski (1993a) An Innovative Region - Baden-Württemberg. In: Gyula Horvath (Ed). Development Strategies in the Alpine-Adriatic Region. Centre for Regional Studies. Hungarian Academy of Sciences Pecs 281-298

Funck R and J Kowalski (1993b) Transnational Networks and Cooperation in the New Europe: Experiences and Prospects in the Upper Rhine Area and Recommendations for Eastern Europe. In: Cappellin R and P W J Batey (Eds). Regional Networks, Border Regions and European Integration. Pion London 205-215

Gomulka S (1986) Soviet Growth Slowdown: Duality, Maturity and Innovation. American Economic Association Papers and Proceedings. 76 2:170-174

Gomulka S (1990) The Theory of Technological Change and Economic Growth. Routledge London

Gorzelak G (1994) Regional Transformation in Central Europe. Paper presented at the 34th European Congress of the RSA Groningen

Hanson P (1981) Trade and Technology in Soviet-Western Relations. MacMillan London

Horvath G (1994) Economic Restructuring and Regional Policy in Eastern-Central Europe. Paper prepared for the 34th European Congress of the RSA Groningen

Jasinski A (1994) R&D and Innovation in Poland in the Transition Period. Economic Systems. 18 2:117-140

Kornai J (1980) Economics of Shortages. North-Holland Amsterdam

Kornai J (1992) The Political Economy of Socialism. Harvard University Press Cambridge Massachusetts

Kowalski J (1983) On the Relevance of the Concept of the Centrally Planned Economies. Jahrbuch für Sozialwissenschaft. Heft 2/83:255-266

Kowalski J (1986) Regional Conflicts in Poland: Spatial Polarization in a Centrally Planned Economy. Environment and Planning A 5:599-619

Kowalski J (1990) Economic Reforms in Centrally Planned Economies and their Consequences for Regional Development. In: K Peschel (Ed). Infrastructure and the Space Economy. Festschrift in Honour of Rolf Funck. Springer-Verlag Berlin/New York 391-402

Kowalski J (1992) Transformation from the Centrally Planned to the Market Economy System. Dimensions of the Problem, Barriers and Consequences. Volkswirtschaftliche Diskussionsbeiträge. Westfäliche Wilhelms-Universität Münster. Beitrag 164

Kravis I B et al (1982) World Product and Income. Johns Hopkins University Press Baltimore

Kux J (1976) A Survey of International Studies of Levels of Labour Productivity in Industry. In: Altmann F A et al (Eds). On Measurement of Factor Productivities: Theoretical Problems and Empirical Results. Vandenhoeck and Ruprecht Göttingen

Lavoi D (1985) Rivalry and Central Planning. Cambridge Massachusetts

Matusiak K (1992) Die technologische Lücke in der Wirtschaft Polens. Ergebnisse einer Befragung. Economic Systems 16 1 April 161-170

Meyer Krahmer F et al (1984) Erfassung Regionaler Innovationsdefizite. Schriftenreihe 06. Raumordnung des Bundesministers für Raumordnung. Bauwesen und Städtebau 06.054 Bonn

Nagy G and E Ruttkay (1994) Regional Dimensions of the Hungarian Economic Transition. Paper presented at the 34th European Congress of the RSA Groningen

Poznanski K (1987) Technology, Competition and the Soviet Block in the World Market. Institute of International Studies. University of California Berkeley

Raiser M and P Nunnenkamp (1993) Output Decline and Recovery in Central Europe. The Role of Incentives before, during and after Privatization. Kiel Working Paper 601. Kiel Institute of World Economics Kiel

Raiser M (1994) Lessons for Whom, from Whom? The Transition from Socialism
in China and Central Eastern Europe Compared. Kiel Working Paper 630. Kiel
Institute of World Economics Kiel

Terrell K (1992) Productivity of Western and Domestic Capital in Polish Industry.
Journal of Comparative Economics 16:494-514

Welfens P (1994) Foreign Direct Investment and Privatization. In: Schipke A and
A M Taylor (Eds). The Economics of Transformation. Springer-Verlag Berlin

Contributors

George Argyrakos
TRENDS
Kondylaki Street 9
Athens 11141
Greece

Carlos R. Arieira
The Institute of Public Policy
George Mason University
Fairfax Virginia 22030-2284
USA

Claudia Azzini
Dipartimento Interateneo Territorio
viale Mattioli 39
10125 Turin
Italy

David Banister
University College London
The Bartlett/Wates House
22 Gordon Street
London WC1H 0QB
England

Cristoforo S. Bertuglia
Dipartimento di Scienze e Tecniche
per i Processi di Insediamento
Politecnico di Torino
viale Mattioli 39
10125 Turin
Italy

Domenico Campisi
Istituto di Analisi dei
Sistemi e Informatica
CNR
viale Manzoni 30
00185 Rome
Italy

Roberta Capello
Dipartimento di Economia e Produzione
Politecnico di Milano
Piazza Leonardo da Vinci 32
20133 Milan
Italy

Evert Jan Davelaar
Bureau Bartels
Katherijnesingel 38
3511 GL Utrecht
The Netherlands

Maryann P. Feldman
Institute for Policy Studies
Johns Hopkins University
Wyman Park Building
3400 N Charles Street
Baltimore MD 21218-2696
USA

Amnon Frenkel
Technion
Faculty of Architecture and
Town Planning
Technion City
Haifa
Israel 32000

Rolf H.Funck
Institute of Economic Policy Research
University of Karlsruhe
Kollegium am Schloss
Bau 4
D-7500 Karlsruhe
Germany

Marina van Geenhuizen
Department of Technology Policy Management
Delft University of Technology
P O Box 5015
2600 GA Delft
The Netherlands

Harry Geerlings
Faculty of Economics
Erasmus University
Burg Oudlaan 50
3062 PA Rotterdam
The Netherlands

Maria Giaoutzi
Department of Geography
National Technical University
Zographou Campus
Athens
Greece

Kingsley E. Haynes
The Institute of Public Policy
George Mason University
Fairfax Virginia 22030-2284
USA

Geoffrey J.D. Hewings
Department of Geography
University of Illinois
103 Observatory
901 S Mathews
Urbana Illinois 61801-3682
USA

Philip R. Israilevich
Department of Geography
University of Illinois
103 Observatory
901 S Mathews
Urbana Illinois 61801-3682
USA

V.P. Jain
School of Computer Systems Sciences
Jawahardal Nehru University
New Delhi 110067
India

Karmeshu
School of Computer Systems Sciences
Jawahardal Nehru University
New Delhi 110067
India

Jan S. Kowalski
University of Karlsruhe
Institute of Economic Policy Research
Kollegium am Schloss
Bau 4
D-7500 Karlsruhe
Germany

Aydan S.Kutay
Institute for Policy Studies
Johns Hopkins University
Wyman Park Building
3400 N Charles Street
Baltimore MD 21218-2696
USA

Heli Koski
Department of Economics
University of Oulu
P O Box 111
90571 Oulu
Finland

Agostino La Bella
Dipartimento di Informatica, Sistemi e Produzione
Università di Roma "Tor Vergata"
via della Ricerca Scientifica
00187 Rome
Italy

435

Silvana Lombardo
Dipartimento di Pianificazione Territoriale ed Urbanistica
Università di Roma "La Sapienza"
via Flaminia 70
00197 Roma
Italy

Mario Lucertini
Dipartimento di Informatica, Sistemi e Produzione
Università di Roma "Tor Vergata"
via della Ricerca Scientifica
00187 Rome
Italy

Paolo Mancuso
Istituto di Analisi dei
Sistemi e Informatica
CNR
viale Manzoni 30
00185 Rome
Italy

Dino Martellato
Dipartimento di Scienze Economiche
Università di Venezia "Ca' Foscari"
DD-3246
30123 Venice
Italy

Alberto Nastasi
Istituto di Analisi dei
Sistemi e Informatica
CNR
viale Manzoni 30
00185 Rome
Italy

Peter Nijkamp
Faculty of Economics
Free University
De Boelelaan 1105
1081 HV Amsterdam
The Netherlands

436

Sylvie Occelli
IRES Piemonte
via Bogino 21
10123 Turin
Italy

Nitin S. Pandit
The Institute of Public Policy
George Mason University
Fairfax Virginia 22030-2284
USA

Gerard Pepping
Faculty of Economics
Free University
De Boelelaan 1105
1081 HV Amsterdam
The Netherlands

Fred Y. Phillips
The Institute of Public Policy
George Mason University
Fairfax Virginia 22030-2284
USA

Jacques Poot
Institute of Socio-Economic Planning
The University of Tsukuba
Tsukuba Ibaraki 305
Japan

Giovanni A. Rabino
Dipartimento di Ingegneria dei
Sistemi Edilizi e Territoriali
Politecnico di Milano
Piazza Leonardo da Vinci 32
20133 Milan
Italy

Piet Rietveld
Faculty of Economics
Free University
De Boelelaan 1105
1081 HV Amsterdam
The Netherlands

Daniel Shefer
Faculty of Architecture and
Town Planning
Technion
Technion City
32000 Haifa
Israel

Graham R. Schindler
The Institute of Public Policy
George Mason University
Fairfax Virginia 22030-2284
USA

Michael Sonis
Department of Geography
Bal Ilan University
52 100 Ramat Gan
Israel

Luis Suarez-Villa
Program in Social Ecology
University of California Irvine
Irvine California 92717
USA

Daniela Telmon
TRADE-OFF
Lungotevere Raffaello Sanzio 5
00153 Rome
Italy

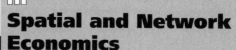